Gerhard Richter

Plant Metabolism

Physiology and Biochemistry
of Primary Metabolism

Translated by David J. Williams

101 Illustrations

Croom Helm London

Georg Thieme Publishers Stuttgart 1978

Prof. Dr. GERHARD RICHTER
Institut für Botanik der Technischen Universität Hannover,
Herrenhäuser Str. 2, 3000 Hannover-Herrenhausen

Dipl.-Biol. DAVID J. WILLIAMS, BSc,
Brückenstr. 47, D-6900 Heidelberg, FRG

Title of the original German edition:
Stoffwechselphysiologie der Pflanzen

Distributed in USA, Canada, Latin America,
United Kingdom and Eire by Croom Helm Ltd,
2–10 St John's Road, London SW 11

1st German edition 1969
2nd German edition 1971
3rd German edition 1976
1st Spanish edition 1972
1st Polish edition 1975

© 1978 Georg Thieme Verlag, Herdweg 63, POB 732, D-7000 Stuttgart 1, FRG
Printed in Germany by Druckhaus Dörr, Inhaber Adam Götz, Ludwigsburg

ISBN 3 13 549601 5 (Thieme)
ISBN 0-85664-955-4 (Croom Helm)

Preface

In recent years intensive experimental work has been done on various aspects of plant physiology, so that it is no easy task for the author of an introductory textbook of general botany to give a short review of the vast amount of new knowledge now available.

Metabolism is one of the most complex branches of plant physiology, and for years teachers and students have been conscious of the lack of an appropriate textbook. For students of biology, biochemistry, microbiology and pharmacy attendance at lectures is usually the only way to gain the necessary knowledge and acquaint oneself with the current state of research. The present textbook is intended to remedy this situation: it is based on lectures on plant physiology and the biochemistry of plant organisms.

For several reasons it seemed best to begin with a discussion of photosynthesis as the typical and central metabolic process of the green plant; this section is the longest in the book. From photosynthesis and chemosynthesis we proceed logically to those areas of reaction concerned with the winning of energy. A chapter on "Water and Ions in Metabolism" rounds off this section. Then follows a discussion of those transformations of matter which require energy from the areas mentioned and elements from the absorbed ions: biosynthesis of the cell's own chemical components. The book ends with a discussion of principles of regulation in plant metabolism. The material to be covered was selected in accordance with this general plan.

Special emphasis has been placed on two aspects of the subject: the close connection between biochemical functions and cellular structures, which is becoming more and more clearly established in molecular biology, and the significance of cross connections and contacts between at first sight unrelated areas of metabolism.

Experience has shown that the description of complicated biochemical processes of reaction is incomplete without a mention of the methods developed to explain and measure them. For this reason, short descriptions of methodological processes have been incorporated into the text wherever necessary. These will help the beginner to understand the more complicated subject matter and acquaint him with the most important aids in biochemical research. As students often have difficulty in grasping the principles and peculiarities of biochemical reactions, the basics are dealt with in an introductory section.

Since the publication of the first German edition much new knowledge has come to light. For the English edition, each chapter has been thoroughly revised and new illustrations, diagrams and formulae have been incorporated wherever necessary.

I am grateful to my students and colleagues, especially Prof. W. Nultsch and Prof. G. Jacobi, for their helpful criticism and suggestions for improvement. Prof. H. Moor, Prof. G. Drews and Prof. W. Wehrmeyer provided the originals of the electronmicrographs, Dr. V. Hemleben read through the manuscript critically, Dr. W. Richter and Mrs. A. Fischer gave valuable assistance, and Mrs. M. Schober played a large part in preparing the manuscript for publication. My special thanks to all these, and to the publishers, Georg Thieme Verlag, for their splendid cooperation in the production of an English edition.

Hannover, May 1978   GERHARD RICHTER

Contents

Introduction

The term *metabolism* encompasses all the conversions of compounds which an organism or any of its cells uses to obtain chemical energy or to synthesize own complex molecular components. Generally, the conversions performed in these processes fulfill the criteria defining chemical reactions. They are almost always accelerated by *biocatalysts,* i.e. *enzymes.* Because of their close connection with living matter they are also termed "biochemical reactions".

The metabolism of the green plant is governed by *autotrophy,* thus differing from the metabolism of most other living organisms. By means of absorbed light energy plants are able to synthesize the organic substrate required for their metabolism from carbon dioxide, water and inorganic salts; oxygen is released in most cases. This primary process of transforming the energy of visible radiation into the energy of chemical bonds is called *photosynthesis.* The second form of autotrophy, *chemosynthesis,* is of less significance; by oxidation of inorganic compounds some organisms obtain sufficient energy to synthesize organic substrate from carbon dioxide and water. They are termed *chemoautotrophic* in contrast to the *photoautotrophic* organisms requiring light.

The metabolism of autotrophic plants stands in contrast to that of all other organisms, which are unable to build their organic substrate in this way. Most bacteria, the fungi, animals and man require exogenous complex organic compounds as vital substrate. These organisms are termed *heterotrophic* and their type of metabolism is called *heterotrophy.*

The existence of all heterotrophic organisms therefore depends on substances produced by autotrophic organisms. These provide all the various complex organic compounds which the heterotrophic organisms require for energy production and biosynthesis. Thus photosynthesis is of fundamental importance for life on earth since plants serve directly or indirectly as food and energy source for most other organisms. Figure 1 illustrates this interdependence. In living matter, organic compounds are subject to a typical cyclic turnover and are connected with oxygen, carbon dioxide and water as reaction partners. From the atmosphere green plants remove carbon dioxide and water which they use to build organic compounds by photosynthesis; oxygen is released. Heterotrophic organisms utilize

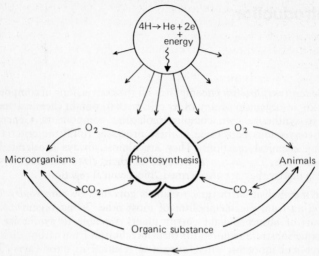

Fig. 1 The central position of photosynthesis in the turnover of living matter in nature.

this organic plant material consuming oxygen at the same time; the end products carbon dioxide and water return to the atmosphere, thus closing the cycle. Any disturbance of this equilibrium in carbon dioxide assimilation and production would have serious consequences for life on earth.

Not only organic material but also biologically utilizable energy passes along definite pathways within living matter. It originates from the radiant sunlight energy set free during the fusion of hydrogen nuclei to helium atoms at very high temperatures. This nuclear fusion releases electrons and relatively large quantities of energy in the form of gamma radiation:

$$4\,H \rightarrow He + 2\,\text{electrons} + \text{energy}$$

The radiation undergoes a complex transformation process and eventually reaches the earth in the form of photons or quanta. Resembling a gigantic nuclear reactor, the sun provides the energy necessary for terrestrial life. Conversion of this energy into the chemical energy of metastable organic carbon compounds is effected via photosynthesis. This is the ultimate source from which green plants and, indirectly, heterotrophic organisms meet their energy requirements (see above).

The flow of energy within living matter is, however, associated with a continuous reduction in quality. The high energy content of radiant energy is converted into the significantly lower one of an organic molecule and eventually into heat, a largely useless form of energy from a biological point of view. Unavoidable losses of energy, manifested as heat, occur in numerous chemical transformations. There is thus a natural downward gradient along which energy flows, which is irreversible; continuous "topping up" from the energy reserve of the sun is required.

From a thermodynamic point of view this dissipation of energy indicates transition of matter from a higher to a lower state of molecular order and a proportionate increase in *entropy**.

Apart from this crucial difference in the metabolism of autotrophic and heterotrophic organisms, there are common features which are often overlooked since the differences are more prominent. Firstly, those reactions which produce energy should be mentioned. Whether itself an independent organism or part of a more highly organized form of life, every cell combusts organic substrates to produce energy in a form which it can utilize. This complex process of *dissimilation* or *respiration* occurs in the same way in autotrophic and heterotrophic organisms. With consumption of oxygen, the fuel glucose is normally degraded to carbon dioxide and water (see below) in a process during which part of the energy released is trapped in "chemical energy equivalents" (Fig. 2). This enables the cell to perform the *biological work* vitally necessary for the maintenance of its integrity. Basically, there are three different forms of work: 1) *chemical* work; 2) uptake and transport of substances (*osmotic* work); and 3) *mechanical* work. We are mainly interested in the chemical work which encompasses the energy-consuming processes underlying the synthesis of cellular material. The most important components of the cell, the *macromolecules* such as proteins, nucleic acids, polysaccharides and also lipids, are formed from relatively small molecular building blocks mediated by specific enzymes. Macromolecular synthesis, giving rise to a net increase of cell substance, is a typical feature of growing cells and organisms.

However, biosynthesis also occurs in cells which are no longer dividing and which are fully differentiated; most of the constituents are subject to continuous renewal, i.e. there is a continuous dynamic turnover. Since degradation and synthesis proceed at equal rates, a "dynamic equilibrium" results, i.e. the cell substance does not in-

---

* *Entropy* is not only a measure of the state of molecular order, but also a measure of the dispersion of energy. Change to a disorganized state indicates an increase in entropy; establishment of a higher state of order indicates a decrease in entropy.

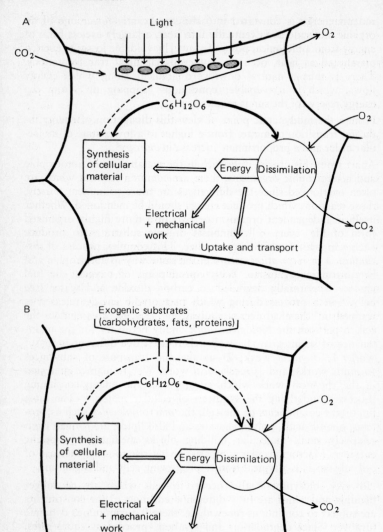

Fig. 2 Differences and mutual features in the metabolism of an autotrophic green plant cell and a heterotrophic animal cell (schematic).

crease in mass. The relative instability of cellular components emphasizes the dynamic element in metabolism.

Considering the synthesis and renewal of several macromolecular cellular components we come to another important feature. The chemical structure of these components enables them to store, transmit and also to express *information*. The information content of these macromolecules is responsible not only for maintaining the type specificity of a cell or organism during growth and multiplication but also for the precise synthesis of biologically active molecules, above all enzymes, the "construction plans" of which are laid down in the genetic material of each cell.

Dissimilation, the process yielding energy, was formerly distinguished as "operating metabolism" from the conversions concerned with the synthesis of cellular substance, the "building metabolism". This distinction is now largely outmoded; the two systems are so closely connected that boundary lines are indiscernible. Dissimilatory processes in autotrophic and heterotrophic organisms produce important precursors for synthesis, while synthetic processes provide substrates for energy production. The continuous renewal of cell components which involves most of the macromolecules and requires an input of energy, also indicates that the obviously contrasting processes of assimilation and dissimilation are closely interlocked.

The common features of autotrophy and heterotrophy are further exemplified by a "cellular division of labour". A close analogy can be seen when comparing a metabolically active cell with a chemical factory. The departments of the latter working together harmoniously under one roof find their counterparts in the various reaction spaces or *compartments* of a cell. In both cases, central regulation guarantees an equilibrium between the supply of raw materials and the output of end products. The production lines in the factory correspond to the reaction pathways of assimilation and dissimilation. Cellular division of labour is reflected in characteristic structures which are again comparable to specialized machines in a factory. The mitochondria are the organelles of energy production, the "power stations" of the cell. The pigment-bearing chloroplasts are the sites of photosynthesis. Protein biosynthesis takes place at the ribosomes. The regulatory and control center, the cell nucleus, contains the genetic material which provides the "construction plans" for the cell. The cytoplasmic membrane assumes control of the influx of products. Finally, the cytoplasm functions not only as a connecting medium between the organelles, but in its own right is an important reaction space in cellular metabolism.

The Laws Governing Metabolic Reactions

The special characteristics of biochemical reactions must be known if complex metabolic events are to be understood. The following discussion is of an introductory nature and cannot replace an exhaustive study of the principles of organic chemistry and biochemistry, provided by the relevant literature.

Chemical Equilibrium and "Steady State"

Knowledge of the laws governing chemical equilibrium and energetics is essential for the understanding of biochemical reactions. Between two reactants A and B a state is reached more or less quickly in which there is no longer any net change of the concentration ratio. The position of this chemical equilibrium is given by $K = \frac{[B]}{[A]}$, where "K" is the thermodynamic equilibrium constant, and the symbols in square brackets are the concentrations of reactant A and reaction product B respectively. Where there are several of these reactants or reaction products, the value of K is given by the product of the end products divided by the product of the reactants. Accordingly, for the conversion of A + B into C + D the equilibrium constant is

$$K = \frac{[C] \times [D]}{[A] \times [B]}.$$

The equilibrium constant K is closely related to another quantity, namely the free energy change (ΔG). This is the gain or loss of energy (in calories) during a particular chemical reaction. Energy is released when the reaction proceeds in the direction of the natural downward energy gradient, i.e. from a higher to a lower energy level. In such an "exergonic reaction", ΔG has a negative sign since the decline of free energy is evaluated in the balance as a loss. In this case, K has a relatively high value; it signifies that the reaction A + B $\rightleftharpoons$ C + D proceeds spontaneously and almost to completion. Finally, when equilibrium has been reached, much C + D but little A + B are present in the reaction mixture; the equilibrium has significantly "shifted to the right". When $\Delta G = 0$, the system does no longer release free energy.

A value of K below 1.0 indicates that there is little tendency for the reaction concerned to proceed spontaneously or approximately to completion. It requires an input of energy to proceed against the natural energy gradient, i.e. from a lower to a higher energy level. In such an "endergonic reaction", ΔG has a positive sign; energy is invested in the system, the free energy change is positive.

The symbol "ΔG°" designates the free energy change under "standard conditions": 1 mole/l of all reactants are dissolved in the reaction mixture; conversion of 1 mole in the reaction is taken as base value; the pH value is zero, i.e. the concentration or "activity" of $H^\oplus$ equals 1; the temperature is 25 °C and the pressure 1 atmosphere. By means of this "standard quantity", the free energy change of a reaction can be calculated for all other concentrations of reactants as "ΔG". The change in free energy of biochemical reactions is sometimes determined for the physiological pH value of 7.0; it is then designated "$\Delta G'$" or "$\Delta G^{\circ\prime}$".

The laws governing chemical equilibrium apply only to *closed systems*. The latter are characterized in physicochemical terms by the fact that they do not exchange either matter or energy with their surroundings. They attain a true thermodynamic equilibrium, time and speed of reaction being without significance. However, a cell or an organism is not a closed system, but an *open system;* matter and energy are exchanged with the surroundings. A state of chemical equilibrium in the form set out above does not occur in the living cell because it would render the cell unable to obtain energy or carry out cellular work. Only a system which proceeds towards equilibrium has the capacity to release energy. Therefore in the living cell or organism, the situation is always such that equilibrium is continually approached but never reached, thus making a continuous supply of energy available for cellular work. Although the concentrations of the chemical components appear to be stationary, they are not usually in true thermodynamic equilibrium, but rather in a "steady state" in which the starting substances enter the system at the same rate as the end products leave it. In between there is a series of intermediate reactions during which the product of each preceding reaction is modified to yield the substrate for the next reaction.

Figure 3 attempts to illustrate these relationships schematically, although a profound explanation is still not available. The individual sub-systems and their components in dynamic steady state equilibrium are represented as vessels with an outflow using different symbols (circles, triangles, squares). The vessels form a descending sequence, and in each a dynamic equilibrium prevails since exactly as much starting material enters as end product leaves. Starting material drops into the uppermost vessel at the same speed with which the end product leaves the last vessel of the system. A photographic snapshot would give the (false) impression that the

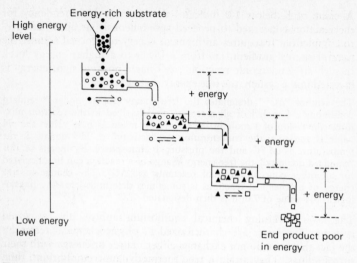

Fig. 3 Hypothetical model of a "dynamic steady state equilibrium" in a biochemical reaction chain. Details in the text.

concentration in each vessel is stationary; a moving film, however, would reveal the (true) state of a dynamic equilibrium.

The first vessel was placed at the highest level to show that, in the cell, starting substances with a relatively high energy content enter the system. Their stepwise degradation to an end product, poor in energy, corresponds to an overall exergonic reaction. This movement from a higher to a lower energy level is symbolized by the natural flow of fluid from one vessel to the next on a level below. The reaction chain proceeds spontaneously towards an energetically determined equilibrium which is, however, never reached. Energy is thereby continuously made available. The state of equilibrium arises when the input to the uppermost vessel ceases; the remaining vessels then run empty very quickly. For the cell or organism, such "running empty" means the end of energy production and consequently death. This catastrophe must be avoided by making available energy-rich substrate – autotrophically or heterotrophically – in order to maintain actively the established state of non-equilibrium. This is a basic attribute of living organisms. A number of important reaction sequences in cellular metabolism have been identified as "open" systems that exist in a dynamic steady state, e.g. the "respiratory chain" (for details see p. 241).

The complex steady state system does not, however, consist exclusively of exergonic reactions; they may be coupled or connected with endergonic processes which proceed by this mechanism (p. 19 f). This *energetic coupling* which plays an important role in metabolism, will take place only if the decline in free energy of the exergonic reaction is larger than the gain in free energy of the endergonic reaction. (For regulation and control of non-equilibrium steady state systems see p. 440).

Enzymes and Coenzymes

If every possible exergonic reaction actually did spontaneously and freely proceed to its equilibrium, the effects on terrestrial life would be fatal: most organic compounds would break down very rapidly to CO_2 and H_2O by reacting with oxygen in the case of equilibrium. However, these compounds are in fact rather stable or *"metastable"* against the action of oxygen within the biologically important temperature range. Although this is a spontaneous type of reaction, i.e. the reaction has a negative standard energy change, no compensation of the state of non-equilibrium takes place. This situation prevails until a reactant is "activated", i.e. when a certain amount of energy, the *activation energy,* is added which will raise the energy content of the molecules. In the chemical laboratory this is effected by heating the reaction mixture, in the cell by the mediation of chemical "energy equivalents" (see below). The "energy diagram" of a reaction (Fig. 4) illustrates this. The activation energy so to speak raises the reactants to a higher energy level so that they can overcome without difficulty the barrier inhibiting the reaction. The resulting "downhill" or "uphill" reaction proceeds quickly to its equilibrium determined by ΔG. From this it follows that both exergonic and endergonic reactions require an input of activation energy.

It is also possible to reduce the activation energy very substantially by adding a *catalyst.* This is a substance which, added in rather small amounts, brings about an acceleration of the reaction but neither appearing among the end products nor being itself permanently changed in the

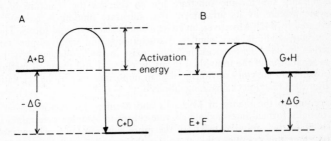

Fig. 4 The free energy change (ΔG) in an exergonic (A) and an endergonic (B) reaction. Additional activation energy must be expended in both cases.

course of the reaction. The catalyst generally forms an unstable intermediate complex with the substrate, this complex decomposing again very quickly, releasing the reaction products (see below). The resulting reduction in activation energy corresponds to a lowering of the barrier preventing the establishment of equilibrium in the non-catalyzed reaction. Although a catalyst causes a reaction to proceed more quickly, it does not generally alter the final equilibrium reached. A special form of catalysis in which the reaction product itself acts as catalyst is termed *autocatalysis*.

The cell uses the principle of *catalysis* too. Its *biocatalysts* are the *enzymes* (also termed "ferments"); they are mostly "proteids". A protein is bound to a non-protein "active" or *prosthetic group* to form a functional unit. When the active group is reversibly bound, it is termed *coenzyme* (see below) and the protein moiety *apoenzyme*.

Enzymes exhibit several important properties: 1. **Specificity of action**; although several ways of conversion of the substrate are possible, there is only one in which the activation energy is lowered sufficiently to allow the reaction to proceed. 2. **Substrate specificity**; of the various compounds which can be converted by the same reaction, only one is reversibly bound to the enzyme and thus activated. In some instances, however, the specificity is restricted to a functional group or a particular type of chemical bond only. Both may occur in different types of compounds. In this case, one speaks of *group specificity*. 3. **Temperature and pH optimum**; an enzyme exerts its full catalytic activity only at a definite pH value and within a limited temperature range. 4. **Activation**; numerous enzymes are only fully active in the presence of ions, e.g. Mg^{2+}, Mn^{2+} or Zn^{2+}. 5. **Inhibition**; the action of many enzymes is often specifically blocked by certain substances (*inhibitors*). This "poisoning" effect can be abolished in some cases by substances which bind the inhibitor. In *competitive inhibition* the molecular structure of the inhibitor is very similar to that of the substrate so that the inhibitor is bound to the enzyme; since the inhibitor is not converted the specific catalytic action of the enzyme is blocked (example p. 224). *Allosteric inhibition* occurs when the inhibitor does not occupy the binding site for the substrate but so changes the structure of the enzyme protein by its attachment that substrate binding is reduced or prevented (p. 442). Moreover, an enzyme may exist in several forms. These *isoenzymes* exhibit very similar characteristics and catalyze the same reaction (p. 443). Differences, however, exist in the molecular structure; thus separation of the individual forms by physical and chemical methods is possible. Isoenzymes obviously play an important role in the regulation of branched reaction chains in metabolism (p. 443).

According to the theory of Michaelis and Menten, the concentration of an initially formed intermediate enzyme-substrate complex determines the velocity of the catalyzed reaction: the more substrate and enzyme molecules enter this complex, the greater is the reaction rate. It is maximal when all molecules of the enzyme and the substrate have formed complexes. This saturation concentration varies for the various substrates and

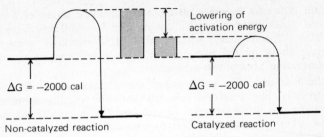

Fig. 5 Lowering of activation energy of a reaction in the presence of a catalyst.

enzymes. A convenient measure for its determination is the *"Michaelis constant"* (K_M). It is equivalent to the substrate concentration (in mole/l) at which the velocity of the catalyzed reaction is half maximal, and equals the dissociation constant of the enzyme-substrate complex. Equal quantities of complex-bound enzyme and of free enzyme are present in the reaction mixture.

By attachment of a substrate molecule the structure of the enzyme molecule is very likely deformed somewhat, the released amount of energy being obviously sufficient to lower the activation energy of the reaction to be catalyzed.

Exactly like catalysts in the chemical laboratory and in technology, enzymes make possible or accelerate the attainment of equilibrium in exergonic and endergonic reactions, but do not influence the point of equilibrium since the free energy change in the reaction is not affected. The energy diagrams (Fig. 5) for a catalyzed and a non-catalyzed reaction illustrate these relationships. Strictly speaking, however, the principles outlined above apply only to "closed systems". In an "open system" an enzyme present in higher concentration may very well influence the point of equilibrium, for instance by accelerating the conversion of a substrate to a rate higher than the rate of supply from the preceding reaction (cf. p. 8), thus shifting the equilibrium of the latter "far to the right".

Coenzyme and Prosthetic Group. It is often difficult to decide whether the active group of an enzyme consists of a coenzyme or of a prosthetic group when only the criterion of weak or tight binding to the protein component is considered. A better distinction is possible from their mode of action. Both coenzyme and prosthetic group are actively involved in the catalyzed reaction and are chemically changed at the end of it. Consequently, they do not act as true catalysts because these are not to be changed by the reaction, per definitionem. Coenzyme and prosthetic group return to their original state of activity only via a second enzymatic reaction. Here a significant difference becomes obvious: the catalytic action of a co-enzyme arises from successive coupling to two different enzyme proteins ("apoenzymes") whereas that of a prosthetic group is brought about by the participation of only one enzyme protein ("holoenzyme") but of two substrates.

In the first case, the active group reacts stoichiometrically, not catalytically, like a second substrate with the starting substrate. Therefore it is better termed a *cosubstrate*. The chemically changed active group subsequently binds to a second enzyme protein and reacts, once again stoichiometrically and as a "cosubstrate", with another substrate compound. Thereby the initially active state of the coenzyme is restored. We consider the mode of action of the most important hydrogen-transferring coenzymes, the "nicotinamide nucleotides" (p. 13 ff), shown schematically in Fig. 6, to be a typical example. Since these coenzymes cause contact between individual enzymes in this way and transfer hydrogen as well as functional groups or molecular fragments from one substrate to another, they represent the true transport systems or *transport metabolites* (Bücher) in cellular metabolism.

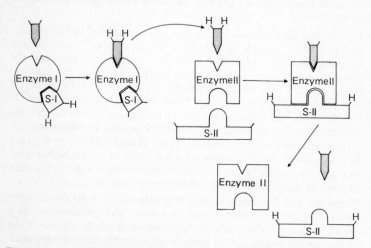

Fig. 6 Mode of action of a "hydrogen transport metabolite" or "cosubstrate". Enzyme I (= "dehydrogenase I") reacts with the substrate S-I and the oxidized transport metabolite. The latter takes over the hydrogen from S-I and binds to enzyme II (= "dehydrogenase II"). This enzyme is specifically adapted to another substrate, S-II, which accordingly accepts the hydrogen from the transport metabolite, thereby converting it into the oxidized form. The transport metabolite and enzyme II now separate from the reduced substrate molecule and are ready to react again.

When the active group is a prosthetic group, however, and is tightly bound to the enzyme protein, this holoenzyme will also react with the starting substrate, the active group being chemically changed. Its initial state of activity is restored – in contrast to a coenzyme – by the reaction with a second substrate, while being bound to the same enzyme protein. This reaction mechanism is typical of another hydrogen-transferring enzyme which contains "flavin adenine dinucleotide" (see below) as a prosthetic group.

Coenzymes participate in numerous metabolic reactions. On the basis of their mode of action, hydrogen-transferring coenzymes (= "coenzymes of the oxidoreductases") and "group-transferring" coenzymes can be distinguished. The name chosen for the latter class is based on the chemical group transferred: methyl group, formyl, carboxyl, acetyl, amino, and phosphate group, C_2-aldehyde as well as sugar molecules. More details about these will be given when we come to the metabolic reactions in which they are involved. Since vitamins are integral constituents of numerous coenzymes and prosthetic groups, their catalytic function is obvious.

The first coenzyme was identified in 1906 when Harden and Young succeeded in dissociating the "zymase", discovered by Buchner, into a protein moiety of high molecular weight and a component of low molecular weight, the "coferment". The isolation of this "cozymase" ("codehydrase I"), today termed "NAD", was finally accomplished by H. von Euler (1931). Warburg and Christian later discovered "codehydrase II", today termed "NADP".

Hydrogen-Transferring Coenzymes. Because of their participation in numerous conversions these coenzymes are of general importance in cellular metabolism.

Nicotinamide Nucleotides. Two compounds contain the nitrogen-containing pyridine derivative nicotinamide (vitamin B_2 complex); they are also termed "pyridine nucleotides" (a "nucleotide" consists of a base, a sugar and phosphoric acid). Since they also contain adenine as second base, they are more accurately termed "dinucleotides". The two coenzymes differ only in the number of phosphoric acid groups bound to the molecule: "nicotinamide adenine dinucleotide" ("NAD") and "nicotinamide adenine dinucleotide phosphate" ("NADP"). As shown by the respective structural formulae, nicotinamide is glycosidically bound to ribose, which is linked via two phosphate groups (= pyrophosphoric acid) to the nucleoside adenosine (the purine base adenine + ribose). NADP contains a third phosphate residue in the 2' position of the ribose in adenosine. Nicotinamide is the actual "active" site in the coenzyme, since here the reversible attachment of hydrogen occurs.

Nicotinamide Adenine Dinucleotide = NAD$^\oplus$ (DPN$^\oplus$)

Nicotinamide Adenine Dinucleotide Phosphate
= NADP$^\oplus$ (TPN$^\oplus$)

In the reduced form and after binding to a new apoenzyme, the coenzyme reacts with a second substrate to which it transfers the hydrogen:

The reversible binding of hydrogen is brought about by the positive charge on the pyridine structure which is characteristic for the oxidized form of the coenzyme. This is expressed in the abbreviation "NAD⊕" and "NADP⊕". The positive charge disappears on taking up two electrons in connection with two H⊕ ions, and the pyridine ring loses its aromatic nature:

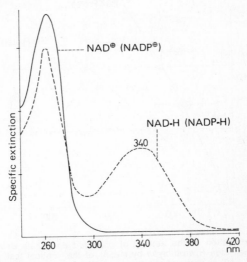

This alteration of the chemical structure gives rise to a characteristic change in the optical properties of the substances: solutions of NAD-H and NADP-H exhibit a distinct absorption maximum at 340 nm (Fig. 7) which is not shown by NAD⊕ or NADP⊕. A sensitive technique of measurement (cf. diagram) of the formation or consumption of reduced coenzyme and hence of the activity of the enzyme involved is based on this change in absorbency. An example of this "optical test" is the method for determining the activity of coenzyme-activated *glucose-6-phosphate dehydrogenase* (Fig. 8).

Fig. 7 Absorption of the oxidized and reduced nicotinamide nucleotides NAD and NADP in the ultraviolet region. The absorption maximum at 340 nm is characteristic for the reduced form.

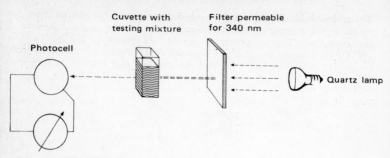

Experimental device used in the "optical test"

It should be noted that this reversible attachment of hydrogen is equivalent to an uptake of electrons since the hydrogen atoms become effective as hydrogen ions and electrons. We shall refer to this important conformity in hydrogen and electron transport in more detail later (p. 21).

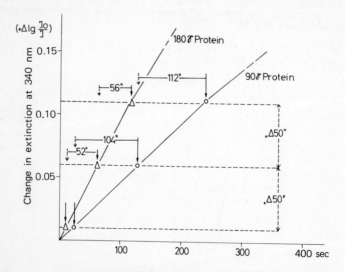

Fig. 8 Determination of the activity of *glucose-6-phosphate dehydrogenase* from the green alga *Hydrodictyon* by means of the "optical test" (according to Richter). "$\Delta 50$" = range of measurement. The formation of NADP–H + H$\oplus$ is measured as increase in extinction at 340 nm. With half the amount of enzyme twice the time is required for formation of the same amount of NADP–H + H$\oplus$.

Flavin Nucleotides. Hydrogen transfer mediated by a prosthetic group characterizes the *yellow enzymes* or "flavoproteins". The name arises from the content of yellow-colored riboflavin (vitamin B_2 complex). This alloxazine derivative has a ribitol (pentahydric alcohol) side chain. As the active group in the enzyme, this "6.7-dimethyl-9-ribityl-isoalloxazine" occurs either in the phosphorylated form as "riboflavin-5'-phosphoric acid" = "flavin mononucleotide", "FMN", or in the extended form as "flavin adenine dinucleotide", "FAD":

Riboflavin-5'-phosphate = Flavin mononucleotide, FMN

The structure of FAD shows that riboflavin and adenosine are linked via two phosphoric acid residues.

Flavin adenine dinucleotide = FAD

Strictly speaking, it is not correct to call the two flavin compounds "nucleotides" since, in contrast to NAD and NADP, they are no N-glycosides of ribose phosphoric acid. However, the names and abbreviations are in general use today.

The reversible binding of hydrogen in FMN and FAD takes place at the nitrogen atoms 1 and 10 of the isoalloxazine ring system (see formula). This reduction may be regarded (as in case of the nicotinamide nucleotides) as an uptake of electrons in connection with $H^{\oplus}$ ions. The concept of electron transfer as an expression of hydrogen transfer (see above) is also valid in this case.

The active state of a flavoprotein, i.e. its oxidized form, is restored when its prosthetic group transfers the bound hydrogen onto a suitable substrate. The enzyme now acts as a "transhydrogenase" or "oxidoreductase". Since its prosthetic group accordingly alternates between the reduced and the oxidized state (depending on the binding or release of hydrogen, or of electrons and hydrogen ions), it constitutes a "redox system" (see p. 21).

Chemical Energy Equivalents and Energy-Coupling

The expression "chemical energy equivalents" of the cell has been mentioned earlier in discussing biocatalysis. It covers compounds with a special structure enabling them to transfer chemically bound energy between energy-supplying and energy-consuming processes. The most important of these compounds is adenosine triphosphate, "ATP". This substance was first discovered in muscle, and later in the different cells of animals and plants as well as in micro-organisms. The chemical structure is known, and has been confirmed by direct chemical synthesis of ATP.

As the structural formula shows, the molecule is composed of three smaller building blocks: the purine base adenine (6-aminopurine), the C_5-sugar β-D-ribose (in furanose form), and three phosphoric acid residues. Adenine and ribose are linked by an N-glycosidic bond between position 9 of the purine ring and position 1' of the ribose. The first phosphoric acid residue is linked by an ester bond to position 5' of the ribose as well as to a second and third phosphate by anhydride bonds, giving rise to a linear polyphosphate structure. Both the anhydride bonds are "energy-rich": their hydrolytic cleavage releasing first the terminal phosphoric acid residue (forming ADP), then the second one (forming AMP), is accompanied by a substantial change in free energy: $\Delta G^o = -7000$ cal*/mole.

An intramolecular bond which releases at least 6000 cal/mole on hydrolysis is indicated in the structural formula as an "energy-rich bond" by the symbol ∼. This energy release may be measured as heat in a calorimeter. But this does not mean that the energy bound in ATP is also active in metabolism as heat. On the contrary, it is utilized by a chemical mechanism (see below).

The free energy of hydrolysis of an energy-rich bond in ATP ($\Delta G^o = -7000$ cal/mole) mentioned above does not normally occur in living cells, since they contain ATP, ADP and inorganic phosphate in concentrations much lower than 1 molal and in nonequimolar quantities. Moreover, in the living cell the equilibrium of the above-mentioned cleavage reaction is affected insofar as ATP for the most part forms complexes with magnesium. Thus the free energy of hydrolysis of ATP to ADP and inorganic phosphate in the cell may be as high as 12 000 cal/mole. However, in order to be consistent in calculations and to make a comparison of results possible, only the change in free energy under standard conditions (= "standard free energy change") is employed.

The formation of ATP from ADP and inorganic phosphate results from "energy coupling" with a strongly exergonic reaction; the energy released in the latter makes the endergonic reaction of ATP synthesis possible. The algebraic sum of the ΔG^o in the two reactions is just negative or exactly zero. This biochemically important principle which is effective in the formation of ATP as well as in an energy-requiring reaction proceeding at the expense of ATP, is illustrated by the following diagram (Fig. 9). The reaction system ATP/ADP is part of a cyclic process coupled with exergonic and endergonic reactions:

---

* 1 calorie is the amount of energy required to raise the temperature of 1 g of water from 14.5° to 15.5 °C.

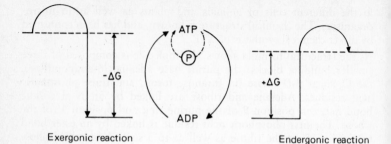

Exergonic reaction   Endergonic reaction

Fig. 9 "Energy coupling"; the role of ATP as the common intermediate in both the energy-supplying ("exergonic") and energy-requiring ("endergonic") reactions of the cell.

An example of an endergonic reaction coupled with ATP can be seen in the conversion of glucose into glucose-6-phosphate which has a higher energy content than free glucose:

α-D-Glucose + ATP ⟶ α-D-Glucose-6-phosphate + ADP

The energy required is supplied by ATP; its terminal phosphate residue is transferred to glucose in an exergonic reaction catalyzed by the enzyme *hexokinase*. Part of the energy released from ATP is retained in the newly formed ester bond of glucose-6-phosphate: about 3000 cal/mole. The energy content of this "phosphate acceptor" is thereby raised and its reactivity increased. The energy diagram (Fig. 10) shows that the glucose molecule has been raised to a higher energy level. The free energy of hydrolysis of glucose-6-phosphate is $\Delta G^\circ = -3300$ cal/mole.

As this reaction shows, ATP has a *high group transfer potential for phosphate*. With the mediation of specific enzymes, the *kinases*, phosphate is attached to different chemical groups (-OH, -COOH, -NH$_2$) in organic compounds. In rare cases two phosphoric acid residues are transferred as pyrophosphate. Relatively more important is the transfer of adenosine monophosphate (AMP) from ATP to an acceptor which is thereby converted to an "activated" compound; pyrophosphate is released. In several reactions the role of

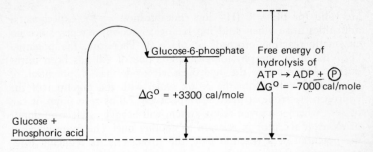

Fig. 10 Energy diagram of the "activation" of glucose by phosphorylation with ATP and *hexokinase*.

ATP is taken over by other energy-rich compounds: GTP, CTP, UTP (for structure, see p. 314 ff) and creatine phosphate.

Electron Transfer in Cellular Metabolism

The mode of action of nicotinamide nucleotides and flavin nucleotides has already taught us that hydrogen transfer in the cell is often synonymous with electron transfer. This is because there are always enough hydrogen ions available in the cell to combine with an "electron acceptor" as it takes up electrons.

Removal of electrons is generally termed *oxidation*. Conversely, a compound is reduced when it takes up electrons. An *electron donor* which has lost an electron is therefore in an oxidized state, while an *electron acceptor* which has taken up an electron is in a reduced state. Both may form a *redox system:* an electron is transferred from a donor molecule (the "reducing agent" or "reductant") to an acceptor molecule (the "oxidizing agent" or "oxidant"). Hereby, the donor is oxidized, the acceptor reduced.

The electron donor is characterized by a particular electron "pressure", and the electron acceptor by a particular electron "affinity". These parameters can be measured as electromotive force or electric potential.

A redox system in which one or several electrons are exchanged in the way described above is characterized by an electric potential, its *redox potential*. This is designated as "E_0" (in volt or millivolt) when the reduced and the oxidized partner are present in the same concentrations. The hydrogen electrode, itself a redox system ($H^{\oplus}$ ions/H_2 molecules), serves as a reference quantity. It represents the zero point of a scale on which any redox system according to its redox potential (E_0) can be arranged. As these standard potentials are valid for pH = 0 ($H^{\oplus}$ ion concentration = 1), experimental difficulties arise when studying biological redox systems, since no enzymatic reaction proceeds at pH = 0. Therefore, the potential "E_0'" based on the physiological pH value of 7.0 has been introduced. Accordingly, the hydrogen electrode has a potential of $E_0' = -0.42$ volt. From the redox scale, with the potential of the hydrogen electrode at pH = 0 respectively 7.0 as zero point, it can be seen whether a given redox system will behave as electron donor or electron acceptor in relation to another one. The more negative the potential of a reaction pair, the stronger is its reducing force and consequently its "electron pressure". Conversely, a strongly positive potential indicates a high oxidative force and high affinity for electrons. Of two redox systems, the pair with the stronger negative potential will reduce the other with the less negative or positive potential.

If individual redox systems are arranged in a series of decreasing electron pressure, starting at the highest point with the most negative system and ending at the lowest point with the most positive one (see diagram), an *electron transport chain* results, so spoken of because electrons do not only follow the natural potential gradient but may also be moved in the opposite direction following the sequence of redox systems. This mechanism of electron transport is effective in several processes in cellular metabolism where co-enzymes and prosthetic groups of specific enzymes constitute the redox systems.

In an electron transport chain transfer of electrons may also change to a transfer of hydrogen. Nevertheless, an electron transfer takes place, however, now in combination with hydrogen ions. The typical "1-electron transfer" is here replaced by a "2-electron transfer".

The change in electron pressure occurring stepwise at each level of an electron transport chain is directly related to the change in free energy. Its magnitude is determined by the difference in the standard redox potentials of the reacting systems: $\Delta G^\circ = -n \cdot F \cdot \Delta E_0$ (n = number of electrons transferred; F = Faraday, 96 500 coulombs = 23 074 cal/volt). Under standard conditions (turnover of 1 mole of substance; valency change = 1; potential difference = 1 volt; pH = 0), the transfer of one mole of electrons corresponds to a change in free energy of 23 kcal; for a 2-electron transfer it amounts to 46 kcal.

When an electron is moved against the natural energy gradient, i.e. in an "uphill" reaction, 23 kcal/mole (or 46 kcal/mole for two electrons) must be supplied to the system; the reaction is endergonic. Conversely, when the electron follows the natural energy gradient

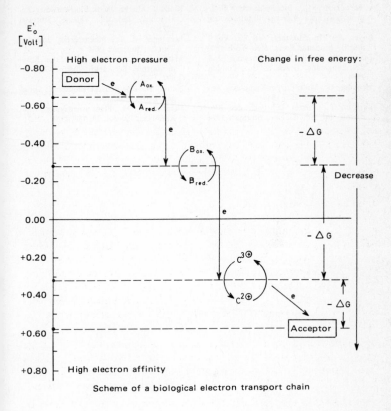

Scheme of a biological electron transport chain

in a "downhill" reaction, the amount of energy mentioned above will be released; it is twice as much in a 2-electron transfer; the reaction is exergonic.

Of course, standard conditions are hardly ever found in living cells. The effective redox potential is determined by the respective concentrations of the oxidized and the reduced reaction partner. Attempts to measure redox potentials in biological systems meet with methodical difficulties. These may be sidestepped by measuring the concentrations of both the reaction partners of a redox system; from these the actual value of the redox potential may be calculated.

References

v. Bertalanffy, L.: Biophysik des Fließ-gleichgewichts. Vieweg, Braunschweig 1953

Boyer, P.: The Enzymes, 3rd Ed. Vol. I and II. Academic Press, New York 1970

Dixon, M., C. E. Webb: Enzymes, 2nd Ed. Longmans, Green & Co., London 1964

Henning, U.: Multi-Enzym-Komplexe. Angew. Chemie 78: 865, 1966

Karlson, P.: Kurzes Lehrbuch der Biochemie, 9th Ed. Thieme, Stuttgart 1974

Klotz, I. M.: Energetik biochemischer Reaktionen, 2nd Ed. Thieme, Stuttgart 1971

Lehninger, A. L.: Bioenergetik, 2nd Ed. Thieme, Stuttgart 1974

Matile, P.: Enzymologie pflanzlicher Kompartimente. Ber. dtsch. botan. Ges. 82: 397, 1969

Netter, H.: Theoretische Biochemie. Springer, Berlin 1959

Shannon, L. M.: Plant isoenzymes. Ann. Rev. Plant Physiol. 19: 187, 1968

Plant Autotrophy

Photosynthesis

Photosynthesis is the name given to the complex process whereby chlorophyll-containing plants synthesize organic substances from carbon dioxide and water in the presence of light. Free oxygen is usually formed. Photosynthesis differs from other light-dependent biological processes insofar as absorbed radiant energy is utilized to effect synthesis of an organic compound with a higher energy content from simple inorganic compounds, like CO_2 and H_2O. This entails an endergonic process (p. 7) and a positive change in free energy. Taking these considerations into account, the following equation for the overall photosynthetic reaction is obtained:

$$6\ CO_2 + 12\ H_2O \xrightarrow[\text{chloroplast}]{\text{light}} C_6H_{12}O_6 + 6\ O_2 + 6\ H_2O$$

$$\Delta G^\circ = +675\ \text{kcal/mole hexose}$$

The elucidation of plant metabolism has occupied researchers since ancient times. It is not surprising that already Aristotle developed certain ideas on the subject. With the passage of centuries, numerous views and hypotheses arose, mostly based on falsely interpreted observations and incorrect conclusions. This is due to the complexity of the photosynthetic process, particularly the involvement of such different reactants and factors as gases, complex organic compounds, mineral salts, water and visible radiation. The correct interpretation of photosynthesis (embodied in the equation above) eventually developed from many individual observations. The diagram below shows the names of scientists who made important contributions leading to the identification of individual reactants and to the recognition of the chemical and energetic aspects of photosynthesis.

Methods. The participation of the reactants appearing in the equation for photosynthesis (see above) can be demonstrated by means of simple experiments.

Carbon dioxide: When a plant is kept in conditions of optimal illumination and water supply, but in a CO_2-free atmosphere from which the small quantities of CO_2 released by respiration are removed by NaOH or KOH, the plant stops growing and wastes away.

Water: Consumption of H_2O in photosynthesis is relatively small compared to the total amount present in the tissues. To demonstrate the participation and the function of water in photosynthesis, an isotope technique has been applied, using isotope-labelled water in which part of the normal ^{16}O is replaced by heavy oxygen (^{18}O).

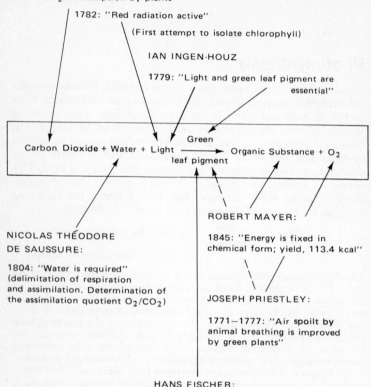

JEAN SENEBIER:

1783: "CO₂ consumption by plants"

1782: "Red radiation active"

(First attempt to isolate chlorophyll)

IAN INGEN-HOUZ

1779: "Light and green leaf pigment are essential"

Carbon Dioxide + Water + Light $\xrightarrow{\text{Green leaf pigment}}$ Organic Substance + O₂

NICOLAS THÉODORE
DE SAUSSURE:

1804: "Water is required"
(delimitation of respiration
and assimilation. Determination of
the assimilation quotient O₂/CO₂)

ROBERT MAYER:

1845: "Energy is fixed in
chemical form; yield, 113.4 kcal"

JOSEPH PRIESTLEY:

1771–1777: "Air spoilt by
animal breathing is improved
by green plants"

HANS FISCHER:

1939: "Structure of chlorophyll elucidated"

Oxygen: In accordance with the phenomenon observed by Bonnet, gas bubbles are collected which originate from illuminated aquatic plants *(Elodea, Fontinalis).* A smouldering wooden chip introduced into the collected gas immediately catches fire, identifying the gas as oxygen. Even clearer are specific chemical tests: alkaline pyrogallol solution turns brown when the gas produced by photosynthesis is bubbled through it. This gas also restores the blue color to a previously bleached solution of indigo carmine. Both reactions are typical for molecular oxygen.

Carbohydrates: During photosynthesis "assimilation starch" is produced which gives rise to an intensive blue color when an illuminated leaf is placed in a solution of KJ and iodine. Starch formation is restricted to the illuminated parts of a leaf ("stencil experiment"!).

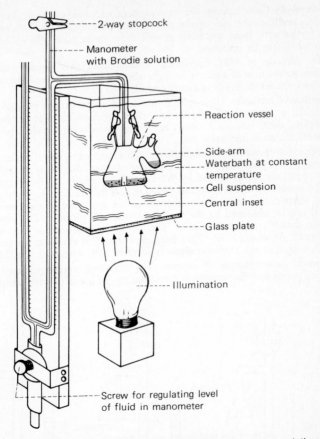

Fig. 11 Manometric determination of photosynthetic oxygen evolution according to Warburg. The U-shaped manometer with one open arm and one closed by a stopcock, contains a solution of defined density, the "Brodie solution". The oxygen produced upon illumination of the algal suspension in the reaction vessel gives rise to an increase in pressure; consequently the Brodie solution will drop in the closed arm and rise in the open one (cf. Fig. 12). CO_2 does not appear in the gas phase since it is dissolved in the medium in which the cells are suspended. From the side-arm of the reaction vessel dissolved substances may be added to the cell suspension by tipping the vessel. The central inset is especially important for measuring respiration; into this inset a cylinder of filter paper soaked with KOH is placed which adsorbs the respiratory CO_2 evolved by the cells in darkness. Thus the negative pressure produced in the reaction vessel exclusively derives from oxygen consumption of the cells. In measuring photosynthetic O_2 evolution it is necessary to determine also the O_2 consumption of the cells due to respiration in a separate dark experiment; the value obtained must be added to that of the "apparent photosynthesis" measured in light.

Chlorophyll: Cells lacking this key pigment of photosynthesis do not produce organic substances under photoautotrophic conditions. This holds true for bacteria, yeast and fungi as well as for those plant cells rendered chlorophyll-free by mutation.

Light: Already in 1882 Engelmann demonstrated that photosynthetic oxygen production in algae is brought about only on illumination with definite qualities of visible radiation (details p. 85 ff).

Quantitative methods. The methods mentioned above cannot be used for quantitative determination of the reactants involved in photosynthesis. Therefore, other suitable methods have been introduced; we shall briefly consider the most important ones.

Oxygen determination by manometry. This method is very useful for measuring the photosynthetic gas exchange of unicellular algae. It utilizes the "respirometer" developed by Warburg (1926) which is a modification of the apparatus of Barcroft and Haldane and that of Brodie. With constant temperature and volume, changes in the gas space of the reaction vessel can be measured by the resulting changes in pressure. The experimental details are given in Fig. 11. The measuring procedure is illustrated in Fig. 12. An example of a "Warburg experiment" is shown in the graph of O_2 evolution of unicellular algae at different light intensities (Fig. 13).

Chemical determination of oxygen: The dissolved oxygen released by aquatic plants into the surrounding medium can be recorded with sufficient

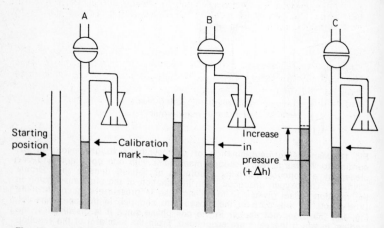

Fig. 12 Registration of a positive pressure change, e.g. resulting from O_2 evolution in the reaction vessel. A: Start of the experiment; the Brodie solution in the closed arm of the manometer is set at the calibration mark. B: Registration of the pressure increase after time t; by turning the clamp screw at the basal reservoir (cf. Fig. 11) the level of the Brodie solution is adjusted to the calibration mark (thus restoring the original volume of the gas phase!). Now the rise of the Brodie solution in the open arm beyond the starting position is recorded with the help of a graduation. The absolute amount of O_2 evolved in the time t is calculated from this pressure increase ("$+\Delta h$").

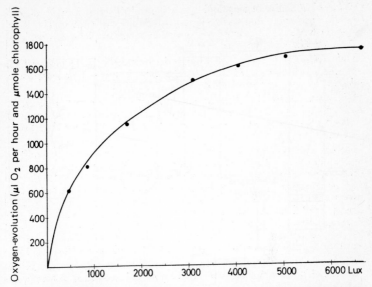

Fig. 13 Photosynthetic oxygen evolution in relation to light intensity by the blue-green alga *Anacystis nidulans*. O_2 determination by the manometric method of Warburg (after Richter).

precision by the titration method of Winkler. The dissolved gas is first trapped as manganese-IV-hydroxide by oxidizing a manganese-II salt in the presence of a strong alkali. Added potassium iodide reacts with the manganese hydroxide in the acidified solution releasing iodine which can be quantitatively determined by titration with thiosulphate. The titre gives the amount of oxygen originally dissolved.

CO$_2$ determination: Manometric and chemical methods for measuring photosynthetic O_2 evolution are unfit for most higher plants. The uptake of CO_2 is therefore preferred as criterion of photosynthetic activity. Methods have been developed to record CO_2 uptake from a closed experimental chamber into which the test plants are placed.

Infrared Absorption Recorder: This often used procedure is based on the characteristic absorption of infrared radiation by CO_2. The apparatus can detect 0.0001 % CO_2 in the air. Its construction and mode of action are shown in Fig. 14.

Hot-wire method: Changes in the CO_2 content of air affect the electrical conductivity of a heated wire in such a way that it can be used for quantitative measurement of the gas. A decrease in the CO_2 content is indicated by an increase in the conductivity of the heated wire.

Measurement of incorporation of radioactive $^{14}CO_2$: The recent introduction of labelled CO_2, the molecules of which contain unstable radioactive carbon isotope ^{14}C instead of normal carbon (^{12}C) has permitted

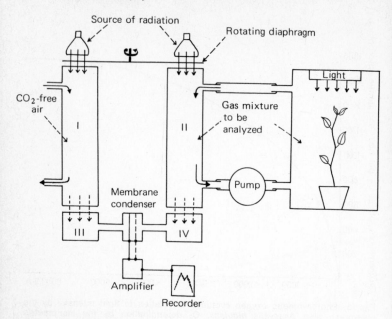

Fig. 14 Working principle of an "Infrared Absorption Recorder". The procedure is based on the characteristic absorption of infrared radiation by CO_2. Two separate chambers ("I" and "II"), each equipped with a source for infrared radiation on top, receive radiation in turns by means of a rotating diaphragm. CO_2-free air serves as a control in chamber I, while the gas mixture from the illuminated plant to be analyzed circulates through chamber II. The radiation having passed through I and II strikes the two chamber sections III and IV, which both contain pure CO_2 and are separated by a metal diaphragm. Together with a metal plate this diaphragm forms an electric condenser. Depending on the CO_2 content, more infrared radiation is absorbed in II than in I, i.e. IV receives less radiation than III. Accordingly, the CO_2 in IV is heated less by specific absorption than that in III. A pressure difference between the two chambers develops which forces the metal plate to move towards the lower pressure. This movement is amplified electronically and recorded. The photosynthetic CO_2 consumption of the test plant can be read directly from the resulting graph.

the development of a very sensitive and specific method of CO_2 measurement. Details are given on p. 142 ff.

The Pigments of Photosynthesis

Light, a form of electromagnetic radiation of natural or artificial origin, is visible to the human eye. In the total spectrum of electromagnetic radiation, visible light falls into the relatively small zone between wavelengths of 400 and 760 nm (4000 to 7600 Å). Below

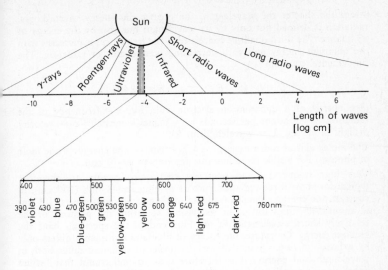

Fig. 15 The zone of visible light within the spectrum of electromagnetic radiation.

this range it is bordered by the ultraviolet, above by the long-wave infrared region (Fig. 15). This "white light" can be divided by a prism into the spectral colors red, orange, yellow, green, blue-green, blue and violet; every color corresponds to definite wavelengths. Recombination of all these colors of the spectrum produces white light. The impression of color is thus created when certain wavelengths of visible light are missing; the eye only registers the rest or the difference. Like the other forms of electromagnetic radiation, light travels with a velocity of 3×10^{10} cm/sec (= 300 000 km/sec).

Several of the properties of light, e.g. diffraction and scattering, seem to indicate that it exists in the form of waves. On the other hand, other properties indicate that visible light is corpuscular, e.g. the "photoelectric effect": an emission of electrons can be brought about by light striking the surface of certain substances such as selenium. This is why the latter is used in photocells where the electrons ejected by light set up an electrical current serving as a measure for light intensity. The blackening of a sensitized photographic emulsion by invading light is another phenomenon indicating the corpuscular nature of light.

Due to these controversial properties light is now thought to consist of waves of very small particles, the *photons,* which are the smallest units of electromagnetic radiation fit for absorption.

They are also termed *quanta.* The energy of a quantum of light is not constant, however, but inversely proportional to the wavelength of radiation: the shorter the wavelength, the greater the energy. Accordingly, radiation is defined not only by its wavelength but also by the energy of its quanta. The third quantity concerned is the frequency; it increases with the quantum energy and with decreasing wavelengths of radiation. The fundamental equation is:

$$E = h \cdot \nu; \quad E = \frac{c}{\lambda}; \quad E = 12\,403 : \lambda$$

(E = energy of a quantum in electron volt, eV; ν = frequency of the radiation in vibrations per second; h = Planck's constant; c = velocity of light in cm/sec; λ = wavelength in A).

The molar unit of radiant energy is 1 Einstein, i.e. the energy of one mole of photons. For visible radiation, this amounts to 40–70 kcal.

Very high frequencies and relatively energy-rich quanta are therefore typical for short-wave radiation such as gamma and X-rays (10^{-9} cm). In contrast, the energy content of quanta of long-wave radio waves (10^4 cm) is about 10^{13} times lower. In the visible spectrum the photons of the relatively short wavelengths at the blue end of the spectrum have the greatest energy. Of course, the energy of blue and red quanta differs only by a factor of two. Quanta of various forms of radiation differ both in wavelength and energy. It is therefore easy to understand that "white light" does not consist of energetically identical quanta, but represents a natural mixture of all the colored quanta between ultraviolet and infrared.

We use the human eye to differentiate and evaluate the different forms of visible radiation. However, it is known that the eyes of other organisms have a different range of optical perception. Thus bees and other insects perceive the ultraviolet light which is invisible for us. Nevertheless, there is a common feature in the absorption of radiation by animals and plants insofar as both are equipped with *photoreceptors* or *pigments* which "swallow", i.e. absorb certain portions of radiation, while they are "permeable" to others. These substances are usually colored complex organic compounds the structure of which is responsible for the ability to absorb light. The pigments hence act as absorbers of visible light of defined wavelengths, making it available for a photochemical process. The non-absorbed (= "transmitted") portions of the radiation are photochemically inactive.

This fundamental law, that radiation is only photochemically active when its photons are absorbed by appropriate photoreceptors or pigments, was first recognized and formulated in 1872 by Grotthus and Draper.

Of course, the absorption range of animal and plant pigments varies. The latter generally absorb throughout the visible spectrum and also in the ultraviolet and infrared region. There is also a difference in the utilization of the absorbed radiant energy. In photosynthesis it is converted into chemical energy through the mediation mainly of chlorophyll; this kind of energy conversion does not occur either in the visual process of man and of animals nor in various light reactions of plants.

In order to elucidate the chemical structure and the mode of action of photosynthetic pigments they must first be extracted from the cells or organisms and purified. A series of methods have been developed for this purpose the most important ones of which will be discussed below.

Isolation of pigments

Pigments are most readily extracted from leaves by grinding the leaves in organic solvents (such as acetone or alcohol) containing a small quantity of water. A green-colored solution is obtained after filtration. When a drop of this solution is put on filter paper, diffusion gives rise to concentric green and yellow zones. These indicate that the green leaf extract is not homogenous, but evidently a mixture of green and yellow pigments.

This separation on paper ("capillary analysis") was already used by Goppelsroeder towards the end of the last century for isolating leaf pigments.

Another procedure, first used by Tswett in 1906 for the separation of pigments, has become very important. He ran a leaf extract through a column of calcium carbonate in a glass tube, fractionating it into green and yellow zones. Tswett identified the two green components as chlorophyll a and chlorophyll b, and the yellow ones as "xanthophyll" (today called lutein) and carotene. Later a better separation of the leaf pigments was obtained by using columns of starch or confectioner's sugar. The pigments are divided in such a way that yellowish-green chlorophyll b at the top, then bluish-green chlorophyll a and finally the yellow xanthophylls accumulate in distinct zones. Another yellowish-orange colored

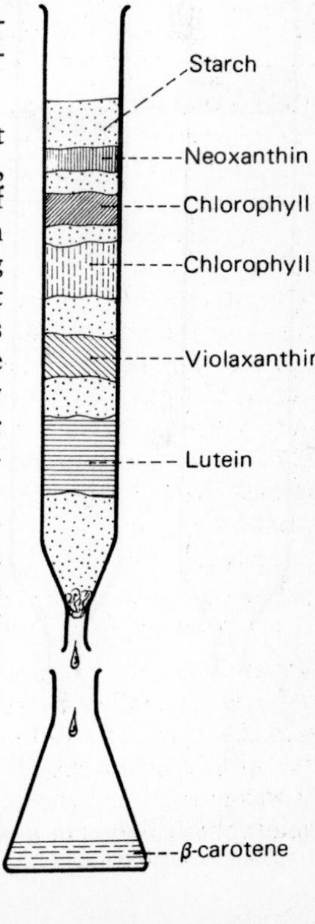

Fig. 16 Chromatography of leaf pigments on a starch column. β-carotene is not absorbed and leaves the column during the "development".

pigment, the β-carotene, is not adsorbed, and travels through the column with the organic solvent (Fig. 16).

The xanthophylls are represented mostly by lutein, which is usually accompanied by violaxanthin and neoxanthin and in many cases also by zeaxanthin. Traces of other xanthophylls may constitute up to 1 % of the total amount.

The pigment composition of leaves described above is also typical for many species of the *Chlorophyceae* and *Euglenophyceae*, though smaller amounts of other pigments may occasionally occur, e.g. α-carotene in some forms of *Chlorophyceae*, or siphonaxanthin, siphonein and also lutein epoxide in many of the siphonaceous green algae. Among the *Euglenophyceae* diadinoxanthin is an important additional xanthophyll occurring together with diatoxanthin,

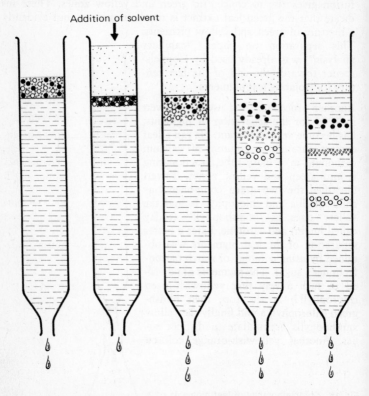

Addition of solvent

Fig. 17 Fractionation of a mixture of three different substances by means of column chromatography.

β-cryptoxanthin as well as echinenone and its hydroxy-derivatives, the latter probably as a constituent of the eyespot together with other keto-carotinoids.

Tswett succeeded not only in discovering both the chlorophylls and other pigments in leaves, but also in developing a valuable separation procedure. Referring to the successful isolation of pigments, he named it "chromatography".

This method was later consistently improved for analytical and preparatory purposes. It serves to separate complicated mixtures of chemically related compounds, including non-colored compounds. Currently routine *adsorption chromatography* is based on the original procedure of Tswett:

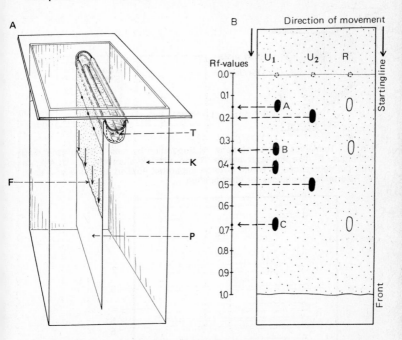

Fig. 18 Paper chromatography. A: Air-tight tank with trough (T) containing the solvent mixture into which the paper (P) dips; the mixture to be separated had been previously applied to the paper in small spots. The liquid moves slowly downwards and forms a front (F). The atmosphere within the tank (K) is saturated with solvent vapor. B: paper chromatogram after development. The two mixtures of unknown composition (U_1, U_2) have been separated into several components. Of these A, B and C are identified by reference substances (R). The position of every spot is given by the "R_f value" which is the quotient:

$$\frac{\text{distance from starting point to substance spot}}{\text{distance from starting point to solvent front}}$$

the adsorbents are packed in vertical glass tubes and the substances to be isolated are added in appropriate solvents at the top of the columns (Fig. 17). The individual components are separated into distinct zones on moving through the column as a result of their different affinities to the adsorbent. They can be directly identified on the basis of their position and color, or can be further processed after elution from the adsorbent. In another kind of column chromatography, the carrier material is first soaked with a fluid, usually water. The substances to be separated are dissolved in an organic solvent which mixes with water only to a limited extent, and then added on to the column. Since by this method the separation of substances results from partition between the aqueous phase and

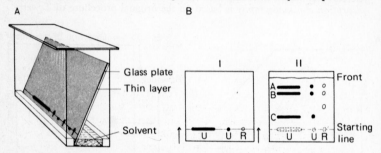

Fig. 19 Thin-layer chromatography. A: Separation by means of an ascending technique. B: The thin-layer before (I) and after development (II). U: Mixture with the substances A, B and C to be separated applied as a line or a spot. R: Reference substances for identification.

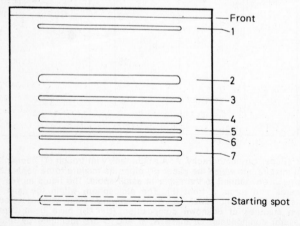

Fig. 20 Thin-layer chromatography of a pigment extract from the green alga *Chlorella pyrenoidosa*. 1, α- and β-carotene; 2, chlorophyll a; 3, chlorophyll b; 4, lutein (+ zeaxanthin); 5, antheraxanthin; 6, violaxanthin; 7, neoxanthin (after Hager and Meyer-Bertenrath).

the organic solvent, it is referred to as *partition chromatography*. Two further important separation techniques for small quantities of substances belong to this type of chromatography: *paper chromatography* (cf. Fig. 18) and *thin-layer chromatography* (cf. Fig. 19). Both techniques represent developments of the methods mentioned above. In the first procedure, the carrier material is filter paper, in the second one a glass plate is evenly spread with a thin layer of cellulose, gypsum, kieselgur, or similar material. Apart from the minor technical equipment required, a further advantage of these two methods is the small amount of material required (5 μg or less!). They are also suitable for separating extracts of plant pigments (cf. Figs. 20 and 30). Both procedures have become indispensable in modern biochemical research; we shall meet them quite frequently in the chapters to come.

Pigment Structure and Absorption of Radiation

The ability of a compound to absorb light is determined by the chemical structure of its molecules. When only very small portions of visible radiation are absorbed, a substance appears colorless. Water is such a substance; molecules of a dye dissolved in it, however, absorb certain wavelengths of the incident radiation, others pass unimpeded through the solution, reach our eye and are perceived as a characteristic color. Leaf pigments in a colorless organic solvent react in much the same way as the molecules of the water-soluble dye.

Absorption of visible radiation and hence the color of an organic compound depend in many cases on the presence of "conjugate double bonds" in its molecules, i.e. a regular alternation of single and double bonds between carbon atoms, in some cases including nitrogen atoms:

$$-C=C-C=C-C=C- \quad \text{or} \quad -C=C-C=N-C=C-C=N-C=$$

All photosynthetically active pigments contain a system of conjugate double bonds. This common feature offers a good opportunity for describing their chemical structure and its significance in the absorption of radiation. Later we shall discuss in detail the molecular basis of the absorption process (p. 98 ff).

It should be mentioned that unsaturated groups in the molecule such as –CH=N–, –N=N–, –N=O or =C=S may also contribute to the color of a compound. They are termed "chromophores".

Chlorophylls

Chlorophylls are chemically classified as tetrapyrrole compounds. The phycocyanins and phycoerythrins, additional pigments of the blue-green and red algae, belong to the same group.

The nitrogen-containing building block of these compounds, pyrrole, has a cyclic structure (see below). The two carbon atoms adjacent to the nitrogen are designated "α", the two others "β"; each has one free valency.

When four of these pyrrole rings are linked by methine bridges (–CH =) they form the closed ring structure of porphin*. This cyclic tetrapyrrole is the structure underlying all porphyrins, a group of biologically important compounds comprising the chlorophylls as well as the heme-containing compounds such as hemoglobin, and the cytochromes, which are involved in biological oxidation (p. 237 ff).

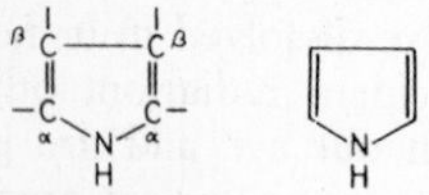

Pyrrole

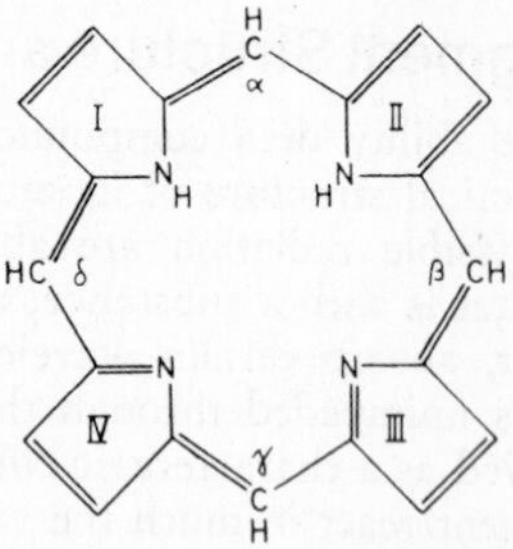

Porphin skeleton

In addition to this cyclic form a linear arrangement of four pyrrole rings exists in nature. It is present in the structure of phycocyanins and phycoerythrins indicating a structural similarity of these compounds to the bile pigments. Phycocyanins and phycoerythrins are therefore also termed "phycobilins" (p. 48).

Let us now consider again the structure of porphin, a compound which does not occur in nature. Since the pyrrole rings I–IV are linked by unsaturated carbon bonds, the whole molecule has a regular series of single and double bonds between carbon and nitrogen atoms. This accumulation of "conjugate double bonds" – there is a total of 11 in porphin – has an important consequence: porphins and the porphyrins derived from them (including the chlorophylls) are intensively colored substances.

The eight β-atoms of the pyrrole rings in porphyrin compounds are linked to different kinds of side chains. An important property of these compounds is their capacity to bind metal atoms within

---

* There are three different forms of this structure, one of which is shown.

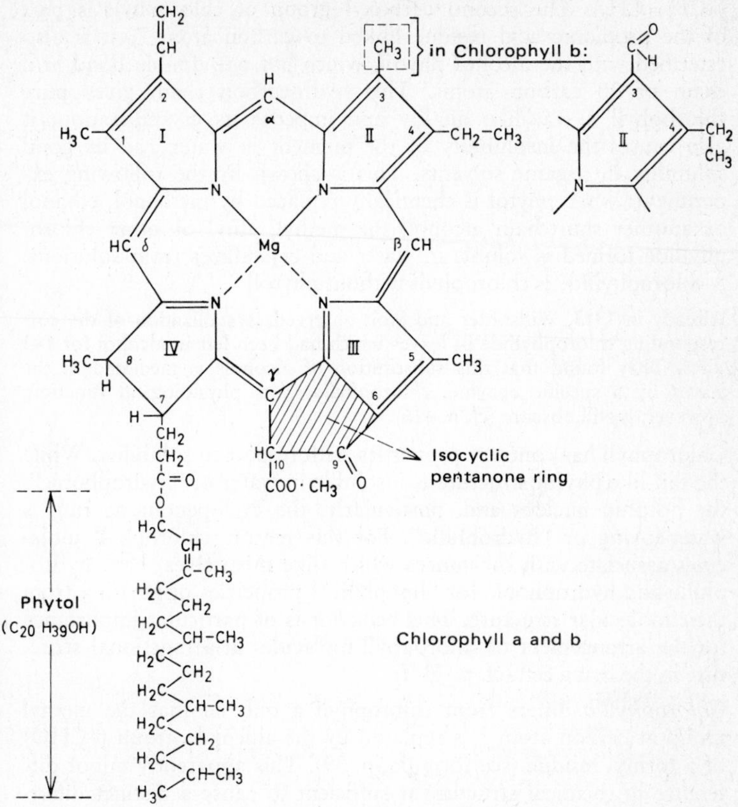

Chlorophyll a and b

Phytol ($C_{20}H_{39}OH$)

Isocyclic pentanone ring

their molecular structure. In the chlorophylls, only magnesium acts as a central atom bound in a complex way. All other biologically important compounds of this class contain iron instead of magnesium (exception: vitamin B_{12} contains cobalt).

The chlorophylls differ in several other points from the rest of the porphyrins: 1. To the carbon atoms 7 and 8 two additional hydrogen atoms are attached (see formula); chlorophylls are hence derivatives of the dihydroporphin (for consequences of this additional hydrogenation with respect to the absorbency of chlorophyll, see p. 45 f). 2. The pyrrole ring III is linked to another isocyclic structure, the cyclopentanone ring, the carboxyl group of which is esterified with a methyl group. 3. Attached to the β-carbon atoms the following side chains occur: in positions 1, 3, 5 and 8, methyl groups; in position 4, an ethyl group; in position 2, a vinyl group

$(-CH=CH_2)$. The second carboxyl group of chlorophyll is part of the propionic acid residue linked to carbon atom 7; it is also esterified with the alcohol phytol, which has one double bond in a chain of 20 carbon atoms. This hydrocarbon chain gives pure chlorophyll a wax-like quality and impedes its crystallization; it also causes the insolubility of the pigment in water and its good solubility in organic solvents. This is shown by the following experiment: when phytol is chemically replaced by methanol, ethanol or another shortchain alcohol, the methyl, ethyl or other chlorophyllide formed is soluble in water and crystallizes from solutions. A chlorophyllide is chlorophyll without phytol.

Already in 1913, Willstätter and Stoll observed crystallization of the corresponding chlorophyllide in leaves which had been left in alcohol for 1–2 days. They found that this substitution of alcohols is mediated in the leaves by a specific enzyme, *chlorophyllase*. Its physiological function, however, is still obscure (cf. p. 426).

Chlorophyll has contrary properties with respect to solubility. While the tail-like phytol structure is insoluble in water or "hydrophobic", the porphin nucleus and, particularly, the cyclopentanone ring is water-loving or "hydrophilic". For this reason, chlorophyll molecules associate with substances which (like themselves) have hydrophilic and hydrophobic (or "lipophilic") properties originating from their molecular structure. This behavior is of particular importance for the arrangement of chlorophyll molecules in a functional structure in the living cell (cf. p. 73 ff).

Chlorophyll b differs from chlorophyll a only in that the methyl group at carbon atom 3 is replaced by the aldehyde group $(-CHO)$ of a formyl residue (see formula, p. 39). This apparently minor difference in chemical structure is sufficient to cause a distinct difference in color and in the absorption spectrum (p. 43).

Chlorophyll c which (together with chlorophyll a) is typical of the "brown" line of algae *(Phaeophyceae, Bacillariophyceae, Xanthophyceae, Cryptophyceae)* is obviously a chlorophyllide since the phytol tail is replaced by a side chain $-CH=CH-COOH$; moreover, a double bond exists between the carbon atoms 7 and 8. Chlorophyll c probably occurs in two forms: "c_1" and "c_2", the latter having a vinyl group in position 4. While "c_2" is practically always present, "c_1" may be absent in certain species of the main "brown" algal groups mentioned above.

Chlorophyll d is found in small quantities in pigment extracts of numerous red algae species. Recent findings indicate that it is probably a true pigment and not an artifact resulting from degradation of chlorophyll a.

On the other hand, we are well informed on the structure of *bacteriochlorophyll a,* the typical pigment of phototrophic bacteria. Two important differences clearly set this pigment apart from the chlorophylls discussed above: 1. The positions 3 and 4 of ring II bear two further hydrogen atoms, thereby eliminating the double bond between these two β-carbon atoms; 2. the vinyl side-chain on ring I is replaced by an acetyl group (–CO–CH$_3$). The presence of two additional hydrogen atoms identifies bacteriochlorophyll a as a derivative of tetrahydroporphin which is chemically defined as 2-acetyl-desvinyl-3,4-dihydrochlorophyll. The pigment thus contains more hydrogen than the chlorophylls a and b.

The form without additional hydrogen, i.e. a true porphin derivative, is *protochlorophyll,* the biosynthetic precursor of chlorophyll a. Ring IV contains a double bond between carbon atoms 7 and 8 (see formula).

Bacteriochlorophyll a

This is eliminated during conversion to chlorophyll a by the addition of two hydrogen atoms. In etiolated seedlings this hydrogenation step is triggered off by light, but in conifers and algae it occurs in darkness too. These processes will be treated in more detail when discussing chlorophyll biosynthesis (p. 425 f).

Protochlorophyll

The Absorption Spectra of the Chlorophylls. Now let us consider the absorption of visible radiation by the chlorophylls and the dependence of this process on the pigment structure already described.

A precise description of the specific absorption of a substance is only possible when one is familiar with its *absorption spectrum*. To obtain this the substance is dissolved and the absorption is determined for every wavelength in a spectrophotometer. When the values obtained are plotted against wavelength, a characteristic graph is obtained: an absorbed portion of visible radiation gives rise to a maximum, a transmitted one to a minimum – both coinciding with the wavelengths involved. However, in many cases "extinction" rather than absorption is used to set up an absorption spectrum. By the Bouguer-Lambert-Beer law, extinction is defined as $E = \lg \dfrac{I_0}{I}$, where "I_0" is the intensity of the incident light and "I" is the intensity of the light which passes through the solution.

In the absorption spectra of both chlorophyll a and chlorophyll b (Fig. 21), the two absorption maxima show that red and blue radiation is rather strongly absorbed, green and dark-red, however, little if at all. Accordingly, the latter spectral regions coincide with distinct minima in the curve. Dilute chlorophyll solutions are green, while more concentrated ones or thick layers are dark-red. In the latter most of the green radiation is absorbed as well, thus closing the "green gap" in the spectrum. For the same reason, several leaves laid on top of each other can only be penetrated by long-wave red radiation.

Chlorophyll a is blue-green and chlorophyll b yellow-green. The difference is also reflected in the absorption spectrum: with chlorophyll b, both absorption maxima have clearly shifted into the green region; the red peak to shorter, the blue one to longer wavelengths.

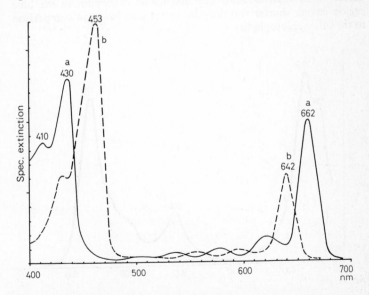

Fig. 21 Absorption spectra of chlorophyll a and chlorophyll b in ether (after Smith). The numbers indicate the position of the absorption maxima.

At the same time, a change in the height of the two absorption maxima is detectable; this varies according to the individual chlorophylls (cf. also p. 45 f).

It should be noted that the position of the two absorption maxima of a chlorophyll is not absolute. It depends on the properties of the solvent in which the measurements are carried out. With an increase in the refractive index of the solvent the absorption maxima will shift to longer wavelengths. The polarity of the solvent also plays a role.

The apparently small difference in chemical structure of the two chlorophylls thus influences the color, respectively the capacity to absorb visible radiation in a far-reaching way. Hence the optical properties of substances, which are structurally very similar, are often more suitable for identification purposes than the small differences in the molecular structure, the analysis of which often entails considerable methodical difficulties.

In bacteriochlorophyll, mainly the presence of additional hydrogen atoms is responsible for the significant shift of the red absorption maximum into the long-wave region of the spectrum (Fig. 22). Photoautotrophic bacteria are thereby enabled to absorb long-wave radiation which green plants are unable to utilize. This radiation is perceived by man as heat. The absorption maximum in the blue region covers shorter wavelengths and is also lower in comparison to the other chlorophylls.

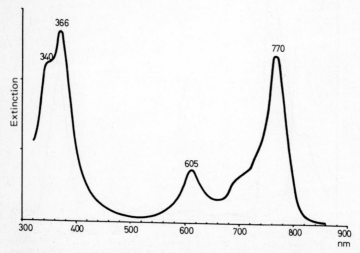

Fig. 22 Absorption spectrum of bacteriochlorophyll a in methanol (after Goedheer).

A further chlorophyll, isolated from *Rhodopseudomonas*, absorbs red radiation of still longer wavelengths. It is termed *bacteriochlorophyll b* in contrast to the typical green pigment of phototrophic bacteria, *bacteriochlorophyll a*. Other forms have been discovered in the green photobacterium *Chlorobium*, *bacteriochlorophyll c* and *d*, which according to the position of their red absorption bands are also termed *chlorobium chlorophyll-650* and *chlorobium chlorophyll-660*. They differ slightly from bacteriochlorophyll a: the long hydrocarbon side chain phytol is replaced by the alcohol farnesol (cf. p. 432); the additional hydrogen in positions 3 and 4 is absent, as is the –CO–CH$_3$ group of the isocyclic pentanone ring; in position 2 the acetyl group is replaced by the side chain –CHOH–CH$_3$.

When one compares the position of the absorption maxima for red and yellow (Fig. 23) of protochlorophyll, chlorophyll a and bacteriochlorophyll a, which differ in the number of hydrogen atoms attached to the pyrrole rings and in the form of the side chains, an

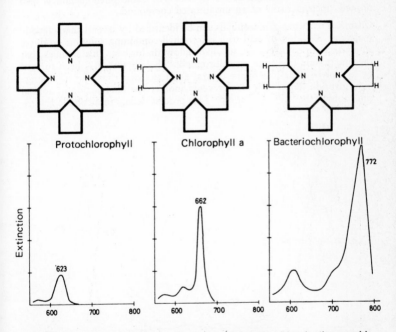

Fig. 23   The effect of increasing reduction of pyrrole rings in the porphin skeleton on position and size of the red absorption bands of individual chlorophylls (after Thomas). Protochlorophyll without additional H-atoms, chlorophyll a with two (ring IV!) and bacteriochlorophyll a with four H-atoms (ring IV and III!). The amount of each pigment in solution (ether) had been adjusted in such a way that the absorption in the blue region gave the same value for each.

interesting rule becomes obvious: not only does the height of the red absorption peak increase with the number of hydrogen-containing pyrrole rings, but its position also shifts to longer wavelengths. The absorption maximum in the yellow region, however, is largely unaffected by the chemical changes in the pyrrole rings; its height and position change only insignificantly.

In Fig. 23 the system of conjugate double bonds is not represented as a system of alternating single and double bonds, but by a heavy line. This signifies that porphyrin molecules are "resonance hybrids" which, strictly speaking, cannot be represented by a fixed alternating series of single and double bonds. An unsaturated compound may exist in various forms with different positions of the double bond(s). However, such a resonance hybrid does not constitute a simple mixture of two or several of these forms, but a definite structure in which the properties of the individual forms are combined. In chlorophyll, there are probably three forms contributing to the structure of the hybrid. The same phenomenon is also termed "mesomerism" of an unsaturated compound.

Chemical reactions. Chlorophylls can be identified by several characteristic reactions. Addition of acid to a chlorophyll-containing solution produces an immediate color change to olive-yellow or brown. This signifies that the magnesium-free form of chlorophyll, "pheophytin", has been formed by hydrolysis. The loss of the central atom causes a characteristic change in absorption (Fig. 24). It also prevents another specific reaction of normal chlorophyll, the "allomerization". This process occurs in alcoholic chloro-

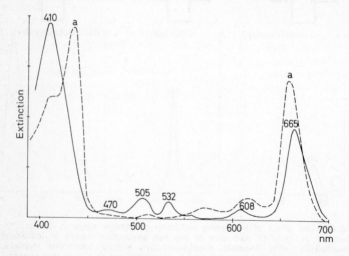

Fig. 24 Absorption spectrum of pheophytin a in ether (after Zscheile and Comar). The absorption of chlorophyll a in ether (broken line) is shown for comparison.

phyll solutions left standing in air and involves the attachment of molecular oxygen to the cyclopentanone ring:

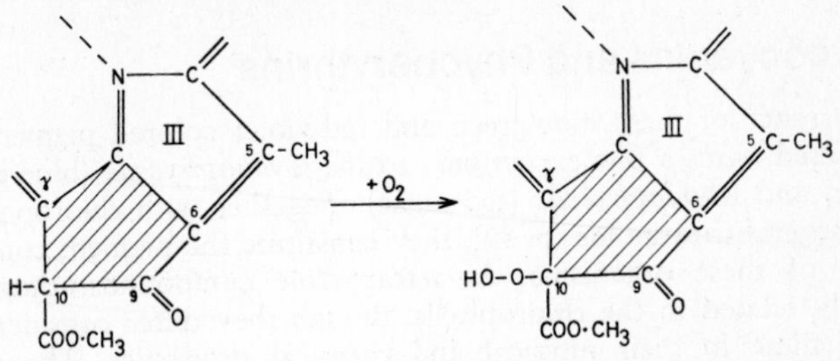

However, allomerized chlorophyll is not a homogeneous substance. On the contrary, at least three fractions of green-colored oxidation products develop which react chemically in a rather similar way despite differences in their structure. In each the crucial reaction has taken place at the cyclopentanone ring. This is indicated by the negative result of the "phase test" (discovered by Molisch in 1896): On layering a 30% methanolic KOH solution on top of a chlorophyll solution in ether a yellowish-brown ring forms at the interphase of the two solvents. A chemically unchanged cyclopentanone ring and a normal configuration at the carbon atom 10 of the chlorophyll or its derivative being tested are essential for a positive outcome of the phase test. After allomerization, neither requirement is met, so the test is negative. On the other hand, the characteristic colored ring appears with both pheophytin and protochlorophyll. The test is positive in the sense that the tested chlorophyll compound had not undergone an allomerization, nor was the hydrogen atom at the carbon atom 10 missing.

The brown color of the ring in the phase test is probably the result of an enolization of the carbonyl group on C-atom 9 involving an initial transfer of the hydrogen atom from position 10 to the oxygen atom at position 9:

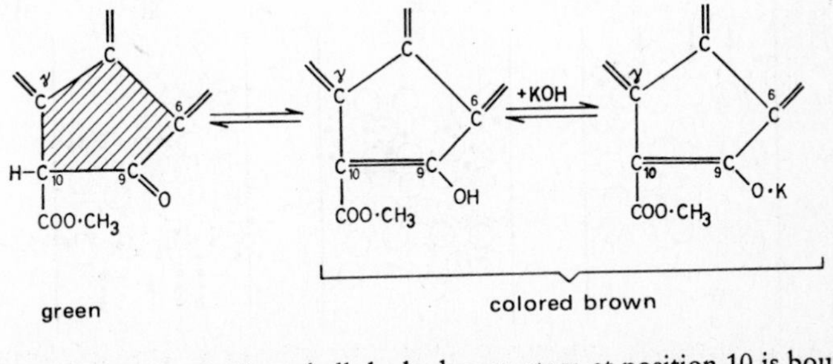

green   colored brown

Since in allomerized chlorophyll the hydrogen atom at position 10 is bound via oxygen, such an enolization cannot occur and the phase test is negative.

Unlike a series of other colored pigments, e.g. methylene blue, chlorophyll is not reduced to a reversible colorless ("leuko"-) form in the absence of oxygen.

Phycocyanins and Phycoerythrins

Occurrence of these blue-green and red-violet colored pigments is restricted (with a few exceptions) to the *Cyanophyceae* (blue-green algae) and *Rhodophyceae* (red algae). Together with chlorophyll a and several carotinoids (p. 59), they constitute the pigment complement of these organisms. As tetrapyrrole compounds, they are closely related to the chlorophylls, though they differ considerably from these in their physical and chemical properties. The most important difference is their solubility in water, which originates from their molecular constitution. They consist of a protein component of high molecular weight and of the "chromophore" (= "prosthetic") group responsible for their color. Because of the structural similarity of this group to the bile pigments, it is termed "phycocyanobilin" and "phycoerythrobilin", respectively. The water-soluble pigments of the blue-green and red algae are hence "chromoproteids" (cf. p. 382). The whole complex is generally termed "phycobiliproteid" or "biliproteid".

Because of their solubility in water, the phycobiliproteids cannot be extracted with organic solvents. Cells are therefore disrupted in an aqueous buffer solution, cell fragments removed by centrifugation of the homogenate and the intensely colored supernatant subjected to "gel filtration" in order to isolate the individual pigments.

In "gel filtration", a method of chromatography, molecules are fractionated according to their size. The mixture of substances in an aqueous solution is layered on top of a column consisting of swollen gel particles. These are dextrans, crosslinked linear polysaccharide macromolecules

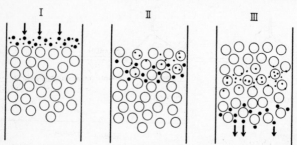

Fig. 25 The individual phases (I—III) during gel filtration of molecules differing in size.

forming a three-dimensional network (cf. p. 174 ff). The mixture of substances is then washed through the column with buffer solution. The resulting separation effect primarily depends on the pore size of the gel particles (Fig. 25): molecules smaller than the pores will enter the gel particles; with decreasing molecular size, the number of accessible pores increases and the rate of elution for these molecules will be retarded. Molecules greater in size than the largest pores cannot enter the gel particles, and therefore pass them by. Since these molecules migrate through the column very rapidly, they form the undermost zone and are the first to be eluted. Conversely, the smallest molecules leave the column as the last fraction. This kind of separation is referred to as a "molecular sieve". Since the pore size of the gel particles, the "exclusion boundary", is known, this separation technique is also very useful for determining the molecular weight of the compounds isolated.

Figure 26 shows the separation of an aqueous cell extract of the blue-green alga *Anacystis nidulans* by means of gel filtration. While the yellow carotinoids and chlorophyll a accumulate in the lower part of the column, the blue phycocyanin occupies its upper part (where the red-violet phycoerythrins will also appear when the extract comes from an organism containing both biliproteids; see below).

The presence of chlorophyll and carotinoids in an aqueous cell extract, as demonstrated by gel filtration, seems to contradict the insolubility of these pigments in water. However, in this case, free pigments are not involved; they rather form a water-soluble complex with protein. A part of the chloroplast pigments occurs in the same form (cf. p. 62).

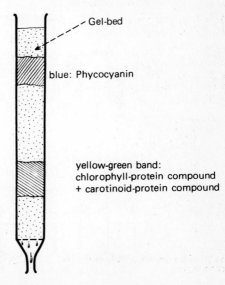

Gel-bed

blue: Phycocyanin

yellow-green band:
chlorophyll-protein compound
+ carotinoid-protein compound

Fig. 26 Gel filtration of an aqueous pigment extract from cells of the blue-green alga *Anacystis nidulans*.

The binding between the chromophore group and the protein in the phycobiliproteids is very stable – in contrast to the above-mentioned complexes formed with protein by chlorophylls and carotinoids. In the latter, the pigment component is readily removed with an organic solvent, but the phycobiliproteids must be subjected to hydrolysis in order to make the phycobilin soluble in organic solvents. Hydrolysis, however, may easily result in secondary chemical changes. Formerly, such artifacts were mistaken for natural components of the biliproteids. This and other methodical difficulties explain why the chemical structure of plant biliproteids is still not completely clarified.

The basic structure consists of four pyrrole rings linked by methine bridges but not forming a closed ring as in chlorophyll. Therefore no central magnesium atom is bound. To the carbon 1' and 8' positions of the pyrrole rings oxygen is attached. The positions 4 and 5 bear propionic acid residues.

Hydrolysis in the presence of methanol, short-term cleavage with HCl or proteolytic degradation (p. 396) set free a blue pigment from C-phycocyanin (below) which may be identical with the native phycocyanobilin. Elucidation of its structure is nearly complete (see below). With identical methods a phycoerythrobilin has been isolated from phycoerythrins which has the following structure:

Phycocyanobilin

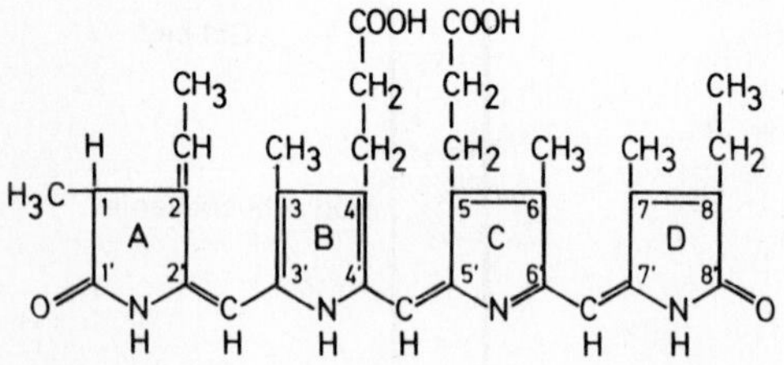

Phycoerythrobilin

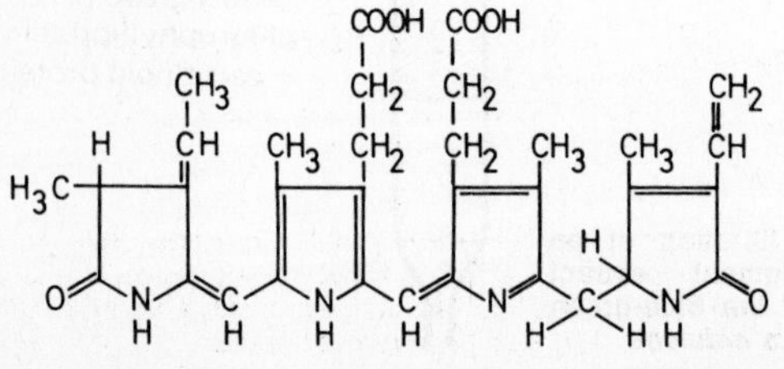

These chromophoric groups are probably linked to protein by the two propionic acid residues as well as by the oxygen function and the nitrogen of ring A.

Recent investigations into the molecular structure of phycocyanins and phycoerythrins indicate that these chromoproteids are composed of subunits. Phycocyanins from blue-green algae appear to contain 2 protein subunits with molecular weights between 14 000 and 17 000, each being attached to one molecule of phycocyanobilin. Together they form the basic unit which tends to aggregate in vitro depending on certain factors of the milieu (ionic strength, pH value, temperature). Thus molecular weights of more than 200 000 result. The same holds true for phycoerythrins whose basic unit has a molecular weight of about 40 000 and probably consists of two protein subunits. From these, high molecular weight aggregates (> 200 000) may arise in vitro. Whether one of these is identical with the native form is still unknown (see also p. 66).

Absorption of Radiation by Phycobiliproteids. The formulae show that the pyrrole rings and methine bridges constitute a system of conjugate double bonds comprising carbon and nitrogen atoms. This is responsible for the characteristic color of the phycobilins, i.e. for partial absorption of visible radiation. Phycocyanobilins absorb mainly the yellow and orange qualities, phycoerythrobilins the green ones.

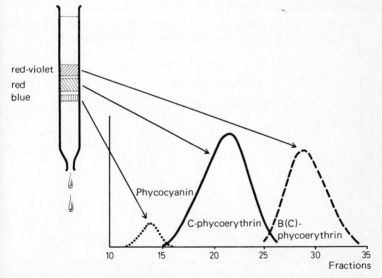

Fig. 27 Isolation of phycobiliproteids by gel filtration. The individual pigments having already accumulated in distinct zones on the column are then successively eluted from it. The elution diagram is obtained by measuring the specific pigment absorption in each fraction collected.

As in the case of chlorophyll, resonance hybrids or mesomeric systems (cf. p. 46) are present.

Phycobiliproteids can be identified by means of their optical properties after isolation by gel filtration (Fig. 27). Phycocyanins and phycoerythrins frequently occur together in the cell, though the quantitative ratio may vary considerably, as indicated by the coloring of the various organisms. Obviously the growth conditions also influence the synthesis and therefore the quantity of these pigments.

C-phycocyanin, the characteristic pigment of blue-green algae, has a distinct absorption maximum at 618 nm (Fig. 28). In contrast, R-phycocyanin absorbs maximally at 552 nm and 615 nm, respectively; this pigment also occurs in some species of the red algae. In allophycocyanin, which has been found in small amounts in many blue-green and red algae, the absorption maximum has shifted to the long-wave red region.

The letters "C" (for *Cyanophyceae*) and "R" (for *Rhodophyceae*) designate a particular type of pigment, rather than its origin.

R-phycoerythrin, the most important pigment of red algae, and B-phycoerythrin absorb between 540 nm and 570 nm (Fig. 28). R-phycocyanin occurs as well in several of these organisms.

Possession of phycobiliproteids enables blue-green and red algae to absorb those qualities of visible radiation which cannot be utilized

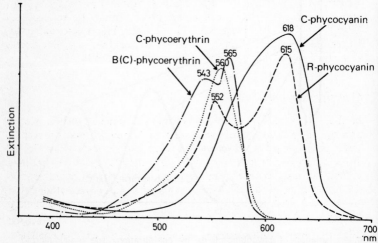

Fig. 28   The absorption spectra of phycocyanins (after ÓhEocha) and phycoerythrins (after Nultsch).

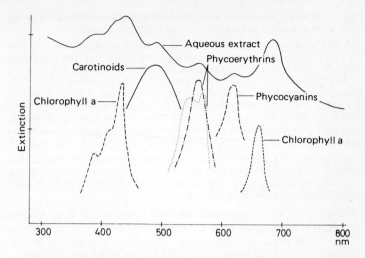

Fig. 29 Absorption spectrum of an aqueous extract from cells of the blue-green alga *Phormidium uncinatum* (after Nultsch). The pigments present are indicated by their main absorption maxima. For chlorophyll a a correction must be applied since the maximum shown is typical for an ether solution only; in aqueous extracts a shift to longer wavelengths will take place.

by green plants and green algae for lack of suitable pigments. The corresponding absorption gap in the green region is narrowed or largely closed in the blue-green and red algae. This is shown by the absorption spectrum of an aqueous cell extract from the blue-green alga *Phormidium uncinatum* (Fig. 29; see also Fig. 49).

Carotinoids

In the yellow to red colored carotinoid pigments, carbon atoms participate exclusively in the system of conjugate double bonds, since these compounds consist of a long hydrocarbon chain. They are hence insoluble in water but readily soluble in fat and fat-dissolving agents, thus behaving like the alcohol phytol bound in the chlorophyll molecule. They are subdivided into *carotenes,* poly-unsaturated hydrocarbons, and *xanthophylls,* oxygen-containing derivatives of carotenes.

The carotinoids involved in photosynthesis, the "primary carotinoids", are distinguished from the "secondary carotinoids" which occur in multiple forms, particularly in flowers and fruits as constituents of chromoplasts, but also in heterotrophic organisms such as bacteria, yeast and fungi. Secondary carotinoids may also develop in photosynthetically active organisms as a result of inadequate mineral salt nutrition (p. 289).

Another classification of the carotinoids is based on the observation that they are distributed unequally on shaking with two mutually immiscible organic solvents (e.g. petroleum ether and 90 %/o methanol). A few of the yellow pigments are "epiphasic" and remain in the petroleum ether, while others are "hypophasic", remaining in the methanol. The molecules of the latter have OH- and CHO-groups as oxygen-containing substituents which are absent in epiphasic pigments. However, the number of oxygen-containing groups (which generally determines the hydrophilicity of a substance) is in any case too small to render carotinoids soluble in water.

For the isolation of carotinoids, column chromatography (p. 33) and recently also thin-layer chromatography (p. 37) are used with great success. Figure 30 shows the separation of the carotinoids from isolated chloroplasts using the latter method.

The skeleton of a carotinoid molecule consists of forty carbon atoms. It is made up of eight building blocks of a chemical structure called "isoprene", an unsaturated C_5-hydrocarbon with a methyl group as side chain (formula p. 430). These units are arranged to

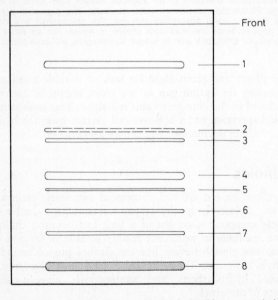

Fig. 30 Carotinoids of isolated chloroplasts (spinach) as revealed by thin-layer chromatography. Under the experimental conditions applied the chlorophylls remain at the origin. 1, β-carotene; 2 lutein-5,6-epoxide; 3, violaxanthin; 4, lutein; 5, antheraxanthin; 6, neoxanthin; 7, zeaxanthin; 8, chlorophyll a and b (after Hager and Meyer-Bertenrath).

form a chain-like central structure with fourteen carbon atoms, seven conjugate double bonds and four methyl groups as side chains (formula A). Attached at one or both ends there is an open or closed ring structure (B, C). With two terminal rings present, these may be identical or different from each other.

The formulae A–C show clearly that the linear arrangement of iso-prene units forms a relatively long system of conjugate double bonds. It usually consists of 11 or more double bonds in the skeleton and the end groups. This molecular construction implies properties with which we are already familiar from the chlorophylls and phycobilins: all carotinoids are intensely colored substances. They absorb in particular the blue and violet portions of visible radiation. The absorption spectra of the individual pigments accordingly show a maximum between 380 nm and 550 nm usually consisting of a main peak and two lower adjacent peaks (Fig. 31).

With increasing numbers of conjugate double bonds, the absorption maxima tend to shift to longer wavelengths, though their absolute position depends on other structural properties of the molecule and the type of solvent used.

For *rhodopin* and *spirilloxanthin,* the typical carotinoids of the "purple bacteria" (p. 132), the absorption maximum extends into the green region (Fig. 31). Since the red radiation as well as a part

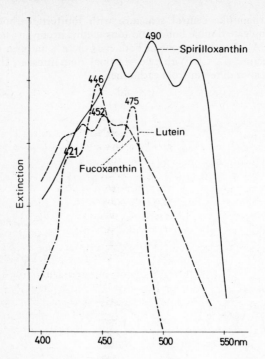

Fig. 31 Absorption spectra of lutein (in ethanol), and spirilloxanthin and fucoxanthin (in hexane).

of the blue radiation are absorbed by bacteriochlorophyll, a rose to purple color (name!) for these organisms results.

Fucoxanthin is responsible for the brown color of brown algae and diatoms. Since it is present in relatively large quantities, it masks the other main pigments – chlorophyll a, β-carotene, violaxanthin, as well as diatoxanthin and diadinoxanthin.

The absorption spectrum of a total pigment extract from leaves differs from that of the chlorophylls a and b only in that the absorption maximum in the blue region is broader; for this the carotinoids are responsible. They insignificantly narrow the "green gap" in chlorophyll absorption. Pigment extracts from green algae and *Euglena gracilis* absorb in much the same way.

Because of their double bonds, the carotinoids exhibit cis-trans isomerism (see formula). Naturally occurring carotinoids mainly show trans-configuration. Cis-isomers have only been reported in leaf extracts; they are given the prefix "neo".

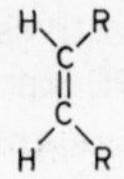

cis compound 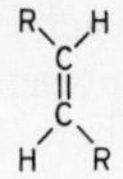   trans compound

Structure of the Primary Carotinoids. *Carotenes* are isomeric compounds of the formula $C_{40}H_{56}$. They differ in the terminal shape of their molecular structure. In the α-carotenes and β-carotenes both the ends consist of closed rings, in γ-carotene, however, only one end is cyclic ("ionone ring") thus eliminating the terminal double bond. In β-carotene the double bond of each ring is conjugated with the last one in the straight chain skeleton; in α-carotene this holds true for one ring only; in the other one the double bond has shifted beyond the next carbon atom. By this typical "α-ionone" structure α-carotene differs from β-carotene which has a "β-ionone" structure.

The structure of β-carotene (with conventional numbering of the carbon atoms) shows that it is a symmetrical and hence optically inactive molecule (see below) that can be cleaved to two identical parts. This symmetrical cleavage of β-carotene is physiologically important since it produces two molecules of vitamin A. β-carotene is provitamin A. In contrast, α-carotene is optically active because of an asymmetrically substituted carbon atom: the plane of polarized light is rotated to the right ("+") by passing through a solution of the pigment (method, p. 354).

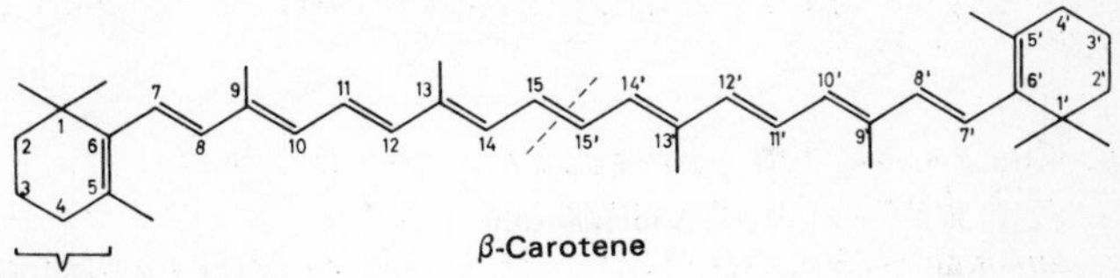

β-Carotene

β-Ionone structure

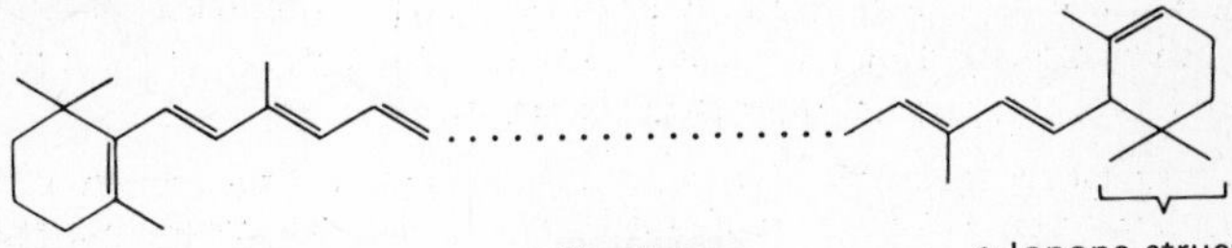

α-Carotene   α-Ionone structure

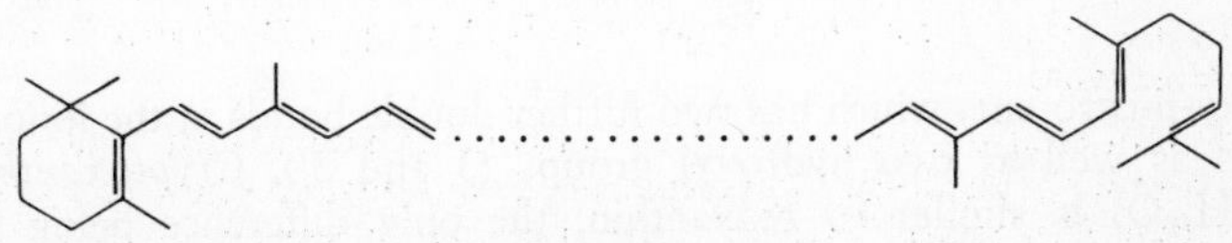

γ-Carotene

Xanthophylls are oxygen-containing derivatives of the carotenes. The oxygen may be present in a hydroxyl, keto, epoxy, carboxyl or methoxy group. There is consequently a greater variety in the chemical structure compared with the carotenes. Our treatment of the xanthophylls is restricted to those compounds which are active as "accessory pigments" in photosynthesis.

The "accessory pigments" comprise chlorophyll b, the carotinoids, phyco-cyanins and phycoerythrins. This traditional distinction is justified by the modern concept of photosynthesis whereby two light reactions occur (cf. p. 105 f) in which the pigments participate in different ways.

Lutein ($C_{40}H_{56}O_2$) is a derivative of α-carotene. Since each terminal ring carries an additional hydroxyl group, the compound is chemically 3,3'-dihydroxy-α-carotene. *Violaxanthin* ($C_{40}H_{56}O_4$) has closed ionone rings, each showing an epoxy group (5–6 and 5'–6') and a hydroxyl group (3 and 3'). All the double bonds present belong to the conjugate system. This also applies to *zeaxanthin* (3,3'-dihy-

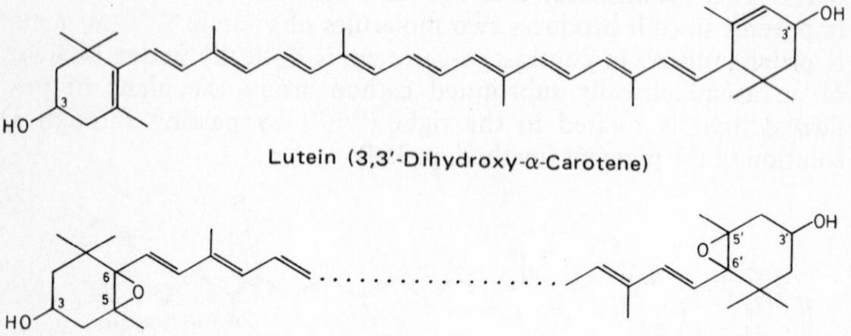

Lutein (3,3'-Dihydroxy-α-Carotene)

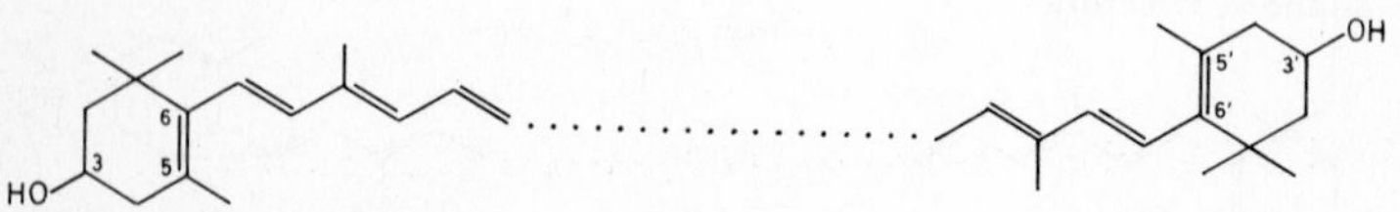

Violaxanthin

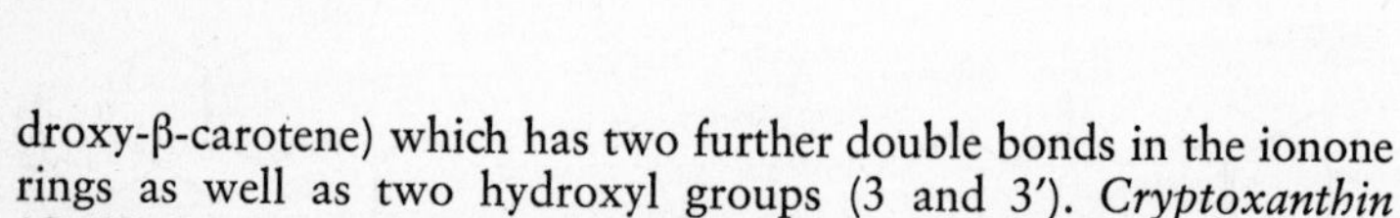

Zeaxanthin

droxy-β-carotene) which has two further double bonds in the ionone rings as well as two hydroxyl groups (3 and 3'). *Cryptoxanthin* ($C_{40}H_{56}O$) is similar to zeaxanthin, the only difference being its lack of the hydroxyl group in position 3'.

Neoxanthin ($C_{40}H_{56}O_4$) may be derived from α-carotene and has, in addition to two hydroxyl groups (3 and 3'), another one in position 6' as well as an epoxy group (5–6).

In *echinenone* ($C_{40}H_{54}O$; "myxoxanthin"), a specific carotinoid of blue-green algae, oxygen is present at position 4 in a keto group.

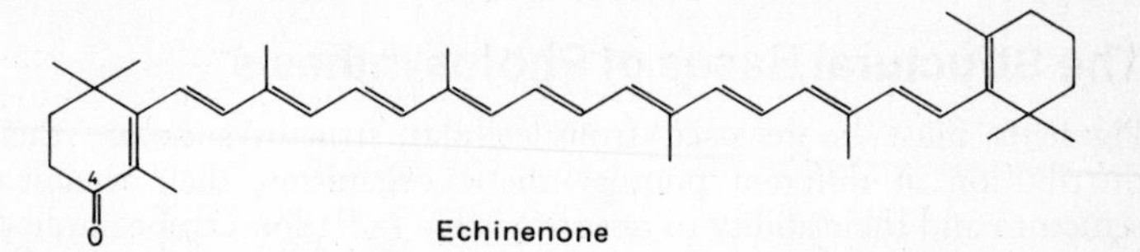

Echinenone

The structure of some other blue-green algae carotinoids, like *myxoxanthophyll*, *oscillaxanthin* and *aphanizophyll*, is unique insofar as they occur in the cells bound to various sugars as "carotinoid-glycosides" (cf. p. 167 f). *Myxoxanthophyll,* a derivative of γ-carotene, is attached to rhamnose (p. 186) or a hexose sugar:

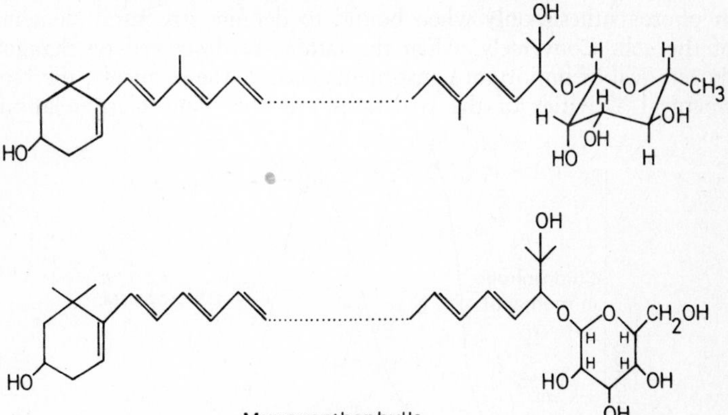

Myxoxanthophylls

The chemical structure of *fucoxanthin* has also been elucidated.

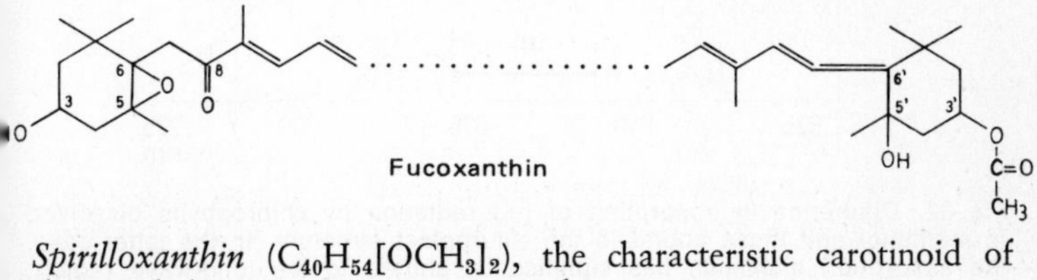

Fucoxanthin

Spirilloxanthin ($C_{40}H_{54}[OCH_3]_2$), the characteristic carotinoid of the purple bacteria, contains two methoxy groups:

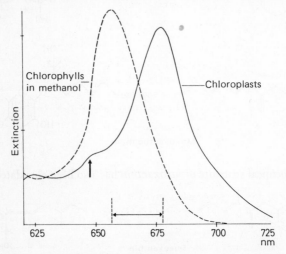

Spirilloxanthin

The Structural Bases of Photosynthesis

Pigments must be released from cellular structures before their distribution in different photosynthetic organisms, their chemical structure and their ability to absorb visible radiation can be studied. However, their physical state and their properties will inevitably change by their isolation and transfer into organic solvents; thus only limited information is to be expected regarding their distribution and properties in vivo. There are good reasons to believe that they cannot react in this isolated state as they do in photosynthesis. The chlorophylls do indeed show a few characteristic photochemical reactions (p. 102) when in solution, though they evidently function in photosynthesis only when bound to definite structural elements of the cell. Conversely, when the latter are destroyed or changed by removal of important constituents such as the pigments, the biochemical activities of the structures are lost. This close relation-

Fig. 32 Difference in absorption of red radiation by chlorophylls dissolved in methanol and those bound to the chloroplast structure. In the latter case, the absorption maximum has significantly shifted to the long-wave region; an additional maximum appears as a shoulder in the short wave-region (↑).

ship between structure and function, which we shall encounter very frequently, is fundamental to the complex reaction mechanisms of photosynthesis.

Two observations indicate a binding of pigments to cellular structures: 1. the difference in absorption of isolated and of cell-bound pigments, and 2. the occurrence of pigment-protein complexes.

As shown by comparison of the absorption spectra of isolated and of chloroplast-bound chlorophyll (Fig. 32), the red absorption maximum of the latter has significantly shifted to longer wavelengths; moreover, an additional shoulder appears in the short-wave region. Even more prominent is the shift of the absorption maximum in phototrophic bacteria (cf. p. 70).

When the absorption of intact photosynthetic organisms is compared with that of extracted chlorophyll a (Fig. 33), the results are

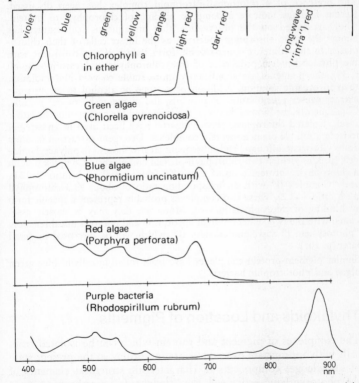

Fig. 33 Comparison of light absorption by chlorophyll a in solution and by various photosynthetic organisms.

of a similar kind. Moreover, the various absorption spectra show clearly how the individual groups of organisms differ in their ability to utilize visible radiation. This behavior, which is not without consequences on photosynthesis (cf. p. 85 ff), is brought about by their different pigment complement.

Since chlorophyll in vivo and in organic solution is chemically the same compound, the difference in absorption must result from the structural binding of the pigment in the cell. While chlorophyll in organic solution has a monomolecular distribution, it is apparently bound in a complex way in vivo; the nature of this binding is still unknown. It certainly does not involve simple aggregations of chlorophyll (colloidal aqueous solution; concentrates in lipids or crystals) since these exhibit either too small or too large a shift of the red absorption maximum.

The presence of a chlorophyll-carotinoid complex has been discussed. Recent findings indicate a complex binding of chlorophyll to protein. They are based on the observation that an aqueous leaf extract has almost the same absorption spectrum as the intact cells or their chloroplasts. In this complex water-soluble form (which is not consistent with the physicochemical properties of free chlorophylls and carotinoids; cf. p. 33), the pigments are significantly more stable to acid and radiation than in organic solution. Addition of acetone or alcohol to an aqueous extract causes precipitation of proteins, the pigments becoming soluble. Consequently, the binding between the latter and the protein must be rather weak. Isolated chloroplasts (Beta vulgaris) have been used in an attempt to isolate soluble pigment-protein complexes. Detergent extraction (sodium dodecylbenzenesulfonate) and subsequent electrophoresis in polyacrylamide gels (p. 336) yielded two chlorophyll-protein complexes: "Complex I" with a chlorophyll : protein ratio of about 8.8 (chlorophyll a : b ratio = 12.0) and "Complex II" with a corresponding ratio of about 13.3 chlorophyll a : b ratio = 1.2). These two complexes probably represent a specific form of binding of chlorophylls in vivo. Moreover, they may be closely associated with the two centers of light absorption and photochemical reaction, "photosystem I" and "photosystem II", which will be discussed in detail later (p. 105).

Similar pigment-protein complexes have been found in cells of blue-green algae and phototrophic bacteria.

Thylakoids and Location of Pigments

The complexes of pigment and protein which can be isolated from cells with appropriate procedures are (according to the present state of knowledge) components of characteristic structural elements of the photosynthetic active cell, the "thylakoids", membrane systems which occur either free in the cytoplasm (phototrophic bacteria and blue-green algae) or within chromatophores or chloroplasts (algae and higher plants). Although the morphology and the formation of thylakoids have been intensively studied, we do not know yet how the pigments are distributed on their membranes and how they are arranged in their molecular structure. Because of methodical difficulties, we depend largely on indirect findings and hypothetical models.

Phototrophic Bacteria and Blue-Green Algae. The members of these two groups of organisms have no plastids although their cells contain photosynthetic pigments. The idea that these pigments may be present in the cytoplasm without binding to specific structures had to be revised since membrane structures have been found in these cells by means of electron microscopy. Their morphology identifies these structures as thylakoids. In contrast to chloroplasts and algal chromatophores, they are not surrounded by a membrane, but lie free in the cytoplasm.

First evidence for the association of photosynthetic pigments with cytoplasmic structures in blue-green algae came from microphotography of living cells using red radiation (600 nm). This technique produces a good contrast because of the specific absorption by chlorophyll. In some cases, however, the contrast became rather distinctive within restricted regions of the chromatoplasm (see below). Accordingly, characteristic distribution patterns for various blue-green algae resulted. In recent experiments these patterns could easily be traced back to the membrane systems of thylakoids in the chromatoplasm. From this observation a preferential if not exclusive localization of chlorophyll a on the thylakoids has been postulated.

In phototrophic bacteria, the thylakoids are either bubble-shaped "vesicles" (ϕ approx. 500 Å) or flattened double membranes. The first type was found in the green sulfur bacterium *Chlorobium (Chlorobacteriaceae)*[*], in the purple sulfur bacteria *Chromatium* and *Thiospirillum (Thiorhodaceae)* and the purple non-sulfur bacteria *Rhodospirillum* and *Rhodopseudomonas (Athiorhodaceae)*. In the state of active photosynthesis the vesicles occupy mostly the peripheral cytoplasmic region of the cell (Fig. 34). The inner space of the vesicles often communicates with that between cell wall and cytoplasmic membrane; this connection originates from the mode of formation of the thylakoids (see below). The attachment of the individual vesicles to a common membrane structure has not been demonstrated so far.

The second type of thylakoid, obviously formed by extreme flattening of the vesicle, was discovered in *Rhodospirillum molischianum*

---

* for classification of phototrophic bacteria cf. p. 132.

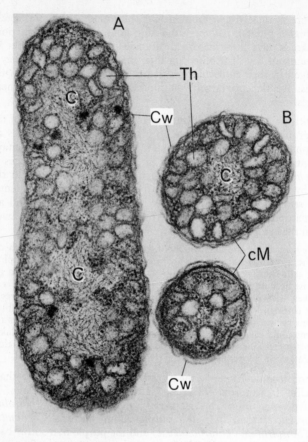

Fig. 34 Vesicular thylakoids in the cells of *Rhodospirillum rubrum*. Electron micrographs of ultrathin sections. A, longitudinal section; B, cross-section; C, nuclear equivalent; cM, cytoplasmic membrane; Th, thylakoids; Cw, cell wall. (Magnification 50 000 x; by courtesy of Dr. R. Marx).

and *Rhodomicrobium vannielii (Athiorhodaceae)*. Generally, the cells contain stacked double membranes partly attached to the cytoplasmic membrane (Fig. 35). The longitudinal axis of the thylakoids runs parallel to that of the cell.

The thylakoids of the blue-green algae are very similar to this second type. In a cross-section through a trichome of *Oscillatoria rubescens,* they are found as stacks near the cell wall in a radial alignment. The double membranes with a distinct inner space show a thickness of about 300 Å. The distance between them is about 1000 Å.

In a longitudinal section, the predominantly axial arrangement and extension of the thylakoids between the end walls (2–4 μ in length) is apparent. In *Oscillatoria chalybea*, they extend parallel to the cell surface; accordingly a cross-section shows a series of double membranes in a concentric arrangement within the cytoplasm adjacent to the cell wall. This is typical for *Anacystis nidulans*, with an average of thirteen concentric double membranes embedded in the

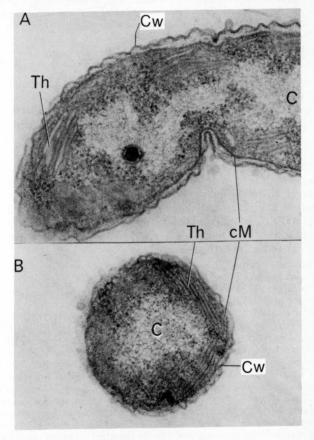

Fig. 35 Lamellar thylakoids in the cells of *Rhodospirillum molischianum*. Arrangement in stacks at the periphery of the cells. A, part of a longitudinal section; B, cross-section. Abbreviations as in Fig. 34.

chromatoplasm (light-microscopic investigations in blue-green algae have revealed that a colorless "centroplasm" can frequently be distinguished from a pigment-bearing "chromatoplasm" in which the thylakoids are localized). A less regular arrangement of thylakoids has been observed in *Nostoc* and *Anabaena*.

Results of recent investigations indicate that phycocyanins and phycoerythrins form polyaggregates (cf. p. 51), the "phycobilisomes". These rod-shaped, round or disc-like particles are linked to the outer surface of the thylakoids (p. 74 ff). Their shape probably varies with the prevailing chromoproteid. Phycobilisomes have been found in several species of blue-green algae, e.g. *Synechococcus lividus,* and obviously consist of C-phycocyanin; it is still unknown whether they also contain C-phycoerythrin or allophycocyanin in other species.

Phycobilisomes also occur in the plastids of *Rhodophyceae* (= "rhodoplasts") where they contain mostly R-phycoerythrin and B-phycoerythrin.

The arrangement of thylakoids described previously is certainly not strictly species-specific and depends on the physiological state of the cell as well. It is known, for instance, that the thylakoids of blue-green algae disappear with aging of the trichome cells. After their disintegration, thylakoids must be formed de-novo in the hormogonia.

The fine structure of the thylakoids in phototrophic bacteria and blue-green algae as well as in chloroplasts can be well observed by means of a new technique of preparation termed "freeze etching": Linked to the inner and outer surface of a thylakoid membrane there are almost spherical structures (p. 75 ff). Their function in the imbedding of pigments into the thylakoid membrane must be shown by further investigations.

The procedure termed "freeze-etching" (see diagram, Fig. 36) is a recently developed method by which, in combination with electron microscopy, it is possible to obtain a three-dimensional display of submicroscopic structures of the living cell. Instead of chemical fixation, rapid cooling of the cells to -190 °C (with liquid propane) is used to preserve their structures. Sectioning with the ultramicrotome is carried out at about -100 °C in such a way that the surface of the deep-frozen object is gently planed; thus the different layers splinter off along native interfaces. In favorable locations, the internal fine structures of the cell are laid bare in relief by these fracture planes. Details, however, are only revealed by the subsequent actual freeze etching. By controlled freeze drying, the masking film of ice is carefully sublimed off the surface of the section to a depth of about 100 Å without deforming the structures. In high vacuum a platinum-carbon layer 20–30 Å thick is vapor-deposited onto the structured surface. The resulting cast can be detached relatively easily, freed from cell residues and examined under the electron microscope. In this way one obtains a reproduction of a three-dimensional replica comprising all the fine structures of the fractured surface.

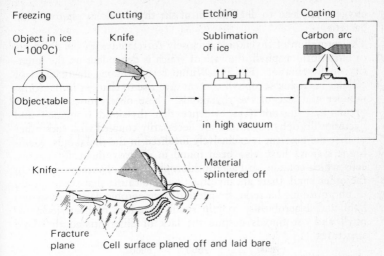

Fig. 36 Sequence of operations in freeze-etching and deep-etching techniques (after Moor). For details see text.

Formation of Thylakoids. Major conformity between phototrophic bacteria and blue-green algae also exists in respect of the formation of thylakoids. They originate from protrusions ("invaginations") of the cytoplasmic membrane into the cytoplasm, the outer surface becoming the inner surface of the thylakoid formed (Fig. 37). These structures enlarge and fill mostly the peripheral regions of the cytoplasm along the end and longitudinal walls of the cell. Aging of the cells, and the influence of external factors, particularly light and

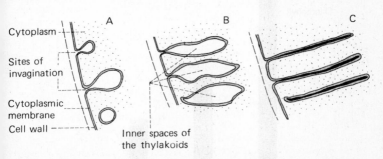

Fig. 37 Thylakoid formation (schematic) in *Rhodospirillum molischianum* (after Hickman and Frenkel). A, invagination of the cytoplasmic membrane into the cytoplasm; B, enlargement by formation of intrathylakoid spaces; C, flattening of thylakoids with elimination of the inner spaces.

oxygen, can lead to degradation of the thylakoids in phototrophic bacteria.

The formation of thylakoids is closely correlated with the synthesis of bacteriochlorophyll, the rate of which is dependent on the intensity of illumination (Fig. 38). Within certain limits, fluctuations of cellular bacteriochlorophyll content give rise to changes in the form, number and size of the thylakoids. These observations must not prematurely be taken as evidence that thylakoid formation and bacteriochlorophyll synthesis are necessarily coupled with each other or that the pigment is exclusively localized in the thylakoids. Membrane systems have also been found in photosynthetically inactive cells which contain no bacteriochlorophyll or only traces of it. On the other hand, there are investigations indicating that some species of phototrophic bacteria, transferred from aerobic dark culture into light and anaerobiosis, contain small quantities of bacteriochlorophyll and carotinoids despite the lack of differentiated thylakoid structures.

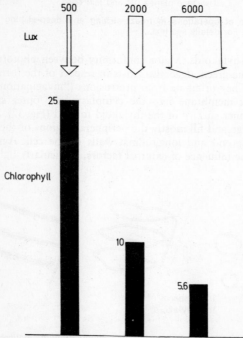

Fig. 38   Effect of light intensity on the bacteriochlorophyll content (μg/mg cell protein) in *Rhodospirillum rubrum* (after Cohen-Bazire).

Since the thylakoids of the *Athiorhodaceae* are obviously not homogenous in origin and function, they have recently been subdivided by several authors into "intracytoplasmic membranes" (ICM) and "cytoplasmic membranes" (CM). Because of their mode of formation, both together constitute a coherent system. The ICM are believed to contain mainly the photosynthetic structures (p. 134), the CM mainly the enzymes of the respiratory chain (p. 235 ff). On transfer from photosynthesis to aerobic dark metabolism, the ICM partly disintegrate and partly acquire other functions by incorporating enzymes of the respiratory chain.

Further findings concerning the problem of the cellular localization of pigments in phototrophic bacteria were obtained from experiments designed to isolate photosynthetically active structures. In these experiments, the cells are gently disrupted and individual particle fractions separated from the homogenate by fractional centrifugation. The pigment content of these fractions is determined and related to the protein moiety. In phototrophic bacteria, particles with the relatively highest pigment content were termed "chromatophores" or "chromatophore material". These terms are unfortunate misnomers because they are also applied to thylakoid structures in the intact cell, although the relationship between these pigment-containing particles and the typical in-vivo structures of the thylakoids is still obscure (see below). On the other hand, the biochemical capacities of the particles, especially in connection with photosynthesis, have been well studied.

"Chromatophores". From cells of *Rhodospirillum rubrum,* pigment-containing particles have been isolated which were largely free from cell-wall material after centrifugation in a sucrose gradient*. It is present in considerable amounts in crude particle preparations. The homogeneity of these "chromatophores" has been established by the electronmicrograph: They are mainly round, disc- or saucer-shaped structures which in ultra-thin sections still clearly show a surrounding membrane but no inner fine structure. The diameter is about 600 Å; hence their size corresponds to that of the vesicular thylakoids, which measure 500 Å to 1000 Å in *Rhodospirillum rubrum.*

A stable and relatively homogeneous fraction of chromatophores with a diameter of about 300 Å and a particle weight of about 15 millions was isolated from cells of *Chromatium* ("strain D").

Experiments of the same kind with *Rhodospirillum molischianum* illustrate that conformity of size alone does not prove the identity of thylakoids and chromatophores. Although the flattened type of thylakoids (p. 63 f) is present in vivo, the isolated chromatophores have the same form as those isolated from *Rhodospirillum rubrum,* i.e. roundish discs with a diameter of 1200 Å to 1500 Å. However, it is not only in this organism

---

* The technique of "gradient centrifugation" will be thoroughly discussed in another context (p. 94 f; "chloroplast isolation").

that the chromatophores have no counterpart in the intact cell. In the green sulfur bacterium *Chlorobium,* no vesicular or flattened thylakoids can be demonstrated in ultrathin sections. Nevertheless, pigment-containing particles have been isolated from the cell homogenate. They have a diameter of 100 Å and a particle weight of about 1.5 million. Highly purified preparations are free from nucleic acids and are homogeneous in the ultracentrifuge (p. 335 f).

There are a number of indications that chromatophores are probably artifacts arising from the disruption of thylakoid membranes. Cytoplasmic membranes of normal bacteria give rise to particles of various size in a similar way when the cells are disrupted by different methods (ultrasonics!). It is true that a chromatophore fraction is far more homogeneous than those membrane particles. The reason for this discrepancy may be the more regular disruption of the pigment-containing structures following only preformed sites, resulting in more homogeneous fragments compared to those from the rather randomly occurring disruption of the cytoplasmic membranes.

The absorption spectrum of a suspension of purified chromatophores from *Rhodospirillum rubrum* shows several absorption maxima; a high one in the long-wave red region at 880 nm, a lower one at 800 nm as well as two further peaks at 588 and 375 in the short-wave region. These are characteristic of the in vivo absorption of bacteriochlorophyll a. A peak at 504 nm with shoulders at 545 and 470 nm is brought about by the carotenoids. Several absorption bands in the long-wave red region are also exhibited by chromatophore preparations from *Rhodopseudomonas spheroides* (Fig. 39). The main maximum is at 850 nm with a shoulder at 880 nm; an adjoining peak appears at 800 nm.

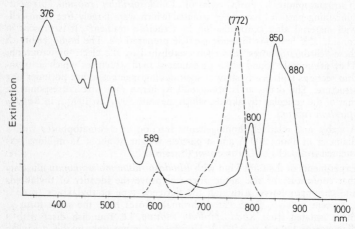

Fig. 39 Absorption spectrum of a chromatophore preparation from *Rhodopseudomonas spheroides* (after Crounse, Sistrom and Nemser). For comparison the red absorption of extracted bacteriochlorophyll (in methanol; broken line) is shown.

These peaks in the infrared region of the absorption spectrum probably depend upon the presence of various complexes which bacteriochlorophyll forms in vivo, particularly with proteins and carotinoids. Analogous complexes of pigments and protein in higher plants have already been discussed (p. 62).

Green Algae and Higher Plants. It has been known for a long time that the chromatophores of algae and the chloroplasts of higher plants are the organelles of photosynthesis. The photosynthetically active pigments are concentrated in them and bound to specific structures.

These structures were detected with the microscope soon after the discovery of chloroplasts. It was observed that the latter are not homogeneously green in color, but contain darker inclusions which were termed "grana". Definite proof of a fine structure was, however, only attained with the aid of the electron microscope when the technique of preparing ultrathin sections of cells after suitable fixation had been improved. It was found that in chloroplasts the biologically important principle of lamellar microstructure is materialized, which is also typical for mitochondria (cf. p. 203 f).

In algal chromatophores and in chloroplasts, the thylakoids are components of a system of higher structural organization. These organelles are independent reaction spaces, separated from the cytoplasm by a double membrane, and thus contributing to cellular "compartmentation". The inner cavity is filled with a semi-fluid material termed "stroma" or "matrix" in which the thylakoids are embedded. The arrangement of the thylakoids in the chromatophores of algae and green flagellates differs from that in chloroplasts of higher plants. The former organelles are "homogenous" in structure: the thylakoids extend quite regularly throughout their whole length, thus forming a rather homogeneous lamellar structure. They are often closely appressed against each other, so that single stacks of double membranes are formed. The pyrenoids are also interlaced by thylakoids although their number is reduced in these areas. Similar observations have been made with the green flagellate *Euglena gracilis*.

In contrast, the chloroplasts of higher plants belong to the so-called "grana type". Here a regular and dense parallel arrangement only exists in the restricted areas of the grana; in contrast, the thylakoids of the stroma are loosely stacked (Fig. 40).

The chloroplasts of the bundle-sheet cells in leaves of the so-called "C_4 plants" (p. 155 f) are an exception; they lack the typical grana, probably due to their spezialized function.

Recent studies indicate that the grana structure may result from the piling up of disc-shaped membranes one upon another ("overlapping"; Fig. 41).

The diameter of the grana is thought to be subject to certain fluctuations depending on various conditions for the plant concerned. The grana thylakoids are linked to those of the inter-grana region, from which they probably arise by evagination. This results in bridge-like connections so

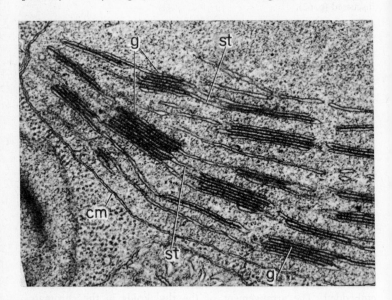

Fig. 40 Arrangement of thylakoids in a chloroplast. Electron micrograph of a thin section through a spinach chloroplast showing a sector. Fixation: glutaraldehyde-OsO₄; magnification: 49 000 (by courtesy of Dr. W. Wehrmeyer). cm, chloroplast membrane; g, thylakoids of the grana region; st, stroma thylakoids.

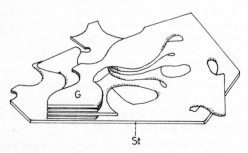

Fig. 41 Three-dimensional model of thylakoid arrangement in the chloroplast (after Wehrmeyer). Two stacks of grana thylakoids are shown of which the one in the foreground (G) has been fractured. They are connected by the stroma thylakoids (St).

that the surface of the thylakoids between the grana appears perforated. This impression is intensified by certain fixation procedures.

In this three-dimensional model of a complex membrane-bound entity, a thylakoid contributes to the construction of several grana, and also connects them. It follows that all the inner spaces of this common structure are probably linked with each other. There is another hypothetical model based on the premise that grana thylakoids originate and develop independently from stroma thylakoids.

Thylakoids are formed by invaginations of the inner membrane which penetrate deeply into the matrix of the plastid. Even in the mature state, a connection may persist between the original membrane and the thylakoids. By extreme flattening of the vesicles formed, the image of a double membrane is given in electron micrographs of ultrathin sections. Differences in the mode of formation and in the arrangement of the thylakoids are responsible for the varied fine structure of chromatophores and chloroplasts discussed above.

Arrangement of the Pigments. In the chromatophores of algae and green flagellates, pigments are probably distributed over the whole surface of the thylakoid membrane. Whether their integration takes place during the development of the thylakoids in the organelles, or afterwards, is an unsolved problem.

On the other hand, in the chloroplasts of higher plants the pigments are thought to be located exclusively in the grana thylakoids. However, this conclusion is based solely on the results of previous investigations of chloroplasts in vivo by fluorescent microscopy and does not unequivocally exclude the location of pigments on the stroma thylakoids. Their molecules may be bound differently from those in the grana and consequently may not exhibit any fluorescence (p. 101 f).

Like in phototrophic bacteria, algae and green flagellates, the role of chlorophyll synthesis in the differentiation of chloroplast microstructure is of considerable interest in this context. Since both processes are difficult to analyze in the normal higher plant, recent studies have concentrated on leaf-pigment mutants the gen-controlled abnormal chloroplast development of which has been investigated in detail. Thus in the ch_8-mutant of *Arabidopsis thaliana,* an inhibition of granum differentiation is manifest, which leads to a 10% reduction in pigment content. In electron micrographs, chloroplasts in the palisade cells of mature leaves of these mutant plants show significant deviations from the normal structure of grana. These may even be completely absent. Apart from the question whether a largely reduced chlorophyll synthesis is the actual cause of incomplete or missing differentiation of the grana, the close correlation between the two processes becomes obvious. As to the thylakoids of the stroma, those appear to be much less affected by the marked reduction in pigment content. They extend throughout the whole length of the chloroplast and are of normal shape. They may contain traces of chlorophyll, as indicated by the positive fluorescence observed in grana-free chloroplasts during blue light irradiation; the emission obviously occurs diffusely from the whole plastid surface. The presence of starch granules is strong evidence that photosynthesis still proceeds in such pigment-deficient chloroplasts.

With an experimental approach similar to the one applied to phototrophic bacteria, pigment location and thylakoid ultrastructure have recently been studied in isolated chloroplasts and fragment fractions (for isolation technique, see p. 92 ff). Despite certain discrepancies, the findings obtained are compatible with the biochemical properties of these isolated structures. Their relation to intermediate processes of photosynthesis are discussed later (p. 106, 124).

The aim of these investigations was first of all to determine the dimensions of the smallest fragment of a pigment-bearing thylakoid in which the characteristic light reactions still proceed. Later on, studies centered on the fundamental problem of the molecular structure of pigment-containing membranes.

On the basis of calculations and experiments with intact cells, the "photosynthetic unit" was thought to contain 250–300 chlorophyll molecules. By definition, this is a physiological unit. Is it perhaps identical with a native thylakoid subunit? The answer to this question should come from studies on the ultrastructure of thylakoid membranes isolated from chloroplasts.

Models of Thylakoid Structure

Different methods, such as electron microscopy including freeze-etching techniques (p. 66 f), X-ray diffraction, polarization optics and immunological techniques have recently been used to study the substructure of thylakoid membranes. Though there is general agreement that these biomembranes represent a "functional membrane", the substructure of which depends strongly on the specific function (cf. p. 267 ff), controversies exist regarding the suitability of the methods applied and regarding the interpretation of the results obtained.

More recent evidence, primarily obtained by freeze-etching techniques, favors a modified model of the originally described three-dimensional arrangement of the thylakoid membrane. It is based on the finding that the fracture plane formed during planing off of the deep-frozen material does not run along the interface between the membrane surface and the surrounding medium (as shown in Fig. 36) but runs within the membrane along the hydrophobic region of the central lipid layer. Consequently, the position of the fracture plane gives rise to two typical fracture faces which will often match. These conclusions are based on the following observations: 1. The morphology of the fracture faces differed greatly from the structure of the adjacent membrane surface; 2. after extraction of the lipids no fracture occurred along internal regions of the membrane; 3. in model experiments with membrane systems rich in lipids the fracture plane always ran along the hydrophobic region.

In accordance with these new findings two major fracture faces are observed which differ greatly in morphology from the external thylakoid surface and often match completely, i.e. they are complementary: small and large particles on the two surfaces interpenetrate each other. A hypothetical model consistent with this view has been developed (after Branton, Park and Pfeifhofer as well as Arntzen and co-workers; Fig. 42, B). The outer surface "I" has a bumpy surface relief which is brought about by numerous larger particles obviously anchored to the central region. In addition other particles are often loosely attached to the outer surface of untreated thylakoid membranes; these particles can be removed by washing with EDTA (= ethylenediaminetetraacetate, a chelating agent) and are considered to function as "coupling factors" (see below) in the process of light-dependent ATP formation in chloroplasts (= "photophosphorylation"; p. 125). The fracture plane running along internal regions of the membrane reveals two major faces: "II", carrying numerous smaller particles (110–120 Å in diameter) and representing a layer adjacent to the outer membrane surface; this layer probably contributes largely to the membrane mass – in contrast to a second layer characterized by surface "III": larger particles (about 175 Å in diameter) are thought to be embedded within this matrix which on the one hand penetrate the layer of smaller particles (surface II accordingly shows a matching pattern with particle-free spaces!), on the other hand protrude distinctly through the inner thylakoid surface into the intrathylakoid space, due to their height (about 90 Å). The occurrence of two classes of particles, as described here, probably holds only for thylakoid membranes within the grana, i.e. where their external surfaces are closely appressed to each other (= "partition region"; Fig. 42, A). In contrast, the membranes of the stroma thylakoids and end membranes of grana stacks are considered to contain only the smaller particles of about 120 Å diameter.

Another model (after Kirk) which attempts to reconcile the data from electron microscopy and those obtained by X-ray diffraction (Fig. 42, C) is based on the assumption that both the particles of about 175 and 120 Å in diameter are embedded within a continuous lipid matrix of the classical bilayer type. Since the thylakoid membrane itself is only 75–90 Å thick it is inevitable that the particles located internally protrude or penetrate the thylakoid surface – provided they are arranged in different layers (Fig. 42, B). It seems certain that these layers are not strictly separated, but interpenetrate each other. The smaller particles appear to occupy positions nearer the outer membrane surface, the larger particles positions nearer the inner surface, obviously penetrating through it, i.e. into the intrathylakoid space. Whether the larger particles correspond to

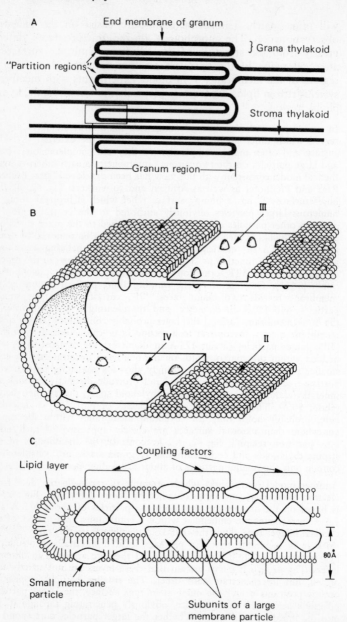

A

End membrane of granum

} Grana thylakoid

"Partition regions"

Stroma thylakoid

Granum region

B

I

III

IV

II

C

Coupling factors

Lipid layer

Small membrane particle

80Å

Subunits of a large membrane particle

the "quantasomes" described earlier in thylakoid membrane preparations with vapor-deposited metal coating, is uncertain; these roundish structures, 160 × 180 Å in size, had been assumed to represent the structural equivalent of the "photosynthetic unit" (p. 74).

The arrangement of the various particles as shown in Fig. 42 C agrees fairly well with the surface reliefs obtained on fracturing thylakoid membranes along their hydrophobic interior (cf. Fig. 42, B). In addition to the particles firmly integrated within the membrane, there are others which are loosely attached to the outer surface (about 100 Å in diameter); they may have a subunit structure. Presumably, they correspond to the "coupling factors" mentioned above.

The elucidation of the thylakoid substructure has not led to clarification of one of the most interesting problems: the position of the photosynthetically active pigments in this membrane structure. Evidence cited previously indicates that chlorophyll and carotinoid molecules may occur loosely bound to proteins as well as to lipids. In each of the models suggested the arrangement of chlorophyll molecules in a protein-lipid-protein membrane is governed by their phycical properties: the hydrophilic Mg-porphyrin ring system will associate with the hydrophilic protein layer, the hydrophobic phytol tail, however, with the hydrophobic lipid layer; the latter will also contain the strictly hydrophobic carotinoids. The porphyrin planes (15 × 15 Å) are thought to be tilted with respect to the membrane plane. However, the existence of such regular monomolecular films of pigment molecules is incompatible with the observed physical properties of intact chloroplasts and chromatophores during light absorption: the expected effects of difluorescence and dichroism did not occur. Moreover, there seems to be no space available for a separate chlorophyll layer in the two membrane models described. Therefore, it appears more plausible to assume that the bulk of the pigments is associated with protein, particularly since chlorophyll-protein complexes have been found in photosynthetically active structures (cf. p. 62). Accordingly, the original assumption of a uniform arrangement of chlorophyll molecules has been discarded in favor of other models which take into consideration the specific assembling of pigment molecules in two distinct "photosystems"

Fig. 42 Hypothetical models of thylakoid membrane structure. A: Arrangement of the grana and stroma thylakoids; B: Three-dimensional diagram of a thylakoid membrane section showing the various fracture faces I—IV and the particles associated with these (cf. text; after Branton, Park and Pfeifhofer; Arntzen and coworkers). C: Model based on the results of electron microscopy and X-ray diffraction (after Kirk). In a continuous lipid bilayer numerous smaller and larger particles are embedded primarily consisting of proteins. The larger ones are thought to consist of four subunits; only two are shown in the diagram. To the outer surface particles of another kind, probably "coupling factors", are attached.

(p. 105 ff). We shall discuss this problem further when we examine these elementary functional units of photosynthesis in detail.

References

Determann, H.: Gelchromatographie. Springer, Berlin 1967

Kirk, J. T. O.: Chloroplast structure and biogenesis. Ann. Rev. Biochem. 40: 161, 1971

Kreutz, W.: Über die Tertiärstruktur des Proteins der Chloroplastenlamellen. Ber. dtsch. botan. Ges. 79: 34, 1966

Linskens, H. F.: Papierchromatographie in der Botanik. Springer, Berlin 1959

Menke, W.: Feinbau und Entwicklung der Plastiden. Ber. dtsch. botan. Ges. 77: 340, 1964

Moor, H.: Beitrag der Gefrierätzmethode zur Aufklärung von Struktur und Funktion der Biomembranen. Ber. dtsch. botan. Ges. 82: 385, 1969

Mühlethaler, K.: Der Feinbau des Photosynthese-Apparates. Umschau 66: 659, 1966

Park, R. B., P. V. Sane: Distribution of function and structure in chloroplast lamellae. Ann. Rev. Plant Physiol. 22: 395, 1971

Sitte, P.: Bau und Feinbau der Pflanzenzelle. Fischer, Jena 1965

Stahl, E.: Dünnschichtchromatographie, 2nd Ed. Springer, Berlin 1967

Wehrmeyer, W.: Morphologie und Morphogenese der Plastiden (Chloroplasten). In: Probleme der biologischen Reduplikation. Springer, Berlin 1966

Studies of the Photosynthetically Active Plant

In the beginning, practical considerations dominated experimental analysis of the highly complex photosynthetic process. Investigations concentrated particularly on the influence of those external factors such as light, temperature and CO_2 supply which significantly affect the production of substances and hence the yield of crop plants. These investigations quite incidentally provided several important results which not only contributed significantly to the understanding of photosynthesis as a fundamental biological process but also served as starting points for further molecular exploration of those reactions which are methodically difficult to get at. We shall discuss the most important findings briefly.

Carbon Dioxide

As a starting substance of photosynthesis, the amount of CO_2 has a significant influence on its output. Of course, a few intermediary reactions of photosynthesis may proceed in its absence, but no synthesis of carbohydrates will take place. The concentration of CO_2 in the air (0.03 %) is not optimal for photosynthesis. When the CO_2 concentration in the gas phase is raised there is an increase of photosynthetic activity up to a limit, the value of which varies with the plant material used. Consequently, in horticultural practice artificial gasing with CO_2 has been applied under constant illumination thus increasing the rate of photosynthesis and hence the yield. The extent to which this procedure is successful with a particular plant species depends on its requirement for light and temperature in the changed conditions of a higher CO_2 supply (see below).

Water

Water is not only a starting substance in photosynthesis, but also participates as a reactant in its transformations and in numerous other metabolic reactions. Because of the enormous importance of water in many vital processes in plants, its uptake, transport and release are treated thoroughly in a separate chapter (see p. 262).

Light

Without light, there is no photosynthesis. As a typical light-requiring process, it is highly dependent on the quantity and (as we shall see later, p. 85 ff) also the quality of the prevailing radiation. Fig. 43 shows that the rate of photosynthesis at first increases linearly with increasing light intensity, then becomes sluggish, and finally reaches a constant value. The curve then runs parallel to the abscissa: the photosynthetic capacity is "saturated" with light. This "saturation point" is reached at different light intensities in individual plants. In so-called "sun plants" relatively high light intensities (Fig. 44, A) are necessary, in "shade plants", however, only rather low light intensities.

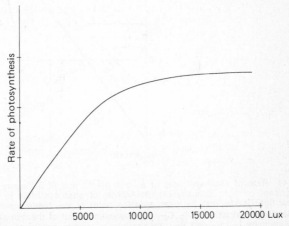

Fig. 43   Effect of various light intensities (in lux) on the rate of photosynthesis.

Another important difference between the two groups is revealed by the shape of the graphs showing photosynthesis rates, particularly when we scrutinize their initial points more closely (Fig. 44, B).

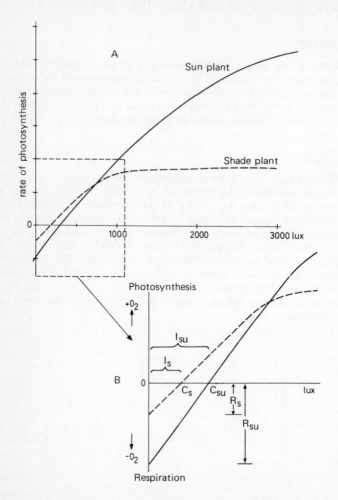

Fig. 44 Rate of photosynthesis of a "sun plant" and of a "shade plant" as dependent on light intensity (A). B: Activity of photosynthesis in the range of low light intensities. C_S: compensation point of the shade plant with corresponding light intensity I_S. C_{SU}: compensation point of the sun plant at the light intensity I_{SU}. R_S and R_{SU}: respiration (= oxygen consumption) of shade and sun plant, respectively. For details, see text.

Neither coincides with zero; both the curves intersect the abscissa, i.e. at points of low light intensities. This point of intersection is termed "compensation point". It specifies the light intensity at which the process of photosynthesis consumes exactly as much carbon dioxide as is produced by the respiration of the plant. The latter is "compensated" by photosynthesis. In sun plants, the compensation point is reached at a significantly higher light intensity than in shade plants. In other words, even at low light intensities, shade plants have a positive metabolic balance and can exist. Sun plants, in contrast, exhibit a negative balance under these conditions. Therefore, light affects the distribution of individual plant species. Even the leaves of a tree top are subject to this rule: "sun leaves" and "shade leaves" can be distinguished; their compensation points are reached at different light intensities. The two types of leaves are easily recognized on the basis of their different anatomical structure.

The gas exchange of numerous plant species is more complex on account of an additional process which releases CO_2 and consumes O_2 with increasing light intensity. It has been termed "photorespiration" since its gas exchange is similar to that in respiration. Actually, photorespiration is closely correlated with photosynthetic CO_2 reduction; that is why we shall discuss its details together with the latter (p. 159).

In the state of light-saturation, photosynthesis has reached an optimal rate in respect of the factor light, but its rate may be further increased by raising the CO_2 concentration in the gas phase. The role of CO_2 as a "limiting factor" in photosynthesis under natural conditions is obvious. At 0.03 %, the carbon dioxide content of air is suboptimal for photosynthesis and, according to the "law of limiting factors", it is the one furthest in minimum. The CO_2 content in the air therefore determines the yield of photosynthesis (cf. p. 78 f).

There is, however, a distinct difference between single plant species in respect of their effectiveness in utilizing the natural supply of carbon dioxide at various light intensities. This effectiveness is particularly high in some plants from tropical and subtropical regions (sugar cane, maize) which bind CO_2 by the "C_4-dicarboxylic acid cycle" (Hatch and Slack; p. 155). In these "C_4 plants", light-saturation in photosynthesis hardly ever occurs, even at highest light intensities, because of the apparently high efficiency of the CO_2-binding mechanism. At medium light intensities they are also often distinctly superior to the "C_3-plants" which incorporate CO_2 by the "Calvin-Cycle" (p. 145).

The photosynthetic utilization of an artificially raised amount of CO_2 in the air depends primarily on the available light intensity (Fig. 45). Another possible limiting factor is the temperature.

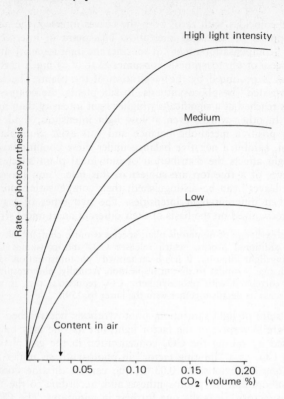

Fig. 45 CO_2 concentration and light intensity as limiting factors in photosynthesis.

Temperature

Figure 46 shows the significance of this factor for photosynthesis at low and high intensities of illumination. In the first case, the photosynthetic rate is practically unchanged – light is the limiting factor. In the second case, it rises considerably within a definite temperature range. When the latter is exceeded, a rapid drop in photosynthetic activity occurs because of heat damage. The curve is a typical "optimum curve", the highest values for higher plants of our latitude lying between 20 ° and 30 ° C. Apart from this there are cases of extreme temperature tolerance: for thermophilic bacteria and thermophilic blue-green algae which live in hot springs at 70 ° to 80 ° C a temperature maximum of 35 ° to 50 ° C is characteristic. This also holds true for higher plants which, owing to adaptation, may live in extremely hot places. Conversely, several species of lichens are able to carry on photosynthesis at temperatures below 0 °C. Even at these low temperatures, they have a positive metabolic balance (see above). From the observed dependence of photosynthesis on the factors light and temperature, an important consequence follows: photosynthesis is not a uniform process, but consists of a light-dependent or "photochemical" reaction complex and a sequence of temperature-dependent or "enzymatic" reactions. The latter are only recognizable in the state of light saturation, where raising of temperature increases the rate of photosynthesis (cf. Fig. 46). At low light intensities temperature has practically no effect on photosynthesis. Evidently, only the photochemical reaction complex is active under these conditions; it is characterized by insensitivity to temperature. This specific feature of all photochemical processes is shown particularly strikingly by a photographic emulsion: the light-dependent formation of silver grains occurs even at extremely low temperatures.

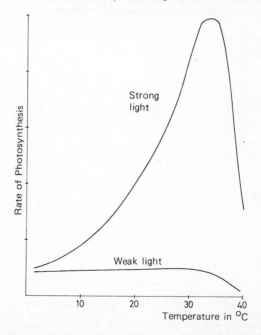

Fig. 46 Effect of temperature on the rate of photosynthesis at low and high light intensities.

The velocity of the whole photosynthetic process is significantly limited when only low light intensities are available for the photo-chemical reaction complex. Thus the latter becomes the bottleneck for the whole reaction (Fig. 47, A), since only a limited supply of precursors is produced for the coupled enzymatic reactions. Conversely, the latter reactions are the bottleneck when the capacity of the photochemical system is fully activated by high light intensities, but the temperature is suboptimal. By raising the temperature, the diameter of the tube is enlarged (speaking figuratively), and an optimal yield of assimilates results (Fig. 47, B).

The different nature of the two photosynthetic subprocesses is also reflected in the size of the "Q_{10}-value". This is a measure of the temperature dependence of a process, namely the increase in reaction rate with a 10 °C rise in temperature. For the photochemical reaction complex of photosynthesis, the Q_{10}-value is about 1.0. This means that a rise in temperature is practically without effect, i.e. the reaction is independent of temperature. The enzymatic reaction complex of photosynthesis has a Q_{10}-value of at least 2.0, signifying that the reaction rate is doubled when the temperature is raised by 10 °C. This subprocess is thereby clearly identified as a sequence of chemical, enzyme-catalyzed reactions.

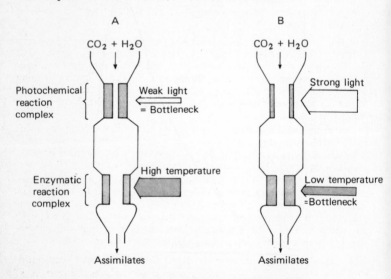

Fig. 47 Effects on the "photochemical" and the "enzymatic" reaction complex of photosynthesis by the factors light and temperature (schematic). A: Low light intensity and optimal temperature: the photochemical reactions become the bottleneck and hence limit the yield of photosynthesis. B: High light intensity and low temperature: now the enzymatic reactions are the bottleneck of the overall process.

The Efficiency of Different Spectral Regions

The importance of light in photosynthesis has only been considered up to now under conditions where white light of defined intensity was applied. The experimental results allow no definite conclusions as to which of the different spectral qualities contained in white light are essential for photosynthesis, and which are ineffective.

The "Action Spectrum". According to the rule cited initially, that radiation can only be effective when it is absorbed, the pigments present determine which spectral regions are effective in photosynthesis. Their relative activity is shown by an *action spectrum* or "activity spectrum". This reveals the dependence of photosynthetic activity (or any other light-requiring process) on radiation of various wavelengths. For each an equal number of quanta is administered. The action spectrum shows even more details if one measures the number of quanta of each wavelength required to obtain the same activity of the light-dependent process. In connection with the absorption spectrum of the pigments present it is possible to identify those light qualities which are effective in photosynthesis. Moreover, the pigment systems responsible for their absorption can be detected.

The first action spectra of photosynthesis were recorded by Engelmann in the years 1882–1884. At first, he used filaments of the green algae *Oedogonium* and *Cladophora* as a test object. He projected a microspectrum onto the cells by means of a prism. To demonstrate formation of oxygen in the illuminated areas of the cells, or of their chromatophores, he used bacteria which react specifically to traces of oxygen by an aerotactic aggregation.

Engelmann observed an intense massing of bacteria in the cell areas struck by red radiation, a less intense one in the range of action of blue radiation, while the green radiation did not cause any aggregation of bacteria. Similar results were obtained with cells of *Euglena viridis*.

In diatoms, brown and red algae, which Engelmann investigated later, a characteristic difference was observed: oxygen production was also induced by green and yellow light. On the basis of these findings, Engelmann assumed that in these organisms pigments other than chlorophylls also participate actively in photosynthesis. In the meantime this conception has been generally accepted and experimentally confirmed.

The action spectra of photosynthesis available today are more precise owing to improved experimental procedures. However, in principle they merely confirm Engelmann's observations of more than 80 years ago. In many cases, the curves of photosynthetic activity largely follow those of pigment absorption. The maxima and minima of the former coincide with the top and bottom values of absorption. Within certain limits it is thus possible to draw con-

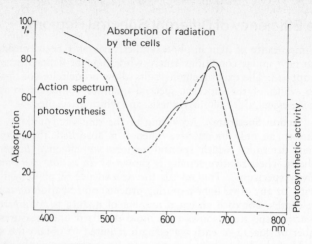

Fig. 48 Action spectrum of photosynthesis in *Chlorella pyrenoidosa,* and the absorption of visible radiation by the cells (after Emerson and Lewis).

clusions regarding pigment effectiveness in the various spectral regions.

In the action spectrum of *Chlorella* (Fig. 48), the maximal photosynthetic activity coincides with the absorption peaks of the chlorophylls in the red and blue spectral regions. In the blue one the maxima of both curves extend into the long-wave region due to the action of the carotinoids.

In diatoms and brown algae, the additional specific carotinoid pigment fucoxanthin gives rise to a definite absorption peak between 430 and 520 nm. It has its counterpart in the curve of photosynthetic activity.

In connection with the total absorption, the action spectra of photosynthesis in blue-green algae indicate unequivocally that the blue and red biliproteids are involved in photosynthesis. In *Phormidium uncinatum,* a filamentous blue-green alga with both phycocyanins and phycoerythrin, the action spectrum of photosynthesis largely follows the absorption curve of the pigments (Fig. 49). The former reaches its highest activity in the main absorption region of the phycoerythrins (between 500 and 560 nm); in the absorption maximum of chlorophyll a and the phycocyanins (610 to 680 nm), photosynthesis is far less active.

The obviously smaller efficiency of chlorophyll a is shown even more strikingly in red algae. For the cells of *Delesseria decipiens*

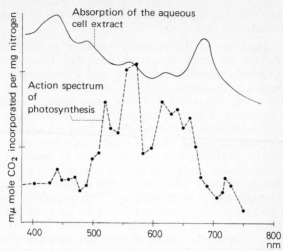

Fig. 49 Action spectrum of photosynthetic $^{14}CO_2$ incorporation in the blue-green alga *Phormidium uncinatum*, and the absorption of visible radiation by an aqueous cell extract (after Nultsch and Richter). Reference quantity of CO_2 incorporation: nitrogen content of the cells.

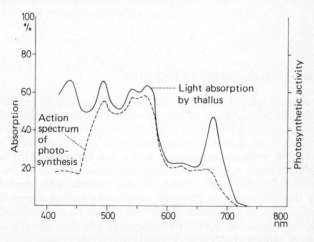

Fig. 50 Action spectrum of photosynthesis of the red alga *Delesseria decipiens*, and the light absorption by the thallus (after Haxo and Blinks).

(Fig. 50), which mainly contain phycoerythrins, the absorption spectrum and the action spectrum of photosynthesis are largely parallel to each other. The superiority of the biliproteids in comparison with chlorophyll a and the carotinoids is demonstrated particularly strikingly.

This superiority of the accessory pigments to chlorophyll a, as revealed by the action spectra obtained with blue-green and red algae, lacked a plausible explanation until a new concept of photosynthesis was developed, based on the participation of two light reactions and, accordingly, of two active pigment systems. Only on cooperation of both an optimal rate of photosynthesis results. An action spectrum of photosynthesis, however, cannot deal with these aspects, since each point of measurement represents the action of only one wavelength of radiation. Only by irradiation with an additional second wavelength will the activation of both the photosystems be achieved. Only then a true estimation of the capacity of photosynthesis is to be expected. This hypothesis has actually been confirmed experimentally. It led to the discovery of the "Emerson effect", which we shall study more closely in another context (p. 91 f). Although the validity of the action spectra of photosynthesis is limited by the new concepts of light-dependent subprocesses in photosynthesis, they confirm in particular the importance of the "accessory pigments". As generally accepted today, the role of the latter consists in transferring absorbed radiant energy to chlorophyll a, which is the site of the actual photochemical reaction in photosynthesis.

Role of the Accessory Pigments

Fluorescence measurements (p. 102 ff) have significantly helped to confirm the participation of the accessory pigments in absorption and transmission of radiant energy. Under appropriate conditions, an excited pigment molecule may transfer its excitation energy to a neighboring molecule provided it is not too far away. This condition is certainly fulfilled in the pigment-containing membranes of photosynthetic structures. Today we may assume with some certainty that the accessory pigments transfer the radiant energy they have absorbed, to a "reaction center" (cf. p. 106). Here, by mediation of chlorophyll a, the actual photochemical reaction, i.e. the conversion of radiant energy into chemical energy, takes place.

The energy transfer from the leaf carotinoids to chlorophyll can be demonstrated by measuring the fluorescence emitted (Fig. 51).

A leaf is exposed to blue and green radiation simultaneously by placing a cuvette containing a chlorophyll solution in the path of the incident white light. The chlorophyll solution largely eliminates all spectral quali-

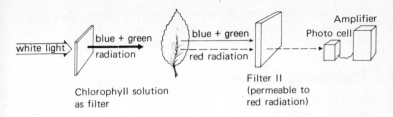

Fig. 51 Experimental device to demonstrate energy transfer of absorbed radiant energy from accessory pigments to chlorophyll which exhibits fluorescence in consequence. For details, see text.

ties except parts of the blue and green ones. Mostly, these blue-green quanta striking the leaf excite the carotenes and xanthophylls. The resulting fluorescence of chlorophyll is demonstrated by a similar experimental device as the one described above. The red radiation emitted is identical with the red fluorescence of pure chlorophyll (cf. p. 102). Energy transfer from chlorophyll b to chlorophyll a is obviously direct and quantitative: even during illumination of leaves at a wavelength which will specifically stimulate chlorophyll b, only chlorophyll a fluorescence is observed. In contrast, carotenes and xanthophylls do not show any fluorescence in vivo; lutein and β-carotene are only slightly active in the transfer of radiant energy, although they are the largest in quantity. They probably play another important role by protecting chlorophyll from photo-oxidation, i.e. light-induced degradation.

Energy transfer from phycocyanins and phycoerythrins to chlorophyll a has also been demonstrated in blue-green and red algae by means of a similar experimental approach. It presumably follows an established sequence when blue radiation is absorbed by the carotinoids: carotinoids → phycoerythrin → phycocyanin → chlorophyll a. In addition, the energy of green and orange quanta absorbed by the phycobiliproteids is also transferred to chlorophyll a. While fluorescence of the biliproteids is only weak in the living cell, it is stronger in isolated pigments. The fluorescent yield is 53 % and 85 % for phycocyanins and phycoerythrins respectively.

Fucoxanthin, the characteristic carotinoid of diatoms and brown algae, is also able to transfer a large part of the energy of absorbed quanta to chlorophyll a.

By means of a slightly modified experiment, it is also possible to measure the rate of photosynthesis under these conditions and to determine simultaneously its relationship to the intensity of fluorescence.

Studies of intact photosynthetic organisms, as described above, have led to two important conclusions: 1. Photosynthesis is a complex process which consists of a photochemical and an enzymatic part.

2. At least two different photoreactions are required for the optimal efficiency of visible radiation in photosynthesis. On these premises, let us now consider these processes at a molecular level.

The Photochemical Reaction Complex

In the introduction photosynthesis was characterized as an endergonic process in which synthesis of organic compounds from carbon dioxide and water is made possible by absorbed radiant energy. Since both the starting substances have a considerably lower energy level than the end products, it is thermodynamically impossible for this process to proceed spontaneously. Carbohydrate and oxygen are only formed when the equilibrium is "shifted" by the input of energy. This is provided by the absorbed radiation owing to the unique property of chlorophyll. An energy diagram of photosynthesis (Fig. 52) shows the energetics of the overall process.

Due to the "activation energy" (cf. p. 9 ff) which must additionally be expended, the energy level of the total reaction is slightly higher than that of the end products. At the same time, this energetic barrier prevents a spontaneous reaction of carbohydrate with O_2 to yield CO_2 and H_2O. Theoretically this should occur because of the high energy content of the two end products. This explains why these and intermediate products of photosynthesis, like most other organic carbon compounds, are "metastable" (p. 9) in the presence of oxygen at normal temperatures. It is easy to understand that metastability must be paid for by an input of additional energy.

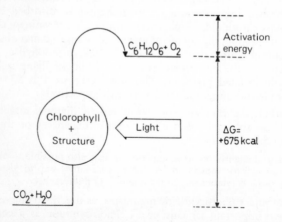

Fig. 52   Energy diagram of photosynthesis.

The energy diagram of photosynthesis shows the change in free energy of the overall process, but gives no information about the complex mechanism of energy conversion which proceeds via many intermediary reactions. Quite a number of these have only been elucidated or at least theoretically recognized in recent years. Although our knowledge suffices to explain the overall process of photosynthesis, for several reactions there is little experimental proof. It may thus be predicted that some of the reactions described below will need revision or will have to be rejected in the near future.

Based upon some experimental key results, the concept that absorbed radiation is effective in photosynthesis in two different photoreactions, has been firmly established in recent years. These reactions occur in two pigment collectives which distinctly differ from each other in composition and physical properties. They are termed *photosystem I* and *photosystem II;* the two photochemical events involved are called *light reaction 1* and *light reaction 2*. Further reaction systems provide the contact between these two photochemical centers. This complex photochemical mechanism furnishes the "energy equivalents" and "reduction equivalents" required for the chemical conversion of carbon dioxide to carbohydrate. The

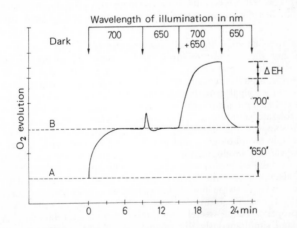

Fig. 53 "Emerson effect" in *Chlorella* (after Myers and French). A: Rate of dark respiration; B: photosynthetic evolution of O_2 during successive short-term illumination (time on abscissa!) with light of 700 nm and 650 nm wave-length, respectively. Δ EH: Increase in the rate of photosynthesis caused by "enhancement" (see text) upon simultaneous illumination with light of 700 nm and 650 nm wavelength; the photosynthetic O_2 evolution measured for each wavelength when applied separately serves as control and is subtracted ("650" and "700", respectively).

reactions involved are also termed the "primary processes" of photosynthesis.

The "Emerson effect" (named after its discoverer) gave the first hint of the cooperation of two light reactions in photosynthesis (Fig. 53).

This effect is based on the observation that the relatively weak photosynthetic oxygen production in cells of *Chlorella* when illuminated with red light (700 nm) is significantly enhanced by simultaneous illumination with red light of shorter wavelengths, i.e. < 670 nm (= "enhancement effect"). As it was also observed in the red alga *Porphyridium* and in the blue-green alga *Anacystis*, Emerson concluded that chlorophyll b in green algae and a biliproteid (phycoerythrin or phycocyanin) in the red and blue-green algae function as an auxiliary pigment. This specific action of two light qualities was subsequently confirmed for other partial processes of photosynthesis.

Investigations of fluorescence, and particularly of short-lived absorption changes of the pigments in vivo (see below) supplied further arguments in favor of the cooperation of two photoreactive systems in photosynthesis. Early suggestions that one is particularly concerned with the oxygen production were confirmed as well as those predictions assuming that both the photoreactions primarily activate and transfer electrons by way of chemically appropriate systems, i.e. redox catalysts.

Methods. The test objects used to elucidate the primary processes in photosynthesis were mainly unicellular green algae and isolated chloroplasts.

Unicellular or Cenobial Algae will grow under controlled conditions of nutrition and illumination relatively easily compared to higher plants because. of their adaptation to water and their uniform cell size. They are, moreover, a physiologically very homogeneous experimental material. The high rate of algal growth, providing sufficient cell material within a few hours, is also an advantage. An important factor is that many species of algae tolerate continuous illumination; only few require a daily dark period. By adapting the illumination period to their life cycle, one obtains a considerable morphological and physiological homogeneity of the cells during the various developmental phases. The conditions of "synchronized culture" are fulfilled in this way. (Fig. 54). The aqueous culture medium used for algae contains the "macroelements" N, P, S, K, Mg and Ca as inorganic salts and also a few "microelements": Fe, B, Cu, Mn, Zn, Mo (cf. p. 281). The sterile culture medium with the inoculated cells is aerated in an appropriate culture vessel with a mixture of carbon dioxide (1–5 %) and air, and simultaneously illuminated. The "light thermostat" shown in Fig. 55 is specially designed to ensure constant temperature too. Besides this simple culture method others have been developed for *Chlorella* which allow synchronized cell growth in adaptation to the life cycle (see above).

Isolated Chloroplasts. Attempts to measure photosynthesis in isolated chloroplasts began at the end of the nineteenth century when Haberlandt

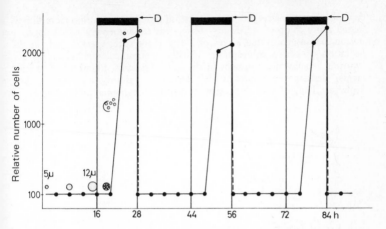

Fig. 54 Cell numbers and sequence of cell division in a long-term synchronous culture of *Chlorella pyrenoidosa* subjected to a light-dark cycle of 16:12 h (after Lorenzen and Senger). Light intensity: 9000 lux. For the first cycle size and shape of a single cell are shown schematically. Cell divisions always occur exactly in the middle of the dark period, provided the culture has been previously diluted ($\leftarrow$ D) to the original relative number of cells (= "100").

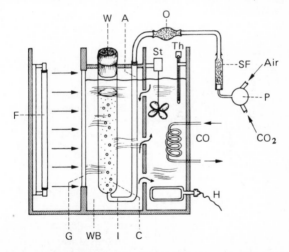

Fig. 55 "Light thermostat" for culturing of algae (schematic). A: Aeration (CO_2 + air); C: Culture tube with nutrient solution and suspended algal cells; CO: Cooling system; F: Fluorescent tubes; G: Glass wall; H: Heating system; I: Inlet for air + CO_2; O: Glass olive with cotton wool wad; P: Pump for air + CO_2 supply; St: Stirrer; SF: Sterile filter for the gas mixture; Th: Contact thermometer for control of heating and cooling; W: Wadding plug; WB: Water-bath for maintaining a constant temperature.

and also Ewart succeeded in demonstrating oxygen evolution from illuminated moss chloroplasts. Many subsequent experiments showed negative results because only the leaves of a few plant species are suited for the isolation of active chloroplasts. Usually, mechanical effects during disruption or substances present in the cells (saponins, tannins and organic acids) destroy the photosynthetic activity of chloroplasts during their isolation. Lack of such destructive constituents explains the preference for

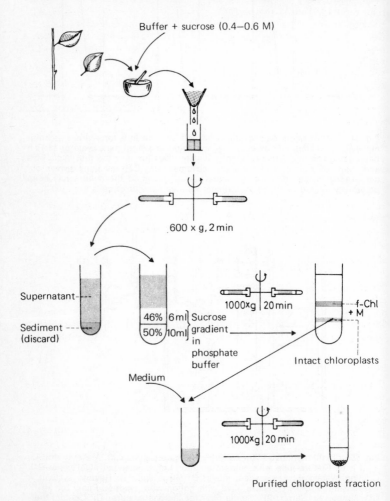

Fig. 56 Procedure for isolation of chloroplasts by means of fractionating centrifugation and sucrose gradient centrifugation.

isolated chloroplasts from spinach, "Swiss chard" (= *Beta vulgaris var. cicla*) and *Phytolacca* as standard objects of photosynthesis research. Apart from a few exceptions *(Acetabularia, Bumilleriopsis)* the isolation of chloroplasts from algal cells appears even more difficult because of the cellular organization; their cell walls especially resist the gentle disruption necessary to obtain intact chloroplasts.

Prior to isolation of the chloroplasts (see scheme, Fig. 56), leaves are rendered largely free from starch by keeping them in darkness. They are then gently disrupted in a buffered medium with addition of sucrose (0.4 to 0.6 molar), sorbitol or mannitol. After passing through a wide-mesh filter the homogenate is subjected to low-speed centrifugation; the resulting supernatant fraction contains many chloroplasts as well as other cellular organelles such as mitochondria, cell nuclei and their respective fragments. For further purification it is layered onto a sucrose gradient in a centrifuge tube which consists, for instance, of two separate layers of sucrose solutions of different concentration, 46% and 50%. During the subsequent "gradient centrifugation"* the intact chloroplasts accumulate in a small deep-green coloured band at about the level of the original boundary between the sucrose solutions, while the other cell organelles and chloroplast fragments appear in other layers of the gradient. After careful removal of the chloroplast band and its suspension in an equal volume of the original medium, the intact chloroplasts are precipitated as a sediment by further centrifugation. The sediment contains about 5% chloroplast fragments and 4–5% mitochondria. Gradients formed from the weakly negative-charged macromolecules of "Ficoll" (10% to 40%) also produced excellent results as far as the isolation of intact chloroplasts from fragments and free thylakoids is concerned.

Besides this method of isolation in hypertonic sucrose solution, there is another one which utilizes hypotonic media, mostly 0.2% NaCl.

Non-aqueous isolation techniques greatly reduce the loss of water-soluble components from the chloroplasts which is inevitable in aqueous isolation media. Since organic solvents (petroleum ether, carbon tetrachloride, hexane) are used during isolation, the chloroplasts obtained are suitable only for certain types of experiments.

---

* The layered sucrose solutions, usually less dense than the particles or macromolecules to be separated, form the gradient. Under the influence of centrifugal acceleration the particles or molecules migrate through the gradient with different speeds, depending on their size, and accumulate in distinct bands (Fig. 57). The gradient merely serves to prevent the spreading of the bands or zones by convection. These bands are thus stable after centrifugation and may be isolated (cf. p. 341). This procedure for separating particles or molecules of different size ("speed method") must not be confused with the "equilibrium density or isopycnic centrifugation", e.g. the "cesium chloride density-gradient centrifugation" discussed later (p. 326 ff), where macromolecules form distinct bands according to their buoyant density.

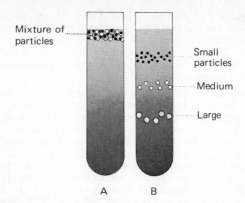

Fig. 57 Technique of gradient centrifugation. A mixture of different particles is layered onto a gradient previously built up from sucrose solutions, usually with increasing density towards the bottom of the centrifuge tube (A). During centrifugation at adequate speed the particles, being different in size, move at different speeds through the solution and accumulate in distinct bands or zones according to their size (B). For isolation of the latter see Fig. 94.

Isolated chloroplasts have recently been broken down to subunit particles ("broken chloroplasts"), grana systems or free thylakoids, on the one hand in order to study their structure, on the other hand to locate the sites of certain partial reactions of photosynthesis. In this context the positions of "photosystem I" and "photosystem II" in the thylakoid membrane are of utmost interest (cf. p. 123 f). Three procedures are applied for disruption of the chloroplasts: ultrasonication, mechanical grinding in a homogenizer, and treatment with digitonin. As a result of high-speed centrifugation or separation on gradients (see above) the thylakoids or their components will accumulate in different more or less homogeneous fractions. Their capacity to carry out primary photosynthetic processes is then tested.

Experimental Analysis of the Primary Processes

① Isolation and characterization of those substances or equivalent structures from chloroplasts which possibly participate in primary processes. A detailed knowledge of their properties permits inferences regarding their participation in hypothetical reaction sequences. Apart from the chlorophylls, carotinoids and biliproteids which are of particular interest in this context, several other compounds have been detected which obviously have a specific function in the photochemical apparatus: cytochromes, plastoquinone, plastocyanin, flavoprotein, ferredoxin and manganese. These will be discussed in detail later.

② Measurement of the amount and synthetic rate of individual reaction products specific for primary processes in relation to external conditions.

Interest focussed very soon on the oxygen released, and later also on "reduction equivalents" and "energy equivalents".

③ Observation and recording of physical changes resulting from light absorption in the living cell. *Fluorescence* (p. 101 ff) became the most important method for demonstrating chlorophyll bound to chloroplasts or other photosynthetically active structures. Excited chlorophyll molecules emit a small part of the absorbed radiant energy (1.5–3 %) as a weak red fluorescence in vivo, too. The intensity and appearance of fluorescence have provided valuable information on the mode of action and distribution of light-absorbing pigments (cf. p. 105 f).

The occurrence of so-called "induction periods" has also been analyzed by means of fluorescence. When green cells are illuminated after having been kept for some time in darkness, the chlorophyll fluorescence that appears shows first a rapid increase to a maximum within about 2 sec, and then declines sharply. One or two slower fluctuations will follow before, after several minutes a steady state is achieved ("Kautsky effect"). Similar reactions, caused by the onset of illumination, are also characteristic for photosynthesis: oxygen evolution and CO_2 uptake also exhibit fluctuations largely parallel to those of fluorescence. A strong flash of light during a dark period directly induces a weak emission of red fluorescent radiation which can be measured with a sensitive photocell. It persists for a remarkably long time, i.e. for several minutes, during which its intensity slowly decreases.

Another method for analyzing the primary processes in algal cells, chloroplasts or their fragments has recently been developed. It registers the conversions involved by direct measurement of the concomitant optical absorption changes. Since these are very small (in the order of magnitude of 0.1 %) and the reactions involved proceed in about 10^{-1} to 10^{-5} sec, highly sensitive recording devices had to be developed. Here, the stimulation of photosynthesis is brought about by short flashes of light (up to 10^{-6} sec) and the whole sequence of absorption changes is recorded between 220 and 800 nm as a function of time. The procedure is called *flashlight photometry*. For separate measurement of a single absorption change the position of which is known, *periodic flashlight photometry* is used. In this procedure, photosynthesis is stimulated by a series of periodic short light flashes and the appearance and disappearance of a specific spectral change is recorded, repeatedly if necessary. The accuracy of registration is thereby considerably increased. Moreover, the sensitivity is about one thousand times higher than in normal flashlight photometry and processes lasting only 10^{-6} sec can be measured. Subsequently, another reversible absorption change with a different kinetics may be registered in the same way; then the next one, and so on, until the complete sequence of spectral changes has finally been recorded.

By applying both these procedures of flashlight photometry, substances could be detected in the photosynthetic active cell or in chloroplasts which are converted within 10^{-6} to 1 sec and which are intimately involved in the primary processes. They will be described in more detail below.

In *flashlight manometry,* the amount of oxygen produced photosynthetically by a light flash is determined by manometry (cf. p. 27 f). Illumination

of the cells with a series of periodic light flashes will improve this method as well. Processes lasting 10^{-4} to 1 sec can be measured.

④ Poisoning of individual photosynthetic reactions by adding to the cells
or isolated chloroplasts (or equivalent organelles) substances with a
specific inhibitory effect. From the resulting changes conclusions on compounds and reaction mechanisms involved may be drawn.

Absorption of Radiation by Atoms and Molecules

Chlorophyll a plays a key role in the photochemical reaction complex of photosynthesis. This is indicated by the fact that it is present
in all photosynthetically active organisms – except phototrophic
bacteria. Before we discuss its properties and mechanism of action,
the main features of photochemical processes will be treated.

Absorption of radiation is an uptake of electromagnetic radiation in the
form of photons or quanta by the atoms or molecules of a substance. The
energy state of the latter is thereby changed. Only quanta of those wavelengths are absorbed the energy of which exactly equals the energy of
transition between two energetically determined "excitation states" in an
atom or molecule. Let us consider the simple case of excitation of an
atom (Fig. 58); the electrons surround the atomic nucleus in discrete

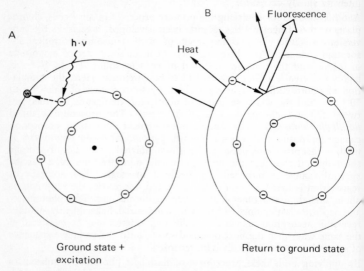

Fig. 58 "Ground state" and "excited state" of an atom. A: By absorption of
a light quantum (h · ν) an electron is raised to an orbital further away from the
atom nucleus, i.e. to a higher energy level. B: On return to the original orbital
(= restoring the "ground state") energy is released either as heat or as fluorescent radiation.

orbitals and at different distances. The electrons nearest to the nucleus possess relatively low energy, while those furthest away possess a comparatively high energy. Displacement of an electron from an orbital near the nucleus to one further away requires input of a definite amount of energy, because a negatively charged particle is being moved away from the nucleus with its high positive charge. This event occurs when a photon strikes an atom and its energy is just sufficient to raise the electron to an outer orbital, i.e. to a higher energy level.

The atom concerned is thereby converted from the "*ground state*" to an "*excited state*". This explains why an atom only absorbs radiation of a few definite wavelengths. Only quanta from these have the exact energy to equal the energy difference between the ground state and an excited state. In other words, the energy of a quantum can be used only as an entity ("all-or-nothing reaction") and not in portions. Since only quanta of one wavelength participate in each electron transition in the atom, an absorption "line" may be observed. Consequently the absorption spectra of atoms usually consist of only a few narrow lines (cf. Fig. 59).

The absorption of quanta is more complex in molecules because due to their size these have different properties as compared to atoms. Here, rotational and vibrational energy of the molecule are closely connected energetically with the electron transition. Both are involved in the absorption event since the electronic configuration and hence the electrostatic field inside and outside the molecule are changed. The atomic nuclei attempt to reach a new, electrostatically more stable state through vibration. The energy required comes from the quantum absorbed. Consequently, the energy of this quantum does not always equal exactly the

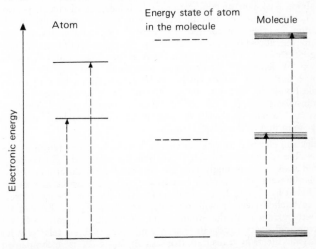

Fig. 59 Excitation states of an atom either free or bound in a molecule. The electron transitions which may occur upon absorption of adequate quanta are indicated by arrows.

amount required for a simple electron transition. Because of the additional excitation of rotation and vibration with their discrete substates, molecules do not exhibit absorption lines, but absorption "bands". There are also differences in the energy levels of individual electron transitions in an atom, depending on whether it is free or bound in a molecule. Figure 59 shows three discrete energy states of an atom which may be reached by absorption of energetically adequate quanta. If the same atom is bound in a molecule, there is a distinct shift in the position of the excitation states (Fig. 59, center). This effect is due to the interference of the other constituents of the molecule with the atom. There are no longer any transitions between the original excitation levels. Transitions are only possible from a level which has broadened to a group or band of transitions as a result of additional excitation of rotation and vibration (Fig. 59, right side).

The "π electrons" are responsible for absorption of visible radiation by molecules consisting mainly of carbon atoms. They contribute to the chemical stability of the molecules by surrounding in an outer orbital two or more atomic nuclei ("π orbitals"). The "σ orbitals" present have a similar function, although their energy level is higher; hence their electrons can be excited by energy-rich ultraviolet radiation only. This is why many organic compounds exclusively absorb this radiation. They are colorless, because the human eye cannot register the extinction of this radiation quality.

When a quantum is absorbed by a molecule, a π electron is raised from the π orbital (= ground state) to a π^* orbital (= excited state). This process is termed a "$\pi \rightarrow \pi^*$ transition". It is characteristic for molecules containing a system of conjugate double bonds. As discussed previously (p. 37), compounds of this molecular structure strongly absorb visible radiation, i.e. they are colored. Yet another electron transition is important in the absorption of radiation by biological pigments. It involves those electrons which do not participate in bond formation in the molecule (= "n electrons"). Their specific orbitals are characteristic of oxygen, nitrogen and sulfur atoms which may be present in a system of conjugate double bonds or linked to such a system. Their role as components of "chromophore groups" has already been mentioned (p. 37). This "$n \rightarrow \pi^*$ transition" occurs relatively seldom; its excited state lasts significantly longer than that of the $\pi \rightarrow \pi^*$ transition.

An electron elevated to a higher energy level will tend to return as rapidly as possible to the ground state, releasing the absorbed light energy. The excited state is accordingly only of extremely short duration. The situation is different when several excitation states are possible in a molecule, designated E_1, E_2 and E_3 in our diagram (Fig. 60). It is far more probable that an excited electron will move between E_3 or E_2 and E_1 than that it will return from E_1 to the ground state. The excitation state of E_3 and E_2 is accordingly very short (10^{-12} sec), while that of E_1 is relatively longer (10^{-9} sec). Electron transition from E_3 or E_2 to E_1 is accompanied by release of energy in the form of heat. In contrast, the transition from E_1 to the ground state lasts long enough to release the original excitation energy either as chemical energy or as radiation. Therefore, in light-

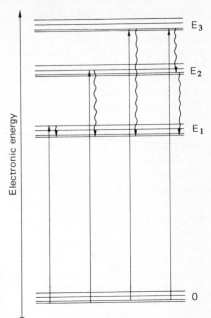

Fig. 60 Hypothetical excitation states (E_1; E_2; E_3) in a molecule. Straight arrows: electron transitions during excitation; curvilinear lines: energy release during electron transition to a lower energy level.

dependent systems, particularly in photosynthesis, the lowest molecular excitation state – corresponding to E_1 – is of utmost importance with respect to the release of chemical energy. The second form of energy release, i.e. the emission of photons caused by electron transition from the excited to the ground state, is termed *fluorescence* (Fig. 58).

Fluorescence. Fluorescence was first observed in fluorspar (fluorcalcium) from which the phenomenon received its name. In addition to a number of other solid materials, numerous organic substances, particularly many dyes, e.g. eosin and fluorescein (hence its name!) fluoresce intensively. Fluorescence of berberine and esculin can easily be demonstrated by putting pieces of twigs from berberis or bark from horse chestnut in water and irradiating the mixture with ultraviolet rays of the 366 nm wavelength. Even traces of esculin dissolved in water are sufficient to produce a bright blue-green fluorescence. In the case of berberine, there is a yellowish-green fluorescence at the points of fracture of the twigs. Fluorescence is therefore a sensitive indicator for very small quantities of substances*; it has become extremely valuable in analyzing the primary photochemical reactions of photosynthesis (p. 88 f and p. 97).

---

* Nowadays, fluorescent substances are added to the paint used for traffic signs; they have a strong luminous effect at night, even when struck by very low intensities of light.

Photochemistry of Chlorophyll

Fluorescence of chlorophyll in organic solution, e.g. in acetone appears upon illumination with short-wave blue light. About 30 % of the blue radiation absorbed is re-emitted as red photons, appearing as an intense red radiation. Dropwise addition of water to the solution results in a gradual "quenching" of the fluorescence. This is because the chlorophyll is converted into a colloidal form of distribution in which the fluorescence is largely abolished. This observation had previously led to the assumption that chlorophyll also occurs in an aqueous colloidal distribution in vivo, since the fluorescence which can be detected in active chloroplasts is relatively faint, amounting to 3 % of the absorbed radiation at the most.

The absorption spectrum of the chlorophylls (Fig. 21) indicates that only red and blue quanta of definite wavelengths are able to excite fluorescence in the molecules. The fluorescent radiation that appears, however, consists of identical quanta for both excitatory light qualities: the "fluorescence spectrum" (Fig. 61) has an identical maximum in the red spectral region after excitation with red or with blue quanta. In terms of wavelength, it differs only slightly from that of chlorophyll absorption. When energy-rich blue quanta are absorbed, red quanta of lower energy are emitted. The same effect is produced by excitation with red quanta, though it is much less pronounced: the wavelength of the fluorescent radiation is now only slightly greater than that of the radiation used for excitation.

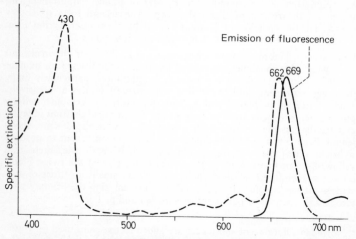

Fig. 61 Absorption spectrum and fluorescence spectrum of chlorophyll a in ether (after Goedheer).

The energetically different excited states in the chlorophyll molecule explain this phenomenon. As shown by the diagram below (Fig. 62), three main states exist: the "1st triplet state" at the lowest energy level, then the "1st singlet state", and finally the "2nd singlet state" at the highest energy level. By absorption of a blue quantum (430 nm) the molecule is promoted to the 2nd singlet state. When a molecule absorbs a red quantum lower in energy (670 nm), it is only raised to the level of the 1st singlet state. The 2nd singlet state has an extremely short lifetime (10^{-12} sec) and a low stability. The reasons for this have already been explained (p. 100 f). The excited electron therefore very rapidly returns to the 1st singlet state (10^{-14}–10^{-13} sec), releasing energy exclusively as heat. Of the original excitation energy of a blue quantum, a portion accordingly dissipates as heat; the remainder is emitted as radiation: the return of the electron from the 1st singlet to the ground state gives rise to emission of a red quantum. This last transition is identical with the reaction resulting from excitation by a red quantum (cf. Fig. 61). For this reason, the same fluorescent radiation is produced by red and by blue quanta. Of course, not every transition from the 1st singlet to the ground state results necessarily in the release of a quantum; the energy may also be released as heat.

It is apparent that the energy which is released from an excited chlorophyll molecule as heat or fluorescent radiation is useless for carrying out chemical work, i.e. formation or cleavage of chemical

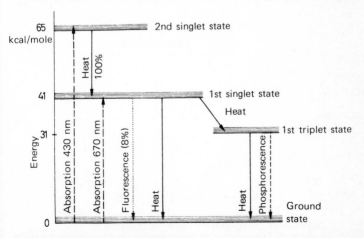

Fig. 62 Excitation states of the chlorophyll molecule upon absorption of a blue quantum (430 nm) and of a red quantum (670 nm), and the events of emission which result from restoring the ground state. Details in text.

bonds. It is lost to photosynthesis. The chloroplasts and other chlorophyll-bearing structures must therefore possess the ability to convert the energy available during a very short-lived state of excitation (about 5×10^{-9} sec) to a more stable form which can be exploited for chemical work.

Since this work is mostly performed by the excitation energy of the 1st singlet state, this excitation state is obviously also of great importance for the photochemical reactions in chloroplasts. However, there are other ways in which this energy may be dissipated, e.g. by a transition to the 1st triplet state which is metastable and has thus a comparatively long life. From there a direct return to the ground state, accompanied by a delayed light emission (= "phosphorescence") is possible; its wavelength is considerably longer than that of the light absorbed. More recent investigations, however, indicate that obviously only a few chlorophyll molecules are in the triplet state during normal photosynthesis.

In recent years studies on the "primary processes" (p. 92) of photosynthesis have shown that the general features and conversions of photochemical reactions do not necessarily also apply to photosynthesis. Some researchers have doubted whether a relatively long-lived state of excitation, i.e. 1st singlet or 2nd triplet (Fig. 62), must be reached at all in order to capture absorbed radiant energy and convert it to chemical energy. The close spatial arrangement of the reactants, particularly of the pigment molecules, in the photosynthetically active structures should permit energy transfer and energy flow (p. 108 f) in a more short-lived state of excitation too.

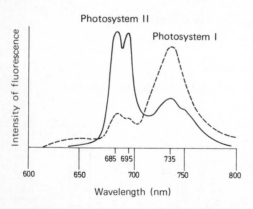

Fig. 63 Detection of "photosystem I" and "photosystem II" in intact chloroplasts by means of fluorescence emission spectra. At low temperatures, e.g. at −196 °C, "photosystem I" fluoresces most strongly at 735 nm, to a lesser extent at 684 nm and 695 nm (broken line). For "photosystem II" peaks appear at 685 nm and 695 nm (solid line).

Studies of fluorescence activity in intact photosynthetic structures have yielded some valuable information regarding the lifetimes of the various states of excitation in pigment molecules. They give an indication of how much time is available for the excitation energy to become effective as chemical energy. Moreover, the registration of fluorescence spectra in intact chloroplasts has led to the detection of two active photosystems in vivo. From the results obtained with this method, which operates at low temperatures, it appears rather certain that the various pigments present participate in two different reaction complexes (Fig. 63). Similar findings were made when measuring the action spectra of photosynthesis in cells of *Chlorella* and *Porphyridium* illuminated with two different light qualities. These findings strongly support the hypothesis of Emerson (cf. p. 91 f) made about 20 years ago.

Coupling of Two Light Reactions

Apart from fluorescence studies on photosynthetically active pigments in vivo (see above), the analyses of their specific absorption spectra at very low temperatures (French) turned out to be a most valuable technique in identifying the two pigment systems. As shown in Fig. 64 the red absorption maximum of chlorophyll a in vivo can be resolved by appropriate methods into several smaller peaks with absorption maxima between 660 and 720 nm. They are specific for several spectral forms of chlorophyll a which are probably bound in complexes with proteins (cf. p. 61 f) or other chloro-

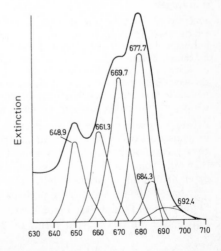

Fig. 64 Absorption spectrum of a chloroplast fraction from *Scenedesmus* containing "photosystem I" and "photosystem II" measured at −196 °C; the red absorption peak is resolved into different smaller absorption peaks specific for the various spectral forms of chlorophyll a and chlorophyll b (after French).

phyll molecules, and absorb differently in the native structure. Corresponding forms were found in algal chromatophores and in particle fractions isolated from them *(Scenedesmus, Chlorella)*. Moreover, relevant evidence for the participation of chlorophyll b, carotinoids and phycobiliproteids in at least one of the two pigment systems was obtained.

The existence of two pigment systems is also substantiated by their successful isolation from chloroplasts and algal chromatophores and subsequent concentrating in two separate particle fractions. Hence, it is commonly accepted that the photosynthetic apparatus in higher plants and in algae is equipped with two active pigment collectives, i.e. "photosystem I" and "photosystem II", in which the "1st light reaction" and the "2nd light reaction" proceed.

In accordance with a recent hypothetical model the pigments of both photosystems are considered to form "photosynthetic units" (cf. p. 74) containing about 300 chlorophyll molecules. Most of these, however, do not take part directly in the photochemical event, but collect the absorbed radiation energy and transfer it to the actual reaction center. This probably consists of only one molecule of a specific form of chlorophyll a which enters into an excited state when energy is conveyed to it: a "high-energy electron" leaves the excited molecule and is led away by an "electron transport chain", i.e. a series of redox catalysts (p. 21 ff). Finally, the electron is either trapped in the stable form of a "reduction equivalent" or conveyed back to the oxidized chlorophyll molecule, making a kind of detour. In the first case, an electron of another origin will restore the electron balance of the excited chlorophyll molecule in the reaction center.

Fast reversible absorption changes of chloroplasts or cells of *Chlorella* which result from application of the "flash-light technique" (p. 97 f) at normal and very low temperatures (–196 °C) and which are now known to be caused by the various forms of chlorophyll a, have provided valuable evidence in analyzing the events in the reaction center. These fast spectral changes result from the changed arrangement of electrons in the excited molecules and permit conclusions to be drawn as to the direct or indirect participation in redox reactions of the pigments concerned.

Before we discuss in detail the electron flow mentioned above and its consequences with respect to energetics, we shall consider the construction of the two photosystems. Of course, we must not forget that only some of the concepts described here are consistent with experimental observations while others are still hypothetical and may have to be revised or even perhaps dismissed in the near future.

"Photosystem I" and "1st Light Reaction". Regarding the distribution of the various in vivo forms of chlorophyll a in the pigment

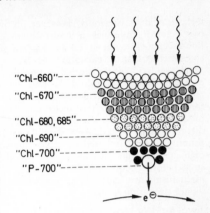

"Chl-660"----
"Chl-670"----
"Chl-680, 685"----
"Chl-690"----
"Chl-700"----
"P-700"----

→ e⊖ →

Fig. 65 Model of the pigment collective (= "trapping center") effective in "photosystem I" (schematic). The site of the actual photochemical reaction is the molecule of "P-700".

collective of "photosystem I", a model has been proposed which is shown in Fig. 65. It includes forms with absorption maxima between 660 and 700 nm ("Chl-660", "Chl-670", "Chl-680", "Chl-690", "Chl-700") which are also arranged in this sequence towards the reaction center, the portions of these forms in the collective decreasing, e.g. one molecule of "Chl-700" per 20 bulk chlorophyll a molecules. At the end of this energy-gathering system a molecule of a distinct form of chlorophyll a, termed "pigment-700" ("P-700") because of its specific absorption in vivo, functions as a reaction center where the photochemical event takes place, i.e. the release of a "high-energy electron" (see above). About 300–500 chlorophyll molecules are concentrated into a "trapping center" for absorbed quanta; one molecule of "P-700" forms the reaction center. The energy will reach the latter by a more or less directed flow through the entire population of pigment molecules. In other words, with the exception of "P-700", all pigment molecules of the system merely serve as "antenna pigments" which "receive" radiation of a particular wavelength according to their specific absorption and transfer it to the reaction center.

According to another version, carotinoids – mainly carotenes – are also among the antenna pigments of "photosystem I", while the xanthophylls find their place in "photosystem II" (cf. p. 116). Some authors now even place chlorophyll b in "photosystem I". This was originally thought to be part of "photosystem II". Certain differences in the pigment distribution of the two photosystems may exist between higher plants and green algae: "Chl$_a$-660" is probably absent in "photosystem I" of the latter.

Details on the mechanism of energy flow in the photosystems are far from clear. That a flow of energy does take place is suggested by the observation that the accessory pigments transfer absorbed light energy so rapidly and efficiently to chlorophyll a that their own fluorescence is practically not stimulated. Generally, however, during an energy transfer between two pigment molecules, a moderate amount of energy is lost as heat or fluorescence. In contrast, the mechanism of the "trapping center" operates so efficiently that the excitation energy, having once passed on to a chlorophyll a molecule, is transferred to a neighboring molecule without emission of a photon or energy loss. An obvious prerequisite is the proximity of another pigment molecule of the same form or of one absorbing at the same or a slightly longer wavelength, with an excitation state of the same or a slightly lower level; in this way an energy transfer in the opposite direction is prevented. As a possible explanation a mechanism has been proposed which is based on the *resonance theory* (Förster): the excited pigment molecule must first subside to the lowest level of the 1st singlet state before energy transfer can occur. Apart from this, other "faster" mechanisms are under discussion which allow energy transfer also on a higher excitation level.

Two models exist concerning the "migration" of energy through the population of pigment molecules in the photosystems (Fig. 66). In the first one, single reaction centers are dispersed at random among a large number of "antenna pigments" which transfer the radiation energy absorbed until it is "trapped" in a reaction center or given off as a photon. In the second one, each reaction center is surrounded and exclusively served by a definite number of about 300 chlorophyll molecules; this allows the absorbed energy to wander at random only within this defined "elementary unit" (cf. p. 106); the energy is either trapped in the reaction center or reemitted.

The Photochemical Reaction in "Photosystem I". The key function of chlorophyll a in photosynthesis may thus be specified as follows: it is only photochemically active in the form of "pigment-700" ("P-700") which acts as a redox system for a 1-electron transfer in the 1st light reaction. "P-700" has a redox potential of about $+0.4$ volt, which is significantly changed in the excited state: its new value of about -0.60 volt is so low that one of its electrons is transferred to a presently unidentified acceptor "Z" which is then reduced. "Z" is also a redox system by nature, with a potential of at least -0.60 volt. This explains why "Z" is a strong reductant after "photoreduction" and therefore exhibits high "electron pressure" (cf. p. 21 f).

A recently discovered redox system of highly negative potential termed "ferredoxin reducing substance" ("FRS") because of its ability to reduce ferredoxin (see below), has been assumed to be possibly identical with "Z". Another version places it near "Z", with "FRS" acting as the first redox catalyst in the electron transport connected with the 1st light reaction.

The energy balance of this endergonic process, in which an electron is moved to a higher energy level against the natural gradient by a potential step of about 1.0 volt; shows that about half of the energy invested by

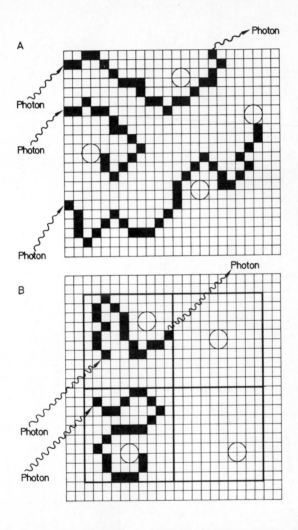

Fig. 66 Models of energy flow in the thylakoid membrane (after Govindjee). A: Absorbed energy travels through the entire mass of "antenna pigment" molecules (squares), either reaching a reaction center (circle) and becoming photochemically active or leaving the system as a re-emitted photon. B: The flow of energy is confined to the region of a "trapping center" comprising about 300 pigment molecules (large outlined square); here too the energy is either trapped in the reaction center or re-emitted.

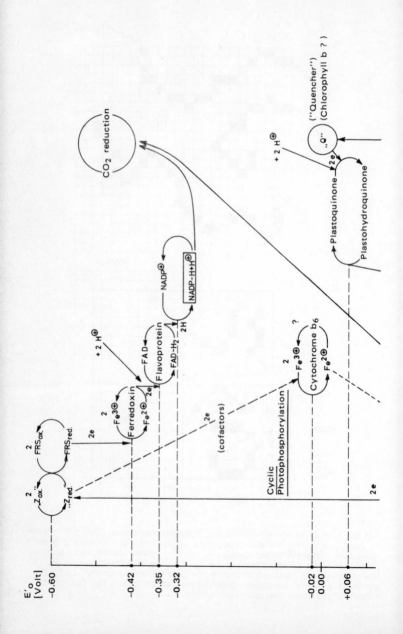

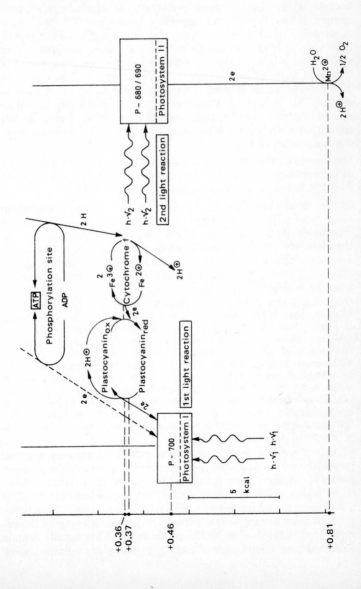

1 mole of red quanta (40 kcal/Einstein) is preserved as chemical energy (a potential difference of 1 volt is equivalent to about 23 kcal; cf. p. 22 f). The rest of the energy is unavoidably lost in stabilizing the reaction product.

Because of its high "electron pressure", the high-energy electron emitted from "P-700" and reversibly bound to "Z" flows in a "downhill" reaction over several steps to a stable end acceptor. Each step or level corresponds to a redox system, the potential of which is slightly more positive than that of the preceding one. The transition from an high-energy to a low-energy level proceeds via intermediate reactions along a chain of redox catalysts of enzymic nature. This biologically important principle of a step-wise reduction of electron pressure and hence of a step-wise decline in free energy has already been commented on (p. 22 f). We shall encounter it still a number of times.

For discussing the details of this light-dependent electron transport in chloroplasts we shall use the schematic presentation on p. 110 f which is based on the assumption of two coupled light reactions.

So far, three redox systems involved in the electron transport in connection with the 1st light reaction have been identified and thoroughly studied. The first, "FRS", must be placed near "Z", as already described on p. 108. The second redox system is ferredoxin, a water-soluble protein with a molecular weight of about 12 000, containing two iron atoms per molecule. It has been crystallized from chloroplasts (spinach) and from algae.

Other ferredoxins of this type (= "type 1") which are very similar to that of chloroplasts and may even replace it in photosynthesis, have been found in bacteria and blue-green algae (= "nonheme iron proteins" or "iron-sulphur proteins"). There are certain differences as to molecular weight and number of iron atoms per mole. All ferredoxins are characterized by an extremely low redox potential of about -0.42 volt which is uncommon for biological systems; it is practically identical with that of the hydrogen electrode ($H_2/H^\oplus$; cf. p. 21 f). Therefore, ferredoxins may act as electron carriers in reactions which release or utilize molecular hydrogen. It is hence plausible that a ferredoxin is closely associated with the enzyme *hydrogenase* (p. 136 f); another one is an essential factor in N_2-binding of several bacterial species (p. 296 ff).

From the reduced ferredoxin, which exhibits a strongly negative potential and an accordingly high electron pressure, the electron passes to another redox system fit for accepting electrons because of its higher potential (about -0.35 volt). It has been identified as a flavoprotein and has been purified from chloroplasts. These flavo-proteins, also termed *yellow enzymes,* represent a group of impor-tant enzymes involved in biological oxidation. They usually contain flavin adenine dinucleotide ("FAD"), in some cases flavin mononucleotide ("FMN") as a prosthetic group (cf. p. 17 f). This acts as a redox system; hydrogen is reversibly attached at two positions of the isoalloxazine system which is part of the active group. As explained previously (p. 18), this reduction may be interpreted as an uptake of electrons coupled with the binding of protons. The concept of electron transport as an expression of hydrogen transfer is also valid in photosynthesis.

The functioning of a flavoprotein, acting as an electron-carrier enzyme in a redox chain, depends upon its return to the oxidized state, by donating its attached hydrogen to another acceptor. These enzymes hence act as *transhydrogenases* or *reductases* (p. 18). The flavoprotein of chloroplasts has been termed *ferredoxin-NADP$^{\oplus}$ reductase* since it catalyzes the reversible transfer of hydrogen or of electrons to an enzyme, the coenzyme of which is nicotinamide adenine dinucleotide phosphate ("NADP", formerly termed "TPN"; redox potential -0.32 volt). The role of this important transport metabolite for hydrogen – together with NAD – has already been discussed (p. 13 ff).

With NADP representing the stable end acceptor for electrons released from excited "P-700" molecules, the electron transport chain connected with the 1st light reaction comes to an end. The reduced form, NADP-H $+$ H$^{\oplus}$, serves as "reduction equivalent", the formation of which had previously been postulated as a substantial result of the light-dependent reaction complex of photosynthesis. Hydrogen bound to this metabolite is required for the subsequent reduction of CO_2. We shall return later (p. 142 ff) to this complex reaction mechanism leading to the formation of carbohydrate. Another product formed in connection with the 1st light reaction and identical with the "energy equivalent" already mentioned, will also be treated later (p. 125 ff).

The biosynthetic pathway of CO_2 reduction leading to glucose was formerly distinguished as pathway of "dark reactions" from the light-dependent reactions of photosynthesis. Meanwhile, we know that quite a few intermediate reaction steps connected with the 1st and the 2nd light reaction may very well proceed in darkness, and that the actual light reaction is restricted to the process of quantum absorption by the various forms of chlorophyll a. Thus the distinction mentioned above is no longer reasonable.

"Photosystem II" and the "2nd Light Reaction". Van Niel has made an important contribution to photosynthesis research with his concept whereby the stable bond between hydrogen and oxygen in the water molecule is broken by radiant energy in the "primary photochemical reaction". While the hydrogen atoms thus released are linked via a series of light-independent intermediary reactions ("dark reactions") to carbon by a bond which is less stable than that with the oxygen in water, the oxygen atoms form oxygen molecules. Thus the hydrogen bound in the metastable carbon compound of the carbohydrate reaches a higher energy level compared with oxygen; an energy potential is established between the two. We may compare its formation with the energy-requiring tightening of a metal spring:

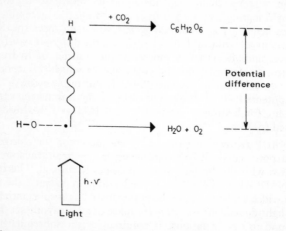

According to the present state of our knowledge, this "photolysis of water" with the release of molecular oxygen is the central event of the 2nd light reaction and takes place in "photosystem II".

Despite a certain superficial similarity, this process has nothing in common with hydrolysis, in which splitting of water occurs when hydrogen and the hydroxyl group form equally strong or stronger bonds with

$$H-OH + R-R' \rightleftharpoons R'-OH + R-H$$

other reactants. The significant difference is that in hydrolysis the free energy change is zero or slightly negative, as shown by the energy diagram (Fig. 67).

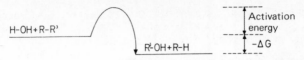

Fig. 67   Energy diagram of hydrolysis, slightly exergonic by nature.

Therefore, hydrolysis as a water-splitting reaction is eliminated from photosynthesis.

There is no chance either that H_2O molecules are cleaved directly into [H] and [OH] radicals by the action of radiation. Such a "photodissociation" requires a relatively high-energy radiation of at least 114 kcal per Einstein[*], corresponding to a wavelength of 260 nm. Since water only absorbs ultraviolet radiation of wavelengths not longer than 190 nm, photodissociation requires this or even a shorter wavelength if no appropriate sensitizer is present. Radiation of these wavelengths and its energy content are not tolerated by living cells; they will soon die. For the same reason, a direct photolysis of CO_2 or of C–H and N–H bonds results exclusively from ultraviolet radiation.

In the course of the history of the earth cleavage of water by high-energy ultraviolet radiation has obviously been of great significance for the synthesis of the first organic compounds. With the appearance of oxygen in the atmosphere, particularly of an ozone layer in the upper atmosphere, the amount of ultraviolet radiation striking the earth's surface was so drastically reduced that direct photodissociation of water came to an end.

The great achievement of photosynthetically active cells consists in having acquired appropriate pigments which enable them to use the biologically far less dangerous qualities of visible radiation instead of high-energy ultraviolet radiation for the cleavage of water. However, the energy of the effective red quanta amounts to about 40 kcal/Einstein only. In contrast to the photodissociation of water by ultraviolet radiation, more than one quantum is required to split one molecule of water. Therefore, there is no chance for a "one-quantum process". We shall discuss this problem of quantum requirement in photosynthesis later (p. 130 f).

The pigment distribution in "photosystem II" (Fig. 68) obviously differs from that of "photosystem I" mainly in that 1) chlorophyll b ("Chl$_b$-650") is one of the active "antenna pigments" and 2) a specific form of chlorophyll a, the "pigment-680/690" (= "P-680/690"), is the active molecule of the reaction center. Presumably absent are those forms of chlorophyll a which absorb between 680 and 700 nm, with the exception perhaps of "Chl$_a$-670" and "Chl$_a$-680" which appear to be constituents common to both pigment systems. Although uncertainty prevails as to the details of the proposed model, several lines of evidence indicate that "P-700" does not participate in "photosystem II" and is only active in "photosystem I".

---

[*] 1 Einstein = 1 mole of quanta, i.e. 6×10^{23} quanta.

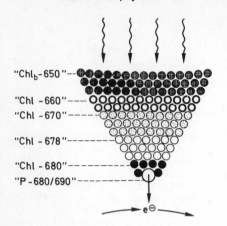

Fig. 68 Model of the pigment collective (= "trapping center") of "photosystem II" (schematic). The reaction center consists of a molecule of "P-680/690".

In other models carotinoid pigments, particularly xanthophylls, have been added to the "antenna pigments" of "photosystem II". In blue-green algae and red algae "Chl$_b$-650" is presumably replaced by phycobiliproteids, in diatoms and brown algae by chlorophyll c (p. 40).

Photolysis of Water. Little is known in detail about the reactions involved in the photochemical splitting of water and the concomitant release of molecular oxygen since the experimental analysis is extremely difficult.

The 2nd light reaction creates high-energy electrons as well, the origin of which is still obscure. It has been assumed that a molecule of "pigment-680/690" is excited by absorption of a light quantum and looses a high-energy electron. The oxidized pigment molecule now restores its negative electron balance at the expense of a $OH^{\ominus}$ ion derived from water, which may become a "hydroxyl radical" ("[OH]"). One molecule of oxygen and two molecules of water are formed from 4 [OH] radicals by a series of yet unknown intermediary reactions in which manganese ions fulfill a specific function. At least four quanta are thus required to release one oxygen molecule and four electrons, respectively. This process is postulated to proceed as follows:

$$4\ H_2O \longrightarrow 4\ H^{\oplus} + 4\ OH^{\ominus}$$
$$4\ OH^{\ominus} \longrightarrow 4\ (OH) + 4\ e$$
$$4\ (OH) \longrightarrow 2\ H_2O + O_2$$
$$\overline{2\ H_2O \longrightarrow O_2 + 4\ H^{\oplus} + 4\ e}$$

Formation of the [OH] radical as a highly oxidized photoproduct must be accompanied by the synthesis of a corresponding reduced equivalent, i.e. the electron emitted must be taken up by an appropriate acceptor. This function is taken over by a yet hypothetical substance "Q" (= "quencher"), a redox system the potential of which was determined indirectly as 0.00 volt. Its identity with a recently discovered substance "C-550" is under discussion. Since the redox potential for the system H_2O/O_2 is known to be $E_0' = +0.815$, the free energy change in the 2nd light reaction can be estimated. The potential step exerted by the emitted electron against the natural energy gradient in order to reach the higher energy level is about 0.8 volt. In its order of magnitude it equals that of the 1st light reaction.

According to another concept of the 2nd light reaction, an excited "P-680/690" molecule removes an electron from an as yet unidentified substance "Y" (potential: $+0.81$ volt) and donates it to a specific acceptor on a higher energy level (about 0.00 volt). The oxidized "$Y^{\oplus}$" is neutralized by taking up an electron from water. The water molecule, on removal of an electron, decomposes to one $H^{\oplus}$ ion and $^1/_4$ O_2 (below).

Theoretically, a quantum with an energy content of 40 kcal/mole would be sufficient to move two electrons simultaneously across the established potential difference; thus two quanta would be sufficient for the release of one molecule of O_2, and thereby also for the 2nd light reaction. However, an expenditure of four quanta per molecule of O_2 is more probable, even if it amounts to a surplus in energy. The latter is very likely utilized for ATP synthesis by the mechanism of "non-cyclic photophosphorylation" (p. 127).

"Hill Reaction". First evidence for the ability of isolated chloroplasts to evolve oxygen was provided by Hill and Scarisbrick (1940). They obtained continuous oxygen evolution during illumination of chloroplast preparations from *Stellaria media* and *Lamium album* in the presence of potassium ferrioxalate and potassium ferricyanide. It ceased when all the Fe^{3+} ions had been converted to Fe^{2+} ions. The amount of oxygen released was equivalent to the number of Fe^{2+} ions formed. This light-induced reduction process

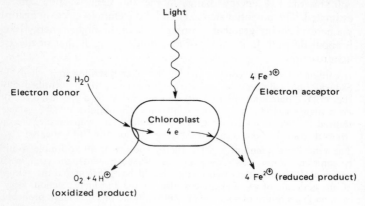

HILL reaction

is not directly dependent on CO_2, the function of which is taken over by artificial hydrogen acceptors or, rather, electron acceptors. Thus the photochemical reaction complex of photosynthesis is separated artificially from the dark reactions of CO_2 reduction. This "Hill reaction" of isolated chloroplasts has been of fundamental significance in photosynthesis research. It has been confirmed many times since and improved upon. For chloroplasts from *Beta* and *Spinacia* ferricyanide alone sufficed as an electron acceptor. The following conversions occur in this most simple Hill reaction:

$$4 \ K_3\left[Fe(CN)_6\right] + 2 \ H_2O + 4 \ K^+ \longrightarrow 4 \ K_4\left[Fe(CN)_6\right] + 4 \ H^+ + O_2$$
$$\Delta G° = +8\,500 \ cal/mole \ (pH \ 7{,}0)$$

It is still unknown which natural redox catalyst functions in the chloroplast as a carrier of the photochemically emitted electron and is actually oxidized by ferricyanide. It is also not certain whether both light reactions participate in the Hill reaction, although an "Emerson effect" has been described (p. 91 f). A number of other compounds have turned out to be suitable "Hill reagents", including such unphysiological substances as quinone and some dyes, particularly 2,6-dichlorophenolindophenol. They are converted quantitatively in the Hill reaction, but block other important photosynthetic processes. Although the Hill reaction had been viewed at first merely as a model reaction or even as an artifact, its physiological significance became obvious when, apart from oxygen evolution, the formation of NADP-H + H⊕ and ATP (p. 127) was demonstrated in isolated chloroplasts during illumination. NADP⊕ is therefore one of those compounds which function as electron acceptors in the Hill reaction.

2nd Light Reaction and Electron Transport. From the excited primary acceptor "Q" the high-energy electron probably passes to a redox system with a potential of $+0.06$ volt which is therefore closely associated with "Q"; it has been identified as the system plastoquinone/plastohydroquinone.

Attached to the aromatic ring in plastoquinone there are two oxygen functions and two methyl groups as well as a side chain comprising 9 isoprene residues (p. 430) with a total of 45 carbon atoms; hence the term "plastoquinone-45":

Plastoquinone (n = 9)

The reversible uptake of electrons by the plastoquinone system takes place with the participation of H^+ ions, in the same way as described for FAD (p. 113). Here also electron transfer is synonymous with hydrogen transfer:

Plastoquinone   Plastohydroquinone

It is very likely that only 10–20 % of the plastoquinones present in a chloroplast participate actively in the 2nd light rection. Besides plastoquinone-45, several other plastoquinones of known and unknown constitution as well as naphthoquinones and tocoquinones are present, the functions of which are unknown. Similar compounds have been detected in cells of the blue-green alga *Anacystis*.

In isolated chloroplasts, "photosystem II" is inactivated upon extraction of plastoquinone; addition of the isolated compound, however, fully restores the photochemical activity.

The plastoquinone system was the first one to be recognized of several electron carriers involved in the transfer of the high-energy electron derived from the 2nd light reaction. Their sequence in the effective electron transport chain is still unclear. They also include compounds from the class of "cytochromes", namely cytochrome f, cytochrome b_6 and cytochrome b_{559} ("b_3"), which have been isolated from chloroplasts. Similar compounds have been identified in phototrophic bacteria, blue-green algae and red algae.

Cytochromes are proteins with enzymatic properties. Their active group contains the same porphin ring system as chlorophyll. In cytochromes, however, the central magnesium atom is replaced by an iron atom. This prosthetic group termed "hemin" identifies the cytochromes as iron-porphyrin proteids. Hemoglobin, the peroxidases and catalases, etc. also belong to this group; for this reason they are termed *hemin enzymes* (for structure, see p. 237 ff). Their function consists in transferring electrons by valence change of the central iron atom which can exist in the divalent "reduced" form as $Fe^{2\oplus}$ or in the trivalent "oxidized" one as $Fe^{3\oplus}$. The latter is easily reduced to $Fe^{2\oplus}$ by uptake of an electron. An appropriate electron acceptor in turn can remove the electron from $Fe^{2\oplus}$ and restore the oxidized state:

$$Fe^{2\oplus} \underset{+e}{\overset{-e}{\rightleftharpoons}} Fe^{3\oplus}$$

This ability to change the valency of iron and the concomitant uptake or release of electrons makes each cytochrome a redox catalyst with a specific redox potential. Both the value of the latter and the specific absorption of visible radiation help to distinguish the individual cytochromes from each other. The spectral features are caused by the system of conjugated double bonds (p. 37) in the porphin structure of the active group.

Just as in chlorophyll and carotinoids, the intense color of cytochromes is brought about by conjugate double bonds. These compounds are most easily identified in the reduced state when their absorption spectra exhibit distinct differences, particularly in the blue and orange spectral regions. These are caused by slight differences in chemical structure, especially in the protein moiety of the individual compounds. As shown by the absorption spectrum of reduced cytochrome c (Fig. 69), characteristic absorption bands exist "α", "β", "γ" and "δ", which are used to classify the cytochromes. Of these, the "γ" or *Soret band* is particularly prominent. By the position of its maximum, cytochrome c can be distinguished from cytochrome f (Fig. 70).

Cytochrome f, which is very similar to cytochrome c of animal cells, is like cytochrome b_6 tightly bound to the chloroplast structure and becomes soluble only after destruction of the latter. Since its native functional activity is thereby abolished, investigations were restricted to direct spectroscopic measurements in intact chloroplasts as well as in small uni-

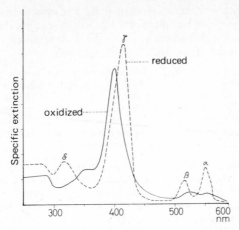

Fig. 69 Absorption spectra of reduced (broken line) and oxidized cytochrome c. The single absorption peaks are marked α, β, γ, δ.

cellular green algae and blue-green algae. Interest focussed mainly on rapid, light-induced absorption changes depending on a shift from the reduced to the oxidized state or vice versa (for the method of "difference spectrophotometry" used, see p. 249).

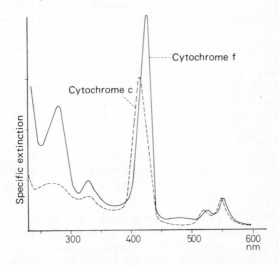

Fig. 70 Absorption spectra of reduced cytochrome c and cytochrome f (after Hill and coworkers).

It is still uncertain whether cytochrome b_6, like cytochrome f, participates in the electron transport of the 2nd light reaction though this compound may well be arranged near the plastoquinone system as indicated by its redox potential of about 0.00 volt.

Cytochrome b_3 (= "cyt. b_{559}"), bound to the native structure, possibly exists in two different forms: one with a low potential (about $+0.05$ volt) and another with a more positive one (about $+0.40$ volt). Since the first one is apparently reduced by "photosystem II", it has recently been placed near "Q" as a direct electron acceptor (see scheme on p. 110 f). In contrast, the high-potential form does not appear to be a constituent of the electron transport chain in connection with "photosystem II".

There is general agreement that another redox catalyst (potential about $+0.37$ volt) accepts the electron from the reduced plastoquinone. It is very likely identical with cytochrome f, while the direct contact with "photosystem I" is apparently made by "plastocyanin", a protein with a molecular weight of about 21 000 which contains copper atoms. This compound, being a "metalloproteid" (p. 382), exhibits a deep blue color in the oxidized state, caused by the copper content. It is rather tightly bound to the structure of chloroplasts and chromatophores. The plastocyanin content of a chloroplast amounts to that of cytochrome f; there are about 500 chlorophyll molecules for each Cu-proteid molecule.

The function of cytochrome f in connection with the 2nd light reaction has also been well confirmed experimentally. Due to its potential of about $+0.37$ volt, which is significantly higher (more positive) than that of plastoquinone or cytochrome b_3, a position near "photosystem I" is assigned to it. However, not cytochrome f but plastocyanin actually links the electron transport chain originating from the 2nd light reaction with "photosystem I". Whatever the details of this coupling between the two light reactions may be, there is general agreement that it is brought about by a tandem connection. Plastocyanin to a certain extent forms the plug connection via which electrons pass to all those molecules of "P-700" that have contributed an electron to reduction of $NADP^{\oplus}$ in the 1st light reaction and could not restore their electron balance by means of the cyclic electron transport coupled with "cyclic photophosphorylation" (see below).

The connection of the two photosystems through an electron-carrying chain results in a flow of electrons from water to NADP since the "P-700" molecule, having returned to the ground state, becomes excited again and supplies a high-energy electron to the acceptor "Z". From here two possible pathways are open to the electron, as discussed previously.

Methods to Locate the Photosystems and the Photosynthetic Electron Transport Chain in the Thylakoid Membrane. In order to elucidate the exact position of a redox catalyst in photosynthetic electron flow, artificial and physiological donors and acceptors have been used with great success. By combining these systems, electrons may be fed in or tapped off from various parts of the transport chain. Thus the position of electron carriers in a definite region of the transport chain as well as their relation to the two photosystems and their possible participation in photophosphorylation can be studied (for details see the following scheme).

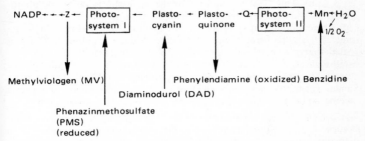

Artificial electron donors and electron acceptors active in photosynthetic electron transport (after Trebst and Hauska)

Inhibitors which specifically block "photosystem I" or "photosystem II" in isolated chloroplasts or chloroplast fragments proved particularly useful. Thus *3(3,4-dichlorophenyl) -1,1 dimethylurea* (= "DCMU") inhibits "photosystem II"; since "photosystem I" is unaffected, its characteristics as well as the coupled electron transport may be investigated. For this purpose an appropriate electron donor must be added to the system, e.g. ascorbic acid. *Dibromothymoquinone* (= "DBMIB") turned out to be a specific inhibitor of "photosystem I"; in its presence only "photosystem II" is active, provided an artificial acceptor for electrons, e.g. potassium ferricyanide has been added.

Finally, the use of "chemical or molecular probes" as well as the application of lipophilic or hydrophilic redox compounds should be mentioned. Substances of the first group, because of their charge and molecule size, will react exclusively with redox systems located on the outer surface of the thylakoid membrane and hence easily accessible. *P-(diazonium)-benzenesulfonic acid* (= "DABS") inhibits a redox catalyst on the acceptor site of "photosystem I". This suggests that the acceptors of the latter are prominently located on the outside of the membrane, i.e. on its matrix side (cf. Fig. 42, p. 76). Studies with artificial acceptors of hydrophilic or lipophilic nature have led to the same conclusion. Both kinds were capable of accepting electrons from "photosystem I"; however, only the lipophilic substances which penetrate the lipid phase of the membrane could serve as electron donors for "photosystem I" (e.g. *2,6-dichlorophenolindophenol*, reduced *methylphenazonium methosulfate*). Obviously its donor site is arranged on the inner surface of the lipid phase, i.e. on the side facing the intrathylakoid space (cf. Fig. 42). The precise position of "photosystem II" is still far from clear; several lines of evidence indicate that the position of the acceptor site is inside the lipid region of the membrane, others support the view that it is located near the outer thylakoid surface, but still in the lipid region.

Therefore, one is tempted to speculate that "photosystem I" is associated with smaller particles (about 110–120 Å in diameter) embedded in the outer surface of the thylakoid membrane, "photosystem II", however, with the larger ones (about 175 Å in diameter) attached to the inner surface (cf. p. 75 ff). This view is supported by the successful isolation of "photosystem I" and "photosystem II" in two separate membrane fractions by gentle fractionation of chloroplasts (for methods see p. 92 ff). The structural analysis, performed with the freeze-etching technique (p. 66 f) revealed that the membrane fraction with "photosystem II" activity consisted mainly of particles about 175 Å in diameter, the one with "photosystem I" activity, however, only of particles 110–120 Å in diameter. Similar fraction experiments yielding the two photosystems in separate membrane fractions have been reported for several algal groups, too.

The concept whereby membranes of the stroma thylakoids and end membranes of grana contain exclusively the smaller particles 110–120 Å in diameter (probably "photosystem I"), whereas membranes of the grana thylakoids in the "partition region" (p. 75) contain in addition the larger ones about 175 Å in diameter (accordingly "photosystem II") awaits experimental proof. Studies with tobacco mutants lacking grana in their chloroplasts and with "C_4 plants" (p. 155), the bundle sheath cells of which contain mostly grana-free chloroplasts, have yielded similar results: in both cases only "photosystem I" activity has been found.

The spatial arrangement and the functional attachment of the two photosystems in the photosynthetically active structures is also indicated by the appearance of characteristic fluorescence changes upon the onset of illumination. The fast short-term increase observed immediately might be explained as indicating that "photosystem II" briefly wastes a certain amount of absorbed radiant energy in fluorescence before it can *spill over* into the weakly fluorescent "photosystem I". This would explain the subsequent decrease of fluorescence and the slower fluctuation until a steady state is reached after several minutes (cf. p. 97).

With the methods described here new insights have been provided into the arrangement of the photosystems and the various electron-carrying catalysts in the thylakoid membrane, and their possible association with already known substructures. Moreover, they have laid the foundation for the concept of a "chemiosmotic mechanism" of photophosphorylation which we shall discuss in detail later (p. 128 f).

Photophosphorylation

Besides the reduction equivalents of NADP-H + H$^\oplus$, "energy equivalents" are required as well for the endergonic process of carbohydrate synthesis. They are formed as the second stable end product of the two light reactions. This process comprises the conversion of radiant energy into chemical energy, which, as mentioned previously, is the most important feature of photosynthesis. Arnon and his coworkers (Berkeley/California) have made valuable contributions towards the clarification of these processes.

We shall first consider the yield of chemical energy coupled with the 1st light reaction.

Cyclic Electron Flow. Apart from the pathway leading to NADP there is a second one open to the high-energy electron promoted to "Z" which finally leads it back to the oxidized "P-700" molecule (cf. reaction scheme p. 110 f). Electron-carrier enzymes, i.e. redox systems ("cofactors") participate here, too; they direct and regulate the "cyclic" electron movement in such a way that a spontaneous return of the emitted electron is prevented.

Without this precaution, the excitation energy invested would be re-emitted as fluorescence or heat, and would not be available for carrying out chemical work. Because of the inserted cofactors, it is released in small portions and thereby conserved in a chemical form. We have encountered the same principle of lowering the energy of the electron gradually, in a series of small steps, in the electron transfer from "Z" to NADP. The high-energy electron resembles a ball which falls from one step down to the next one, setting free a portion of its inherent energy. The highest step would correspond to the redox system "Z" and the lowest one to the oxidized "P-700" molecule. This acts first as a donor, then as an acceptor of the high-energy electron. The members of the electron transport chain are redox systems, the nature and sequence of which are still unknown.

Although the participation of ferredoxin as well as that of cytochrome f and plastoquinone as electron-carriers in the cyclic electron flow is under discussion, it has not been confirmed by the results obtained with algae and chloroplasts. Cytochrome b_6 (= "Cyt. b_{563}"), however, appears to be one of the electron-carrier enzymes in the cyclic electron transport – at least in several algae, since this redox catalyst is oxidized and also reduced by "photosystem I" in cells of *Chlamydomonas* and *Porphyridium*.

The potential gradient traversed by the electron in its circular route back to the oxidized chlorophyll molecule "P-700" amounts to at least 1.0 volt. This strongly exergonic process is a device by which – coupled to a specific endergonic reaction – this energy is to some extent transformed to a chemical form. Here, we meet the biological principle generally underlying the production of chemical energy by the living cell (cf. p. 19 f). In accordance with its function as a specific carrier of chemical energy in the cell, ATP also participates in the energy-conserving step of the 1st light reaction. By coupling with this step ADP is converted to ATP:

$$ADP + \text{inorganic phosphate} \longrightarrow ATP + H_2O$$

$$\Delta G^O = + 7000 \text{ cal/mole}$$

Part of the energy released is conserved in the form of an "energy-rich" bond (p. 19) in the ATP molecule and is available as an "energy equivalent" for the endergonic reaction complex of photosynthetic CO_2 reduction.

As this phosphorylation practically only occurs in light and consumes neither oxygen nor energy-rich substrate, as does ATP synthesis linked to cellular respiration (= "oxidative phosphorylation"; see p. 241 f), it is hence termed "photophosphorylation". Recent studies show that photosynthetic ATP formation may proceed via two different reaction pathways. One of these is identical with the circular ("cyclic") electron transport described above by which the high-energy electron returns to "P-700"; it is termed *cyclic photophosphorylation*.

Cyclic photophosphorylation can be demonstrated in isolated chloroplasts or chloroplast fragments when illuminated under nitrogen in a medium containing ADP, inorganic phosphate, $Mg^{2\oplus}$ and an appropriate redox catalyst. Methylphenazonium methosulfate, vitamin K_3 and anthroquinone sulfonic acid have turned out to be particularly suitable for this purpose. More recently, ferredoxin has been added. An important condition for cyclic photophosphorylation is the blocking of the oxygen-evolving "photosystem II" by means of inhibitors. The most effective one is the herbicide DCMU (p. 123).

Doubts have been raised recently as to whether cyclic photophosphorylation measured in the way described above, is of much importance in higher plants.

Noncyclic Electron Flow. An electron transport chain also forms the basis of the second mechanism of light-dependent phosphorylation. This is distinguished from the first one as *noncyclic photophosphorylation*. In this case, the high-energy electron emitted from "photosystem II" does not return to the oxidized chlorophyll molecule via a circular chain of redox catalysts, but finally reduces $NADP^{\oplus}$ (cf. p. 113). The decline in free energy of the electron transfer between "Q" and cytochrome f is sufficient for the phosphorylation of ADP to ATP. It has been assumed that the phosphorylation site is located between plastoquinone and cytochrome f. There are several indications that this reaction site may also serve cyclic phosphorylation, as shown in the overall reaction scheme (p. 110 f). It is an open question whether cyclic phosphorylation has, in addition to this common site for ATP formation, one of its own. As our reaction scheme shows, a location between ferredoxin and cytochrome b_6 (potential drop of about 0.40 volt) in the course of the cyclic electron flow could be possible. It has not yet been confirmed experimentally.

In isolated chloroplasts stoichiometric amounts of ATP, NADP-H + $H^{\oplus}$ and oxygen were found to be formed in noncyclic photophosphorylation:

$$NADP^{\oplus} + H_2O + ADP + H_2PO_4^{\ominus} \xrightarrow{\text{Light}} NADP\text{-}H + H^{\oplus} + ATP + H_2O + 1/2\ O_2$$

This correlation is also valid when typical Hill reagents (p. 118 f) are used. Since the redox potentials of the individual electron acceptors vary considerably, the amount of bound chemical energy also varies.

Photophosphorylation in vivo has been demonstrated mostly in algal cells (*Ankistrodesmus, Chlorella, Anacystis*) by measuring the ^{32}P incorporation into ATP from added ^{32}P-orthophosphate, as well as by the characteristic changes in the distribution of organic P-compounds resulting from a periodic change of light and dark periods. In absence of CO_2 the photophosphorylation that takes place is thought to be exclusively of the cyclic type; this is indicated by its independence of "photosystem II" and of specific electron acceptors of noncyclic electron flow as well as by the action spectrum (p. 85) obtained. Presumably, cyclic photophosphorylation is not directly involved in the CO_2 reduction of these algae.

The contribution of cyclic and noncyclic photophosphorylation to the total ATP synthesis in light is difficult to estimate. The noncyclic type is rather prominent in those cells which are illuminated in the presence of an appropriate electron acceptor such as CO_2, $NO_3^{\ominus}$ and $SO_4^{2\ominus}$. Noncyclic photophosphorylation obviously also provides the additional portion of ATP which, apart from CO_2 reduction, is required for endergonic reactions within and outside the chloroplasts (cf. p. 158 f).

Mechanisms of Photophosphorylation. Very few details are known of the chemical mechanism of light-dependent ATP synthesis in chloroplasts.

As an alternative to the *chemical mechanism* of photophospho-rylation described above, which requires a coupling of electron transport and phosphorylation, a *chemiosmotic hypothesis* (Jagendorf and Uribe) has been formulated.

The *chemiosmotic hypothesis* is based on a concept originally developed by Mitchell to explain ATP formation in mitochondria (cf. p. 241 ff) and was later modified with respect to the rather similar processes in chloroplasts and bacterial chromatophores. According to this concept, electron transport and ATP synthesis proceed independently from each other in respect of time and site. Originating from a directed flow of $H^\oplus$ ions across the membrane, the energy of the redox reactions taking place establishes an electrochemical proton gradient. In contrast to the chemical mechanism, the membrane as an entity functions as the "high-energy intermediate" (p. 242) or energy carrier. In chloroplasts, the redox systems of the light-induced electron transport in the membrane of the grana thylakoids play a key role. As discussed earlier (cf. p. 18) electron transport may proceed partially with the participation of protons, e.g. from "Q" to plastoquinone or from cytochrome f to plastocyanin. Because of the arrangement of the redox systems in the membrane structure, the required protons are only taken up from the outside, but released exclusively into the intrathylakoid space. A directed photosynthetic electron flow across the thylakoid membrane, resulting from a specific arrangement of the photosystems and the redox catalysts, is shown by the hypothetical model in Fig. 71 A in a schematic way (= *zigzag scheme*). Protons are accumulated in the inner space of the thylakoid (= lowering of the pH value!). The gradient of proton activity or of membrane potential formed is to a certain extent equivalent to the "high-energy intermediate" (see above). It brings about ATP synthesis by membrane-bound enzymes *(ATPases)*, the structural arrangement of which causes a proton to move from the inner compartment to the outer one, while a hydroxyl ion moves in the opposite direction (Fig. 71 B). Since both ions react immediately with the appropriate partner to form water on each side of the membrane, the concentration of the latter at the "phosphorylation site" is kept low and ATP synthesis is thus favored. The participation of high-energy compounds cannot be excluded. When ATP consumption is low and its synthesis is accordingly reduced, the increasing "pressure" of the proton gradient will gradually inhibit the electron transferring reactions involved in its formation. This regulatory effect is suspended with beginning or increasing ATP synthesis. According to the *chemiosmotic hypothesis*, "uncouplers" of photophosphorylation (p. 130) enable protons to migrate through the membrane from the interior to the outside, thus preventing formation of a gradient and subsequent ATP synthesis. This would account for the inhibitory effect of detergents on photophosphorylation.

Several important details have recently been elucidated: 1. On illumination of isolated thylakoids with an unimpaired electron transport, protons are selectively accumulated from the surrounding medium within the inner space of the thylakoids, causing a drop of its pH value by about 3.5 units. 2. An artificial pH gradient induced in darkness also gives rise

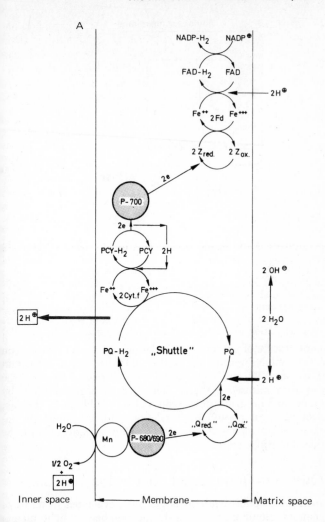

Fig. 71 "Chemiosmotic Hypothesis" of photophosphorylation. A: Proposed mechanism of a directed proton transfer across the thylakoid membrane resulting from specific arrangement of the photosystems and the redox catalysts involved in photosynthetic electron flow (after Trebst and Hauska; schematic). Cyt. f = cytochrome f; Fd = ferredoxin; PCY = plastocyanin; PQ = plastoquinone; Q = quencher. B: ATP synthesis by *ATP-ase* as a result of proton accumulation in the intrathylakoid space giving rise to an electrochemical gradient or membrane potential (after Jagendorf). For details, see text.

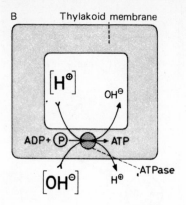

Fig. 71 B

to ATP synthesis. 3. In illuminated chloroplasts and isolated thylakoid membranes from phototrophic bacteria an electrical field is formed via membrane-bound electron flow, yielding a positive charge inside and a negative charge outside the thylakoid. Although other important details await experimental confirmation the *chemiosmotic hypothesis* has provided some elementary facts for the understanding not only of phosphorylation but also of ion uptake and ion exchange, of electron transport and charge separation in chloroplasts and mitochondria.

A common feature of photophosphorylation and the ATP formation coupled to cellular respiration is the inhibitory effect of certain substances on ATP synthesis during the electron transport; the energy is then released as heat. This effect is brought about by $NH_4^{\oplus}$ ions, sodium azide, desaspidin and salicylaldoxime; they act as "uncouplers" of photophosphorylation. Other compounds uncouple the oxidative phosphorylation as well or selectively block the latter, like dinitrophenol (cf. p. 243).

The Quantum Requirement of Photosynthesis

The efficiency of photosynthesis is generally evaluated in terms of its "quantum requirement". This is the number of light quanta required to form one molecule of oxygen (O_2). The expression "quantum yield", i.e. the photosynthetic work done by one quantum, is also used.

On the basis of the concept that the light reactions co-operate by means of a tandem or series connection, four electrons must pass along the whole reaction chain from water to NADP in order that one molecule of oxygen may be released. Two quanta of light per electron are required, giving a theoretical quantum requirement of eight for one molecule of O_2. In fact, most experiments have yielded values of 8–10 quanta. In measuring the quantum requirement of photosynthesis, isolated chloroplasts have been preferred to algal cells and intact leaves, since measurements in the latter are often subject to error due to interference of other processes (respiration, induction periods, growth). The values obtained in isolated chloroplasts were generally a bit higher, amounting to 10–14 quanta per molecule of oxygen.

References

Amesz, J.: The function of plastoquinone in photosynthetic electron transport. Biochim. Biophys. Acta (Amst.) 301: 35, 1973

Arnon, D. I.: Role of ferredoxin in photosynthesis. Naturwissenschaften 56: 295, 1969

Avron, M., J. Neumann: Photophosphorylation in chloroplasts. Ann. Rev. Plant Physiol. 19: 137, 1968

Bendall, D. S., R. Hill: Haem-proteins in photosynthesis. Ann. Rev. Plant Physiol. 19: 167, 1968

Bishop, N. I.: Partial reactions of photosynthesis and photoreduction. Ann. Rev. Plant Physiol. 17: 185, 1966

Bishop, N. I.: Photosynthesis: The electron transport system of green plants. Ann. Rev. Biochem. 40: 197, 1971

Björkmann, O., J. Berry: High-efficiency photosynthesis. Scientific Amer. 229 No 4: 80, 1973

Fork, D. C., J. Amesz: Action spectra and energy transfer in photosynthesis. Ann. Rev. Plant Physiol. 20: 305, 1969

Gregory, R. P. F.: Biochemistry of photosynthesis. Wiley-Interscience. London, New York, Sydney, Toronto 1971

Heath, O. V. S.: Physiologie der Photosynthese. Thieme Verlag, Stuttgart 1972

Hind, G., J. M. Olson: Electron transport pathways in photosynthesis. Ann. Rev. Plant Physiol. 19: 249, 1968

Jagendorf, A. T.: Acid-base transitions and phosphorylation by chloroplasts. Fed. Proc. 26: 1361, 1967

Jagendorf, A. T., E. Uribe: Photophosphorylation and the chemiosmotic hypothesis. In: Energy conversion by the photosynthetic apparatus. Brookhaven Symposia in Biology 19: 215, 1966

Kandler, O., W. Tanner: Die Photo-

assimilation von Glucose als Indikator für die Lichtphosphorylierung in vivo. Ber. dtsch. botan. Ges. 79: 48, 1966

Ke, B.: The primary electron acceptor of photosystem I. Biochim. Biophys. Acta (Amst.) 301: 1, 1973

Kreutz, W.: Neue Untersuchungen zur molekularen Architektur der Thylakoide: Erste Hinweise für eine „Protonenpumpe". Ber. dtsch. botan. Ges. 82: 459, 1969

Pirson, A., H. Lorenzen: Synchronized dividing algae. Ann. Rev. Plant Physiol. 17: 439, 1966

Rabinowitsch, E., Govindjee: Photosynthesis. Wiley & Sons, New York 1969

Robertson, R. N.: Protons, electrons, phosphorylation and active transport. Cambridge Monogr. Biol. 15, 1968

Rumberg, B., U. Siggel: pH-Changes in in the inner phase of the thylakoids during photosynthesis. Naturwissenschaften 56: 130, 1969

Tamiya, H.: Synchronous cultures of algae. Ann. Rev. Plant Physiol. 17: 1, 1966

Trebst, A., H. Bothe: Zur Rolle des Phytoflavins im photosynthetischen Elektronentransport. Ber. dtsch. botan. Ges. 79: 44, 1966

Trebst, A.: Energy conservation in photosynthetic electron transport of chloroplasts. Ann. Rev. Plant Physiol. 25: 423, 1974

Witt, H. T.: Neuere Ergebnisse über die Primärvorgänge der Photosynthese. Umschau 66: 589, 1966

Witt, H. T., B. Rumberg, P. Schmidt-Mende, U. Siggel, B. Skerra, J. Vater, J. Weikard: Über die Analyse der Photosynthese im Blitzlicht. Angew. Chemie 77: 821, 1965

Photosynthesis without Evolution of Oxygen

Bacterial Photosynthesis

As mentioned previously, there are some species of bacteria able to carry out photosynthesis. Bacteriochlorophyll and a few carotinoids absorb the radiation. These pigments are bound to the characteristic thylakoid structures. Bacterial photosynthesis is different insofar as oxygen is not formed as an end product. In fact, bacterial photosynthesis only functions in the absence of oxygen, i.e. in the state of *anaerobiosis*. A further point of difference from photosynthesis of green plants is that light-dependent reduction of carbon dioxide is coupled with oxidation of inorganic or organic substances. In the presence of oxidizable substrate, only a few species of these phototrophic bacteria will tolerate oxygen in darkness.

The current classification of the photosynthetically active bacteria into three groups is based in part on their coloring, in part on their metabolic features:

1. *Green Sulfur Bacteria,* Chlorobacteriaceae *(Chlorobium, Chloropseudomonas)*
2. *Purple Sulfur Bacteria,* Thiorhodaceae *(Chromatium, Thiospirillum)*
3. *Purple Non-Sulfur Bacteria,* Athiorhodaceae *(Rhodospirillum, Rhodopseudomonas, Rhodomicrobium)*

The green sulfur and purple sulfur bacteria are strict anaerobes and grow mainly autotrophically, i.e. in purely inorganic media. Their light-dependent CO_2 reduction is coupled to the oxidation of inorganic sulfur compounds: hydrogen sulfide, sulfite or thiosulfate ("photolithotrophic"). The purple non-sulfur bacteria utilize organic compounds anaerobically not only as hydrogen donors for the reduction of CO_2 in the light, but also include them in their cellular metabolism, often together with CO_2 ("photoorganotrophic"). In darkness, however, organic compounds are heterotrophically metabolized in the presence of oxygen.

The green sulfur bacteria are distinguished from purple bacteria on the basis of their coloring. The former lack the typical aliphatic carotinoids which absorb between 440 and 560 nm, particularly spirilloxanthin (p. 59 f). In the purple bacteria, this pigment masks the green color of bacteriochlorophyll so that a red-violet to brown coloring results. This kind of classification does not reflect the variety of metabolism observed in photosynthetically active bacteria. A few species thus represent intermediate forms between the three groups mentioned above.

As yet there is still no clear concept of the mechanism of bacterial photosynthesis. This is due to the multiplicity of the possible metabolic reactions and the physiological plasticity of these organisms. The findings obtained up to now and their interpretation have given rise to two contrary concepts on the nature of photosynthesis without oxygen formation. According to one the action of light is restricted to formation of "energy equivalents", i.e. of ATP, by photophosphorylation while the formation of "reduction equivalents" is only facultatively connected with it or even proceeds independently of the light reaction. The photochemical apparatus of the phototrophic bacteria would then be constructed in a way very similar to "photosystem I" of the higher plants and algae. The finding that the "Emerson effect" (p. 91 f) cannot be demonstrated in *Rhodospirillum rubrum* is consistent with this first concept. The second interpretation is based on the uniformity of the photosynthetic mechanism in all chlorophyll-bearing organisms. It was modified in phototrophic bacteria, probably in the course of evolution, in that

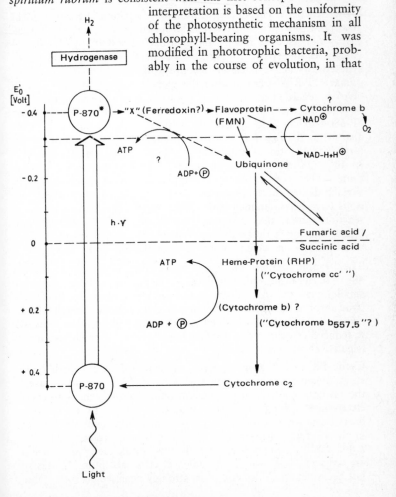

no release of oxygen takes place. The oxidation product from the "photolysis of water" is reduced to water by hydrogen from an inorganic or organic donor – according to the theory of Van Niel – and so disposed of.

The first concept on the nature of bacterial photosynthesis can be illustrated best using the purple non-sulfur bacterium *Rhodospirillum rubrum*. This organism has been studied relatively thoroughly, and represents a kind of "model organism" of phototrophic bacteria.

Light Reaction. The demonstration of short-term light-dependent absorption changes of bacteriochlorophyll in vivo led to the discovery of a special form: "pigment-870" ("P-870") which is obviously characteristic for species of all three groups of photosynthetically active bacteria *(Rhodospirillum, Chromatium, Chlorobium)*. "P-870" (termed after the specific absorption maximum in vivo) is regarded as the actual photochemically active chlorophyll of these bacterial species, and hence corresponds to "P-700" and "P-680/690" in chloroplasts. By analogy with the latter, "P-870" also forms the reaction center of an aggregation of bacteriochlorophyll and carotinoid molecules which act as "antenna pigments" (p. 107), absorbing radiant energy and transferring it to the reaction center. They may be isolated as pigment-protein complexes from thylakoids or "chromatophores" (p. 169 f). Moreover, particle fractions have been obtained which were apparently identical with the reaction center: they contained exclusively "P-870", but no "antenna pigments"; they exhibited the specific absorption of "P-870" of intact cells and were photooxidized on illumination.

The "trapping center" for energy (cf. p. 107 f) is thought to contain 40–50 bacteriochlorophyll molecules; the pigment collective is hence smaller than that of "photosystem I" in chloroplasts (p. 106 f). On light absorption, which can be measured as a weak, reversible bleaching, "P-870" emits a high-energy electron and becomes oxidized. A reduced cytochrome molecule of the "c-type" compensates for its negative electron balance.

Cyclic Electron Transport. First, we shall follow the path of the emitted electron (see reaction scheme on p. 133). As in the case of the 1st light reaction of green plant photosynthesis the high-energy electron, following an energy gradient, returns to the oxidized bacteriochlorophyll molecule via several redox catalysts arranged in an electron transport chain. The nature and arrangement of the redox catalysts which participate in this "cyclic electron transport" are only partly known. The primary electron acceptor "X" has still not been identified. Because of its strongly negative potential (about

–0.4 volt), it could be a ferredoxin (p. 112). A corresponding compound has been extracted and purified from cells of *Chromatium*; it is firmly bound to the thylakoid structure. Cytochrome c_2 and an ubiquinone (structure see below) almost certainly belong to the electron transport chain. The participation of a specific flavoprotein (with flavine mononucleotide, "FMN", as active group, p. 17) which would be located between "X" or ferredoxin and ubiquinone, is as yet uncertain. Ubiquinone obviously helds a key position in the cyclic electron transport: in *Chromatium*, this redox catalyst (potential 0.00 volt) probably establishes the contact with external electron donors (see below).

Ubiquinones are chemically related to the plastoquinones (p. 119 f) of the chloroplasts. As their name indicates, they are widely distributed in nature. This holds particularly true for "coenzyme Q", which has been found in many animal and plant cells (see "respiratory chain", p. 235 ff).

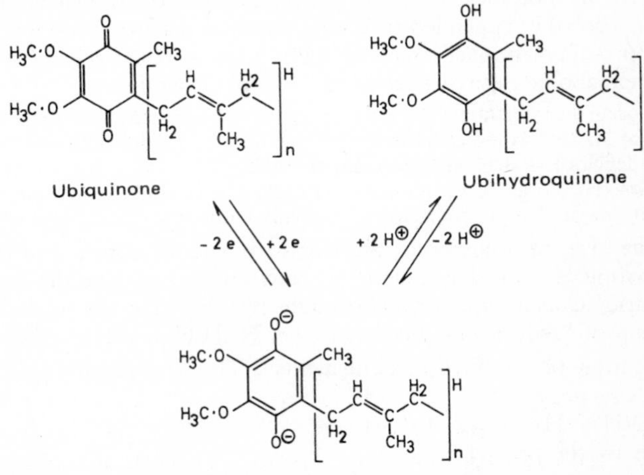

Ubiquinone

Ubihydroquinone

Ubiquinones, like all quinones, consist of an aromatic core and a side chain (structure, see above). In addition to the methyl group in the ortho-position, they have two methoxy groups. The side chain contains a various number of isoprene residues; the compounds are accordingly termed UQ_2, UQ_6, UQ_{10}, etc. The structure of ubiquinones in phototrophic bacteria is still unknown. Hydrogen or electrons are transferred by the same mechanism as in the plastoquinone/plastohydroquinone system (p. 119 f).

Ubiquinone transfers the electron to a "heme protein" ("RHP" = *Rhodospirillum* hemeprotein, "cytochrome cc'''") which is obviously only present in photosynthetically active bacteria. It is eluted from "chromatophores" (p. 69 f) by intensive washing with buffer. The electron transport is thereby blocked, but may be restored by addition of isolated "RHP". Another hemin enzyme, cytochrome b (which is also a native component of chromatophores), is a potential electron acceptor for reduced RHP. The cycle is closed by cytochrome c_2: this redox catalyst transfers an electron to the oxidized molecule of "P-870" and thus balances its electron deficit.

Photophosphorylation. The biologically important principle underlying cyclic electron transport has already been explained (p. 125 f). In the photoreaction of bacterial photosynthesis, too, the drop in potential of about 0.8 volt resulting from the return of the high-energy electron, is used to conserve chemical energy via the ADP/ATP system. The experimental data indicate that two sites of ATP formation exist along the electron transport chain; these are recorded in our scheme (p. 133).

In contrast, the site for the formation of "reduction equivalents" is certainly not located within the cyclic electron transport chain but is probably appended to it as a terminal reaction involving the flavoprotein mentioned above (p. 135). NAD serves as specific transport metabolite for hydrogen in bacterial photosynthesis.

The occurrence of noncyclic electron transport terminated by the reduction of $NAD^{\oplus}$, as indicated in the scheme on p. 133, implies that the electron deficit of "P-870" must be compensated by means of external electron donors (H_2S, organic compounds). However, it is doubtful whether this "light-driven electron transport" is actually employed.

At the level of ubiquinone, an accessory reaction comes into play, consisting of a transfer of electrons between the cycle and the system fumaric acid/succinic acid. This might possibly give rise to the formation of "reduction equivalents", i.e. $NAD\text{-}H + H^{\oplus}$ (see below).

Formation of "Reduction Equivalents". The first possibility for a synthesis of $NAD\text{-}H + H^{\oplus}$ is given by the process of *photoreduction:* molecular hydrogen is activated by the mediation of the enzyme *hydrogenase,* cooperating with ferredoxin and the photochemical system in order to form a bond with the coenzyme:

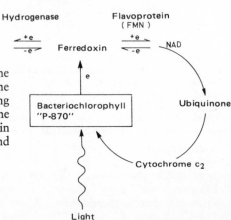

This process has also been demonstrated in unicellular green algae (p. 139 ff).

The *hydrogenase* reaction may also proceed in the opposite direction: when a surplus of NAD-H + H$\oplus$ results from a shortage in reducible substrates (CO_2, $NH_4\oplus$), this reaction, as it were, opens a valve for the "pressure of reductive capacity" which will then leave the cell as molecular hydrogen. Via the *hydrogenase* system the oxidized state of the transport metabolite is restored as well as the normal equilibrium between NAD$\oplus$ and NAD-H + H$\oplus$ in the cell. This light-dependent evolution of hydrogen is termed *photoproduction of H_2* (p. 141).

The second possible formation of NAD-H + H$\oplus$ apparently takes place outside the photochemical reaction complex and utilizes exclusively organic substrates via an anaerobic reaction cycle largely corresponding to the "citric acid cycle" which will be discussed later (p. 219 ff). The carbon skeleton of the organic compounds is fed into cell metabolism (p. 139).

Such a reduction of NAD$\oplus$, for instance by succinic acid, requires the participation of ATP or another energy-rich compound. In this case, the electrons must be moved against the energy gradient over a potential difference of 0.32 volt (succinic acid/fumaric acid: $E_0' = 0.00$ volt; NAD$\oplus$/NAD-H + H$\oplus$: $E_0' = -0.32$ volt). The mechanism of "reversed electron flow"

$$\text{Succinic acid} + \text{NAD}^{\oplus} \xrightarrow{\text{+ ATP or X} \sim} \text{fumaric acid} + \text{NAD} - \text{H} + \text{H}^{\oplus}$$

operating here seems to be a typical feature of bacterial photosynthesis. It was first observed in oxidative phosphorylation (p. 243 f) of animal mitochondria. The physiological significance of feeding electrons over a by-path into the cyclic transport chain lies in the fact that substances with a redox potential more positive than that of NAD$\oplus$ may reduce it if ATP is available. Indirectly, this amounts to a raising of electrons to the energy level of NAD$\oplus$ by radiation energy.

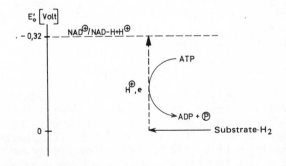

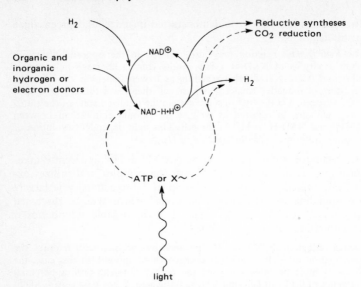

The "exogenous" entry of electrons causes the cyclic electron transport to become an open or "noncyclic" one.

An energy-driven reduction of NAD⊕ could also occur in green sulfur and purple sulfur bacteria if the redox potential of the inorganic substrates serving as hydrogen donors is similarly unfavorable. The scheme above summarizes the pathways of NAD-H⊕ formation which have been discussed.

A further, as yet theoretical possibility of NAD-H + H⊕ formation would consist in a tight coupling of the light reaction and the electron transport depending on it with a hydrogen or electron donor such as hydrogen sulfide or thiosulfate.

The light-dependent oxidation of inorganic sulfur compounds by green sulfur and purple sulfur bacteria was recognized rather early, since the free sulfur formed is deposited in easily discernible droplets in the cells. The chemical conversion taking place is represented by the following overall equation:

$$6\ CO_2 + 12\ H_2S \xrightarrow{\text{light}} C_6H_{12}O_6 + 12\ S + 6\ H_2O$$

Reduction of CO_2 in Light. The conversion of CO_2 to carbohydrate in phototrophic bacteria is apparently brought about by the same or a very similar reaction mechanism as the one in green algae or higher plants, i.e. by the "Calvin cycle", which will be discussed a bit further on (p. 145 ff). The key enzymes, particularly *carboxy-dismutase* (p. 148) have been found in various species of phototrophic bacteria. Pulse labeling experiments with $^{14}CO_2$ (p. 143 f) gave rise to the same rapidly labeled compounds as in green plant photosynthesis. Since NAD assumes the role of a transport metabolite for hydrogen, the *phosphotriose dehydrogenase* (p. 149) of these organisms is specifically activated by NAD-H + H$^{\oplus}$.

In illuminated photoorganotrophic bacteria, in addition to reduction of CO_2 to carbohydrate, a more or less intensive assimilation *(photoassimilation)* of those organic compounds occurs (acids, alcohols) which serve as hydrogen donors. Since various transitions between the two processes are possible, the process of photosynthesis in these organisms becomes rather heterogeneous and thus difficult to elucidate. Moreover, both *photolithotrophic* and *photoorganotrophic* bacteria produce precursors for amino acids and other compounds via a "reductive carboxylic acid cycle" (reversal of the "citric acid cycle"! p. 219). In this process, CO_2 is fixed with consumption of NAD-H + H$^{\oplus}$ and ATP.

Photoreduction in Green Algae

The ability to assimilate CO_2 in light under anaerobic conditions with the utilization of an electron donor other than water is not restricted to phototrophic bacteria. A few algae also carry out photosynthesis without evolving oxygen: molecular hydrogen serves as the electron donor, provided the cells possess an adaptive *hydrogenase* (p. 136 f).

This phenomenon was discovered in 1940 by H. Gaffron in the green algae *Scenedesmus obliquus (strain D₃)* and *Ankistrodesmus sp.* In the meantime, *photoreduction* has also been studied in several species of other algal groups, but it is not certain whether the presence of this enzyme is always associated with the ability to carry out photoreduction. *Hydrogenase* does not seem to occur in higher plants.

Photoreduction only takes place when algae have been previously adapted to anaerobic conditions and their hydrogenase has been activated. This enzyme is inactive in the presence of oxygen formed during photosynthesis. The adaptation is complete when the cells have been kept in darkness for a few hours in a hydrogen atmosphere containing about 4 % CO_2. Upon subsequent illumination, the algal cells take up hydrogen and CO_2. Since no oxygen is released, the gas uptake can be measured directly by the reduction of pressure in a closed manometer vessel. In weak light, or after "stabilization" (see below) this process continues indefinitely. If a certain intensity of illumination is exceeded, a very rapid *"deadaptation"* sets in: the rate of hydrogen uptake decreases fast and normal photosynthesis with evolution of oxygen takes over. The inhibition of hydrogen utilization by the cells and the beginning

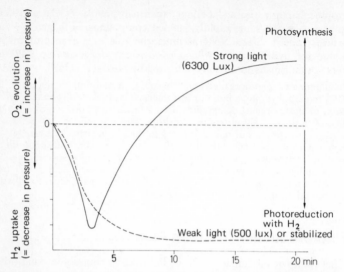

Fig. 72 Photoreduction and photosynthesis of H₂-adapted cells of the green alga *Ankistrodesmus braunii* at different light intensities (after Kessler). Manometric measurement of the gas exchange; decrease in pressure by uptake of hydrogen = photoreduction; increase in pressure by evolution of oxygen = photosynthesis.

evolution of oxygen changes the conditions in the manometer vessel in favor of a constant increase in pressure (Fig. 72).

Photoreduction can be stabilized by experimentally "switching off" the oxygen-evolving system. Accordingly, most inhibitors of oxygen evolution have a stabilizing effect on photoreduction: hydroxylamine, o-phenanthroline and the herbicide DCMU (p. 123). The same effect is achieved by depriving the cells of the trace element manganese, which is an essential factor in the oxygen evolving system (cf. p. 116).

The action of light in photoreduction is probably restricted exclusively to the activation of "photosystem I" while "photosystem II" is – so to speak – by-passed. Electrons respectively hydrogen for the reduction of $NADP^{\oplus}$ are supplied by molecular hydrogen with the mediation of activated *hydrogenase*.

Studies of CO_2 reduction in H₂-adapted algae (using $^{14}CO_2$ or $NaH^{14}CO_3$) have not shown any fundamental difference from the normal aerobic mechanism of the "Calvin cycle" (p. 145 ff).

The *photoproduction of hydrogen* in purple non-sulfur bacteria (p. 137) also has its counterpart in the green algae *Scenedesmus* and *Chlorella*. Their cells must be previously illuminated in an atmosphere with no or relatively little CO_2. However, there appears to be a significant difference with respect to the origin of the hydrogen evolved: in the green algae, it probably derives from the photolysis of water.

Yet another reaction carried out by H_2-adapted algae must be mentioned for the sake of completeness. The introduction of small quantities of molecular oxygen (up to 5 %) into the gas space produces such an increase in CO_2 fixation by hydrogen-adapted cells of *Scenedesmus* in the dark that its rate is comparable to that obtained in photoreduction. The same intermediary products are formed, i.e. those of the *Calvin-Cycle*. In this dark reaction, oxygen assumes the role of an electron acceptor in the *hydrogenase* reaction. The energy released in this moderate "knallgas reaction" (p. 194) is used to form "energy equivalents" which are available for CO_2 reduction. Since the energy equivalents originate from a chemical reaction, this process of CO_2 reduction is termed *chemosynthesis*; we shall be dealing with it later (p. 192 ff).

The following reactions are thus characteristic of hydrogen-adapted green algae:

1. **Photosynthesis**

$$6\ CO_2 + 12\ H_2O \xrightarrow{h \cdot v} C_6H_{12}O_6 + 6\ O_2 + 6\ H_2O$$

2. **Photoreduction**

$$6\ CO_2 + 12\ H_2 \xrightarrow[+\,hydrogenase]{h \cdot v} C_6H_{12}O_6 + 6\ H_2O$$

3. **Photoproduction of hydrogen**

$$2\ H_2O \xrightarrow[-\,CO_2]{h \cdot v} 2\ H_2 + O_2$$

4. **Knallgas reaction and CO_2 fixation in the dark**

$$n\ H_2 + n\ O_2 + n\ CO_2 \xrightarrow[+\,hydrogenase]{-\,light} C_6H_{12}O_6 + n\ H_2O$$

References

Clayton, R. K.: Primary processes in bacterial photosynthesis. Ann. Rev. Biophys. Bioengineering 2: 131, 1973

Gaffron, H.: Carbon dioxide reduction with molecular hydrogen in green algae. Amer. J. Bot. 27: 273, 1940

Gest, H., A. San Pietro, L. P. Vernon: Bacterial photosynthesis. Kettering Res. Lab., Yellow Springs 1963.

Kelly, D. P.: Autotrophy: Concepts of lithotrophic bacteria and their organic metabolism. Ann. Rev. Microbiol. 25: 177, 1971

Kessler, E.: Stoffwechselphysiologische Untersuchungen an Hydrogenase enthaltenden Grünalgen. Planta 49: 435, 1957

Oelze, J., G. Drews: Membranes of photosynthetic bacteria. Biochim. Biophys. Acta (Amst.) 265: 209, 1972

Pfennig, N.: Photosynthetic bacteria. Ann. Rev. Microbiol. 21: 285, 1967

Conversion of CO_2 to Carbohydrate

One reactant, CO_2, has not been discussed up to now. As already indicated, chemical transformation of the second starting substance in photosynthesis is not directly linked to the two light reactions. These supply, apart from oxygen, NADP-H + H$^{\oplus}$ and ATP, which meet the requirement for reductive power and chemical energy in the conversion of CO_2.

Since Wood and Werkman discovered in 1936 that heterotrophic bacteria are able to take up free CO_2 and to incorporate it chemically into their cell substance, doubts were cast on the concept of a photochemical CO_2 transformation on chlorophyll. The discovery of a CO_2-binding reaction ("carboxylation") in bacteria and later in animal cells turned out to be the formal reversion of the cleavage of CO_2 from the carboxyl group of an organic acid (R-COOH $\rightleftharpoons$ R-H + CO_2) which is frequently observed in respiration and fermentation. Since the process is strongly exergonic, the reaction equilibrium is highly unfavorable for the reversal reaction, i.e. the binding of CO_2 in a carboxyl group. Therefore a good yield results only when the reaction is coupled with an energy-supplying reaction, e.g. with the splitting of ATP. The fixation step may be linked to a simultaneous reduction of the compound formed, as in the carboxylation of pyruvic acid by *malic enzyme* (cf. p. 157):

$$CH_3-CO-COOH + CO_2 + NADP-H + H^{\oplus} \rightleftharpoons HOOC-CH_2-CHOH-COOH + NADP^{\oplus}$$

Photosynthetic CO_2 fixation was originally considered to be a similar process of reversible carboxylation, particularly in respect of the participation of energy and reduction equivalents. However, no direct formation of carbohydrate, nor a significant increase in free energy could be expected from such a process. One main difficulty complicating the elucidation of photosynthetic CO_2 reduction was that the common chemical analysis of cellular components could not clarify whether the intermediate compounds demonstrated were specific for photosynthesis or for carbohydrate degradation. Moreover, a detection of the relatively small quantities of the substances involved was nearly impossible. The problem could be successfully tackled only with the application of radioisotopes (see below). The first organisms to be studied were unicellular green algae (p. 92 f) and later phototrophic bacteria as well as low and higher plants.

Use of Isotopes to Elucidate Biochemical Reactions. The chemical conversions which a compound undergoes in a complex reaction process can be tracked when one atom (or several atoms) of its molecules is replaced by the isotopic form of the atom*, particularly if it is radioactive, i.e. emits radiation. The radiation serves as a marker of the molecule or of one of its groups and clearly discloses its route in a reaction, but also in the living cell. The reaction products formed from the molecule contain the isotope and can easily be identified since they emit radiation.

Isotopes of carbon (^{14}C), phosphorus (^{32}P), hydrogen (^{3}H) and nitrogen (^{15}N) are used frequently in biochemical research. ^{15}N is a stable non-radioactive isotope; the other three decay with emission of β-radiation of various strengths.

When a compound is labeled with a radioactive isotope, this is only introduced into a certain percentage of the molecules; the ratio of labeled to unlabeled molecules is defined as "specific activity".

The increasing amounts of isotopes produced in nuclear reactors together with the development of sensitive and exact methods for measuring the different kinds (mass spectrometry for stable isotopes, Geiger-Müller counter and scintillation spectrometer for radioactive isotopes) have made a wide application of the isotopic technique possible.

Elucidation of CO₂ Reduction by Means of Radiocarbon

The long-living carbon isotope ^{14}C turned out to be an ideal tool for tracing the specific reactions in which carbon is bound in photosynthesis and converted to the end product carbohydrate. By using ^{14}C a fundamental difficulty could be avoided, consisting of the fact that the chemical apparatus for conversion of CO_2 consists of organic compounds with the same chemical elements and atomic groupings as the reaction products. Therefore, it is difficult to separate the two by means of conventional chemical analysis; nor would an identification of intermediary products be possible, which only occur in traces in cells with a rapid turnover.

Specific primary and intermediary products in photosynthetic CO_2 reduction are detected by the following technique: Photosynthetically active cells of unicellular green algae are exposed for a short time to CO_2 or $NaHCO_3$, a number of molecules of which contain radiocarbon (^{14}C) as a "tracer". The cells do not distinguish between ^{12}C and ^{14}C since the two forms react identically.

The contact of the cells with ^{14}C must be as short as possible if one wants to record the primary product of CO_2 fixation. Long-term incubation

---

* Isotopes are atoms of an element which have a different atomic weight, but the same nuclear charge and number of electrons. Chemically, they react identically, though their physical behavior is different. Radioactive isotopes are unstable and decay with release of radiation.

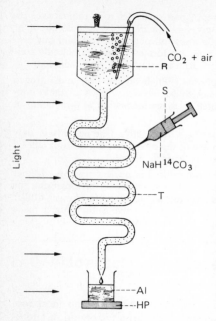

Fig. 73 Experimental system for studying short-term incorporation (0.4–10 sec) of radiocarbon (¹⁴C) during photosynthesis of unicellular algae. $NaH^{14}CO_3$ is injected into the algal suspension which flows from an illuminated reservoir (R) through a transparent plastic tubing (T) until it drops into boiling alcohol (Al). Fixation of ¹⁴C is confined to the period of time required by the algal suspension to pass through the coils of the tubing from the point of injection to the killing solution. The syringe (S) contains the $NaH^{14}CO_3$ solution. HP: Hot-plate (after Calvin and coworkers).

results in a rapid "outflow" of ¹⁴C into a relatively large number of compounds in the cell. With incorporation times of different lengths and by appropriate analytical methods for identification of soluble cellular components, the route of radiocarbon via typical intermediates to the end products can be pursued.

Figures 73 and 74 show two experimental devices for "pulse" labeling experiments of 0.4–15 sec and 30–60 sec duration respectively, using unicellular green algae in the state of optimal photosynthesis. Fixation is terminated by pouring the cells into boiling alcohol. The soluble organic compounds are simultaneously extracted, while the non-fixed $^{14}CO_2$ escapes. The radioactive, i.e. labeled substances in the alcoholic extract are separated by means of two-dimensional paper chromatography (cf. p. 35 ff). To ascertain the position of labeled substances, the developed paper chromatogram is placed in close contact with a sheet of X-ray film and left in the dark for 1–4 weeks ("autoradiography"). By emission of β-radiation, every radiocarbon-containing compound concentrated in a discrete spot changes the sensitive film in such a way that a darkened spot appears after development and fixation. The film sheet, the "autoradiogram", thus gives a precise image of the ¹⁴C distribution on the paper chromatogram (Fig. 75 A, B, C); the individual substances may then be identified.

Precise information on the mechanism of synthesis of ¹⁴C-labeled compounds was obtained when the quantitative distribution of ¹⁴C in the

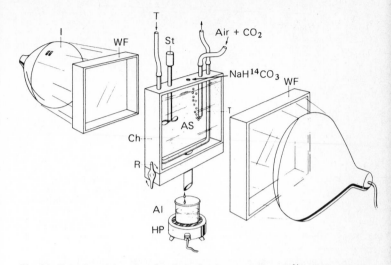

Fig. 74 Experimental system to study the incorporation of ¹⁴C during photosynthesis of unicellular algae. Into a pre-illuminated algal suspension (AS) in an illumination chamber (Ch) NaH¹⁴CO₃ is injected. R, device for removal of samples; by turning the wing-screw the contents of the chamber are instantaneously transferred into the boiling alcohol (Al). HP, hot-plate; I, illumination of the chamber from both sides; St, stirrer; T, system for maintaining constant temperature in the chamber; WF, water filter to eliminate the heat produced by the lamps.

individual carbon atoms of the molecules had been determined by means of chemical and microbiological methods. The carbon chain of the atoms is degraded stepwise into C_1 or C_2 fragments with a chemical structure suitable for measuring their radioactivity.

Mechanism of CO₂ Reduction

The crucial experiments to elucidate photosynthetic CO_2 reduction were carried out with two green algae, *Chlorella pyrenoidosa* and *Scenedesmus obliquus* using radioisotopes.

After 60 sec photosynthesis of *Chlorella* cells in the presence of $^{14}CO_2$, several ¹⁴C-labeled substances appear in the autoradiogram (Fig. 75 C): sugar phosphates, organic acids and amino acids. By a stepwise decrease of the contact time with $^{14}CO_2$, the number of these compounds is reduced; a time of less than 2 sec finally results in only one labeled compound: 3-phosphoglyceric acid ("3-PGA"). Its carboxyl group contains about 95 % of the radiocarbon incor-

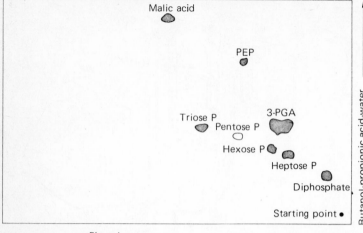

Fig. 75 A Autoradiogram of photosynthetic products obtained from cells of *Chlorella pyrenoidosa* after incubation with NaH^{14}CO$_3$ for 2 sec in light (after Calvin and coworkers). "Two-dimensional" paper chromatography of an alcohol extract (after previous removal of lipids and a concentrating process). The components labeled by ^{14}C produced characteristic black spots when the developed chromatogram had been exposed to X-ray film. For details, see text. FBP: fructose biphosphate; PEP: phosphoenol pyruvic acid; 3-PGA: 3-phosphoglyceric acid; RuBP: ribulose biphosphate; SeBP: sedoheptulose biphosphate; UDPG: uridinediphosphate glucose.

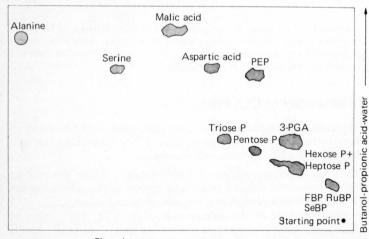

Fig. 75 B As Fig. 75 A, except: 7 sec incubation with NaH^{14}CO$_3$ in light.

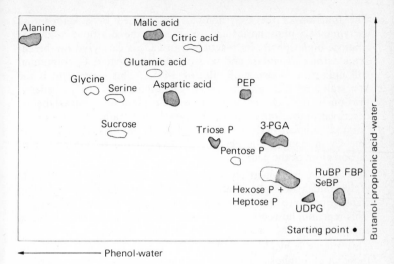

Fig. 75 C As Fig. 75 A, except 60 sec incubation with NaH¹⁴CO₃ in light.

porated (*C in the reaction scheme below). Accordingly, the first reaction step in the reduction of CO₂ is apparently a carboxylation in which CO₂ is incorporated into the COOH group of 3-PGA. The origin of the other two carbon atoms of this compound (discussed as hypothetical "C₂ compound") was obscure until the genuine "CO₂ acceptor" was discovered and the process of CO₂ reduction was recognized as a cyclic process with several enzymatic steps. This process was finally shown to fulfill two important functions: 1) CO₂ is reductively converted to a C_6 sugar, and 2) the specific acceptor for CO₂ is regenerated. The biochemical reaction complex was elucidated by the team of Calvin and Benson using green algae. It is hence termed the "Calvin cycle"*.

Ribulose-1,5-biphosphate   ?   3-phosphoglyceric acid

---

* In 1961 M. Calvin received the Nobel Prize for chemistry.

The function of the CO_2 acceptor is assumed by a ketopentose carrying two phosphoric acid residues in the positions 1 and 5: ribulose biphosphate. Its reaction with CO_2 is catalyzed by the enzyme *carboxydismutase* and leads via an intermediate C_6 compound still unknown to two molecules of 3-PGA. This compound is the first stable reaction product of CO_2 fixation but is not, as believed previously, the CO_2 acceptor. Under the conditions of pulse labeling described above, only one of the two 3-PGA molecules has incorporated radioactivity, and that exclusively in its carboxyl group. The C_6 intermediate is hence split between the second and third carbon atom of the pentose.

The 3-PGA is reductively converted to 3-phosphoglyceraldehyde ("glyceraldehyde-3-phosphate", "G-3-P"), the COOH group being transformed into the CHO group formally by removal of oxygen. This reaction, however, is endergonic and hence requires an input of energy to go in the direction of the aldehyde formation.

This energy is available as ATP, the "energy equivalent"; its synthesis in photophosphorylation reactions has already been described (p. 125 ff). We have also discussed the origin of the hydrogen required, the "reduction equivalent" and its transport metabolite: NADP-H $+ H^{\oplus}$. The reduction of 3-PGA occurs in several steps and requires the participation of two enzymes (cf. scheme). The first one, *phosphoglycerate kinase,* transfers a phosphoric acid residue from ATP to the carboxyl group of 3-PGA. The bond formed is rich in energy since it is an anhydride bond between a carboxyl group and phosphoric acid, analogous to that in ATP. The significance of this "substrate phosphorylation" (cf. p. 20 f) consists

3-Phosphoglyceric acid   1,3-Biphosphoglyceric acid

Substrate-enzyme complex   D-Glyceraldehyde-3-phosphate

———————— Phosphotriose dehydrogenase ————————>

in the "activation" of 3-PGA, i.e. raising it to a higher energy level. The 1,3-biphosphoglyceric acid ("1,3-PGA") formed is thus enabled to react with the SH group of the specific apoenzyme of *phosphotriose dehydrogenase,* releasing the energy-rich bound phosphate. The resulting bond is also energy-rich, since it is a thioester. Hydrogen is now transferred to this acyl-S-enzyme protein from NADP-H $+ H^\oplus$, bringing about a reductive cleavage of the complex to G-3-P and reactivated HS-apoenzyme.

This energy-consuming reduction of the C_3 compound is the most important intermediate reaction of the whole cycle since here CO_2 is converted to the "metastable" carbohydrate compound of higher energy content. It embodies the final form of chemically bound energy originating from absorbed radiant energy. The conversion of triose to hexose which follows is energetically without problems since it takes place on practically the same energy level.

The molecules of the aldotriose G-3-P are mostly converted by enolization to the corresponding ketotriose dihydroxyacetone phosphate (96 %). The rapid establishment of equilibrium in the cell is catalyzed by the enzyme *triosephosphate isomerase,* which also ensures a rapid supply of G-3-P when the relatively small quantity present of this substance is consumed by the subsequent reaction. During this reaction a molecule of G-3-P reacts in an aldol condensation with a molecule of dihydroxyacetone phosphate, forming a

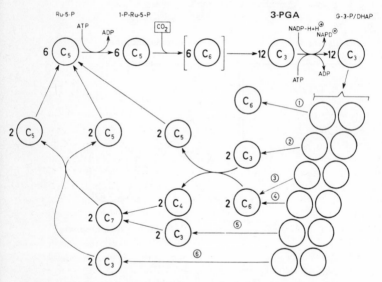

C_6 sugar, fructose-1,6-biphosphate ("FBP") (balance: $C_3 + C_3 = C_6$). The name of the enzyme involved, *aldolase,* indicates the reaction mechanism. The equilibrium strongly favors FBP formation (89 %). The latter compound then loses the phosphoric acid residue at position 1 by the action of a *phosphatase* and is converted to fructose-6-phosphate. The molecules of this compound partly leave the reaction cycle as an end product and enter other reaction pathways of cell metabolism. These will be treated after the discussion of photosynthesis.

The reaction cycle has thus solved its first problem: CO_2 has been converted to a C_6 sugar. In accomplishing its second function, i.e. the regeneration of the specific CO_2 acceptor ribulosebiphosphate

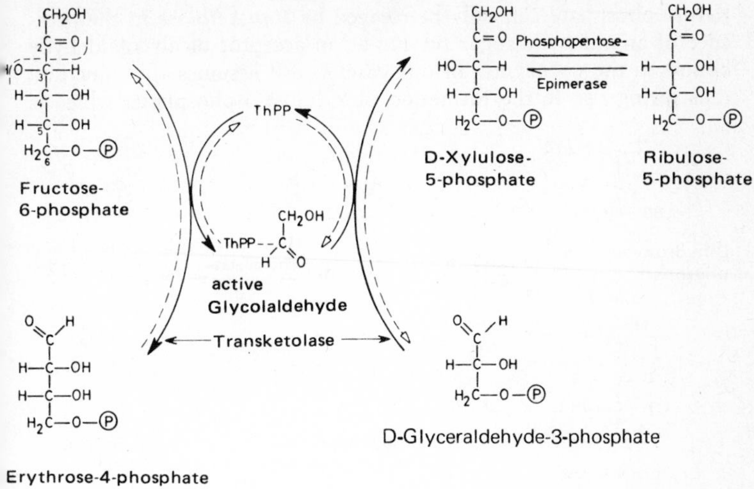

Fructose-6-phosphate

D-Xylulose-5-phosphate

Ribulose-5-phosphate

active Glycolaldehyde

Transketolase

Erythrose-4-phosphate

D-Glyceraldehyde-3-phosphate

("RuBP"), F-6-P and G-3-P serve as starting substances. The reaction sequence (see scheme p. 150) is at first sight confusing. In the beginning F-6-P and G-3-P react with the enzyme *transketolase* in such a way that a C$_4$ sugar phosphate, erythrose-4-phosphate, and a C$_5$ sugar phosphate, xylulose-5-phosphate, are formed. The conversion is brought about by separating the atoms 1 and 2 from F-6-P as a "ketol grouping" (CH$_2$OH-CO-; name of the enzyme!) which is transferred as "active glycolaldehyde" to the coenzyme of the *transketolase*. This coenzyme is identical with thiamine pyrophosphate, an active group which preferentially transfers C$_2$ fragments (see also p. 215 f).

The structure of thiamine (= "aneurin", vitamin B$_1$) consists of a substituted pyrimidine ring linked by a methylene group to a sulfur-containing thiazole ring. Probably the C$_2$ fragment is reversibly bound to carbon atom 2 of the thiazole ring.

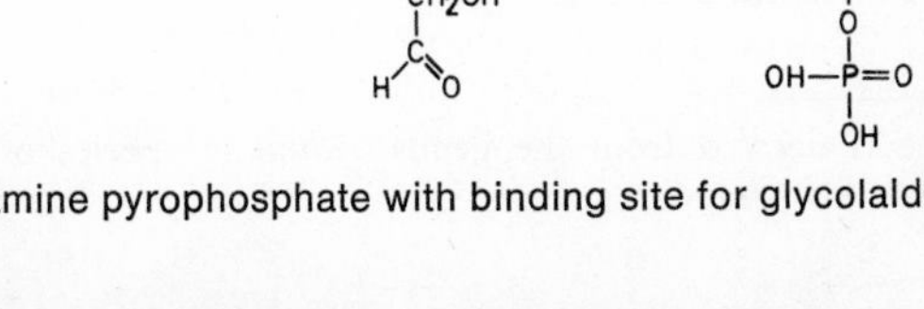

Thiamine pyrophosphate with binding site for glycolaldehyde

Ketose phosphate can only be cleaved by *transketolase* in the presence of an aldehyde sugar serving as an acceptor of glycolaldehyde bound to the coenzyme. In our case, G-3-P assumes this function, thus giving rise to the formation of xylulose-5-phosphate, which is

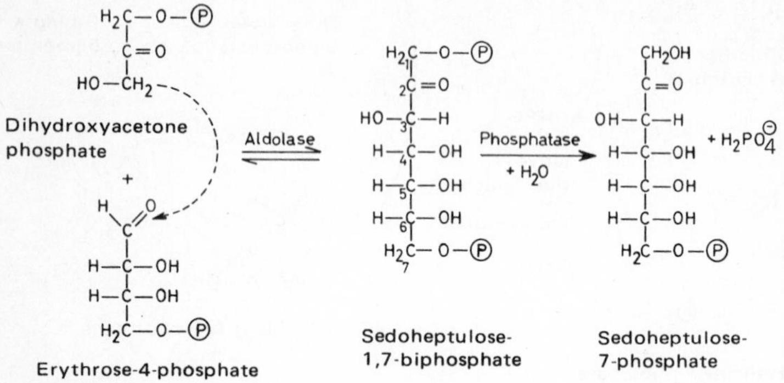

Dihydroxyacetone phosphate

Erythrose-4-phosphate

Aldolase

Phosphatase

Sedoheptulose-1,7-biphosphate

Sedoheptulose-7-phosphate

converted by *phosphopentose epimerase* to ribulose-5-phosphate (Ru-5-P). The latter is the immediate precursor of RuBP (balance: $C_6 + C_3 = C_4 + C_5$). The C_4 fragment left over, erythrose-4-phosphate ("E-4-P"), then reacts with dihydroxyacetone phosphate in a kind of aldol condensation, again catalyzed by *aldolase,* producing sedoheptulose-1,7-biphosphate*, a C_7 sugar (balance: $C_4 + C_3 = C_7$).

Sedoheptulose-7-phosphate ("S-7-P"; see above) is formed from this compound by dephosphorylation catalyzed by a *phosphatase*. S-7-P also serves as substrate for *transketolase:* the first two carbon atoms are removed as a molecule fragment and transferred to a G-3-P molecule after reversible binding to the coenzyme. The reaction produces one molecule of xylulose-5-phosphate and one molecule of ribose-5-phosphate which are both converted by enzymatic epimerization (above) to Ru-5-P (balance: $C_7 + C_3 = C_5 + C_5$). All the intermediates are thus converted to pentoses. In the final phosphorylation step, *phosphoribulokinase* transfers a phosphoric acid residue from ATP to Ru-5-P (for regulation of this reaction, see p. 445). The ribulose-1,5-biphosphate formed is ready to fix CO_2. With the regeneration of the specific CO_2 acceptor completed, the cycle has solved its second problem.

---

* The name is derived from the genus *Sedum* in species of which the heptulose was first found.

The cyclic process of photosynthetic CO_2 reduction acquaints us with an important rule which applies to most sugar transformations in the cell: they occur at the level of their phosphate esters and thus have a higher energy content than the free compounds. The free energy of hydrolysis

Ribulose-5-phosphate   Ribulose-1,5-biphosphate

of the various phosphate ester bonds is of the order $\Delta G° = -2300$ to -5000 cal/mole.

The CO_2 reduction cycle may be summarized as follows (cf scheme, p. 150): since one molecule of hexose monophosphate is formed from 6 molecules of CO_2, 6 molecules of RuBP (C_5) are required. These give rise to the formation of 12 molecules of triose phosphate (C_3), two of which are linked to form one molecule of hexose monophosphate (C_6). The remaining triose phosphate molecules serve exclusively for the regeneration of RuBP, which requires a total of four reaction steps: $C_3 + C_3 \rightarrow C_6$; $C_3 + C_6 \rightarrow C_5 + C_4$; $C_3 + C_4 \rightarrow C_7$; $C_3 + C_7 \rightarrow C_5 + C_5$. From the sum of all these reactions the following balance-sheet results:

$$6\ \text{Ru-5-P} + 6\ \text{ATP} \longrightarrow 6\ \text{1-P-Ru-5-P} + 6\ \text{ADP}$$

$$6\ \text{1-P-Ru-5-P} + 6\ CO_2 + 12\ \text{NADP-H+H}^{\oplus} + 12\ \text{ATP} + 6\ H_2O \longrightarrow$$
$$ \longrightarrow 12\ C_3\text{-P} + 12\ \text{NADP}^{\oplus} + 12\ \text{ADP} + 12\ H_2PO_4^{\ominus}$$

$$2\ C_3\text{-P} \longrightarrow \text{P-}C_6\text{-P}$$

$$\text{P-}C_6\text{-P} \longrightarrow C_6\text{-P} + H_2PO_4^{\ominus}$$

$$10\ C_3\text{-P} \longrightarrow 6\ C_5\text{-P} + 4\ H_2PO_4^{\ominus}$$

$$6\ CO_2 + 12\ \text{NADP-H+H}^{\oplus} + 18\ \text{ATP} + 6\ H_2O \longrightarrow C_6\text{-P} + 18\ \text{ADP} + 17\ H_2PO_4^{\ominus} + 12\ \text{NADP}^{\oplus}$$

Analogous studies on chlorophyll-containing plants from all systematic divisions have yielded results proving that the reactions of the photosynthetic CO_2 reduction detected in *Scenedesmus* and *Chlorella* obviously proceed identically in most photosynthetic organisms.

The reactions involved during the conversion of CO_2 to carbohydrate very likely take place in the stroma or matrix (p. 157 f) of the chloroplast where most of the specific enzymes are located. This would explain why these enzymes are easily lost from chloroplasts during inadequate isolation procedures. According to other concepts, at least *carboxydismutase* is believed to be a membrane-bound enzyme or to form a "multienzyme complex" (p. 406) with other enzymes of the Calvin cycle which is part of the thylakoid membrane.

The *carboxydismutase* found in chloroplasts has been called "fraction 1 protein" as it may comprise up to 50 % of the total soluble leaf protein. The molecular weight is about 525 000; the enzyme protein probably consists of 8 large identical subunits (molecular weight about 55 000 each) and 8–10 small identical subunits (molecular weight about 15 000 each). The large subunit is coded for in the chloroplast genome, the small subunit, however, in the nuclear genome (cf. p. 395 f).

Other Mechanisms of CO₂ Fixation

The C₄-Dicarboxylic Acid Cycle. A number of species of tropical *Gramineae* such as maize, sugar cane and millet, and some members of the genera *Atriplex* and *Amaranthus* show markedly high rates of photosynthesis which are obviously based on a particularly effective CO_2 utilization. In contrast to most other plants, they are able to utilize intensively even an extremely low supply of CO_2. $^{14}CO_2$-fixation experiments showed that they have a special CO_2-binding mechanism: phosphoenolpyruvic acid (PEP) acts as acceptor and is converted to oxaloacetic acid by the enzyme *phosphoenolpyruvate carboxylase* (Hatch and Slack). This reaction also proceeds at very low CO_2 concentrations and is in this respect clearly superior to the typical carboxylation of the Calvin cycle. Oxaloacetic acid is reduced to malic acid by *malate dehydrogenase,* with NADP-H + H$^{\oplus}$ supplying the hydrogen. Because of the participation of these two acids, this pathway of CO_2 fixation is termed the "C₄-dicarboxylic acid cycle", or the "Hatch-Slack-cycle", after its discoverers. The plants mentioned above are hence termed "C₄ plants" as apposed to "C₃ plants" (CO_2 fixation via the Calvin cycle).

In the leaves of C₄ plants there exists a characteristic differentiation of the assimilatory cells into two types (Fig. 76): the mesophyll cells with relatively small, grana-containing chloroplasts, which synthesize malic acid and contain little starch, and the larger parenchyma cells enveloping the leaf vascular bundles like a sheath ("bundle-sheath cells"), which contain chloroplasts mostly without grana structures and store starch. There is probably close physiological contact between the mesophyll cells – which fix CO_2 via the C₄-dicarboxylic acid cycle – and the bundle-sheath cells in which the reactions of the Calvin cycle occur. Malic acid from the mesophyll cells is enzymatically degraded in the bundle-sheath cells to CO_2 and pyruvic acid, simultaneously resulting in the formation of NADP-H + H$^{\oplus}$. The CO_2 released is immediately bound to ribulose-1,5-biphosphate and used for synthesis of hexose phosphate in the Calvin cycle. Having returned to a mesophyll cell, pyruvic acid can be converted to PEP with consumption of ATP (cf. reaction p. 213) and used again as

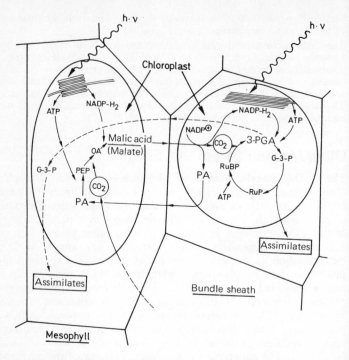

Fig. 76 Translocations and reaction pathways during CO_2 fixation by the "C₄-Dicarboxylic acid cycle" (after Hatch and Slack; schematic). G-3-P, glyceraldehyde-3-phosphate; OA, oxaloacetic acid; PA, pyruvic acid; PEP, phosphoenol pyruvic acid; 3-PGA, 3-phosphoglyceric acid; RuBP, ribulose-1,5-biphosphate; RuP, ribulose-5-phosphate.

carrier molecule for the CO_2 transport into the bundle-sheath cells. At first sight, this appears to be a complicated reaction sequence. However, it places the C_4 plants in a position to utilize even tiny quantities of CO_2 and to convert them to carbohydrate by means of the reduction equivalents, which are always available in sufficient quantity. This is probably an adaptation to the particular conditions of the tropics and subtropics. The low rate of "photorespiration" (p. 159 ff) probably also contributes to the generally high rate of net photosynthesis in these plants.

CO_2 **Fixation in Succulent Plants.** Not only the C_4 plants exhibit CO_2 fixation via PEP, but also a series of succulent or semisucculent plants (*Bryophyllum, Kalanchoe, Sedum, Kleinia, Crassula, Opuntia*). The CO_2 taken up at night through the open stomata in leaves and green shoots is bound to PEP and reduced to malic acid which may then secondarily be converted to other organic acids. Quantitatively, more CO_2 is fixed than is released by the respiration occurring simultaneously. The acids formed are stored in the cell vacuole and cause a strong acidification of the cell sap (pH value about 4.0). During the day, there is a light-dependent transportation of malic acid back to the chloroplasts where – as happens in the C$_4$ plants – it is decarboxylated ("deacidification"; the pH value in the vacuole increases to about 6.0). The CO$_2$ released enters directly into the Calvin cycle.

Besides NADP-H and H$^\oplus$, oxaloacetic acid is possibly first synthesized in chloroplasts and then cleaved by a *phosphoenolpyruvate carboxykinase* to PEP and CO$_2$ with consumption of ATP: Oxaloacetic acid + ATP $\rightleftharpoons$ phosphoenolpyruvic acid + ADP + CO$_2$.

Malic enzyme (p. 142) also appears to participate in this decarboxylation of malic acid.

Since this "diurnal acid rhythm" was first discovered and studied in members of the *Crassulaceae,* it is also referred to as *Crassulaceae acid metabolism.* For succulent plants it constitutes an expedient adaptation to their hot and dry locations. Since the stomata stay mainly closed during the day, loss of water is kept low although a CO$_2$ deficit arises. The latter can, however, be compensated by the nightly "acidification". The substance balance thus remains positive without impairing the water balance. The fact that the reactions involved take place in different reaction compartments of the cell, is of significance for the regulation of the diurnal acid rhythm; this is therefore a complex process in which several factors are involved (p. 439).

Isolated chloroplasts evidently fix ^{14}CO$_2$ in the light and incorporate radiocarbon in a few characteristic intermediates of the CO$_2$ reduction cycle. The amount of this CO$_2$ fixation is small compared with that in intact leaves or algal cells. This is probably due to the heavy loss of stroma substance during isolation of the chloroplasts. A large part of the active enzymes is obviously lost during this process. There are only a few exceptions, e.g. in isolated chloroplasts of the marine green alga *Acetabularia.*

Logic dictates that the chloroplast or its equivalent must possess all the enzymes, coenzymes and cofactors necessary for converting CO$_2$ to carbohydrate. A few of the enzymes involved are also active in other important cellular metabolic processes, e.g. in the "pentose phosphate cycle" (p. 255 ff) where they catalyze the same reactions, though generally in the opposite direction. *Aldolase, triose phosphate isomerase* and *phosphoglycerate kinase* are active in the oxidative glucose degradation of "glycolysis" (p. 208 ff). All the enzymes active in photosynthetic CO$_2$ reduction have been demonstrated in the green cells of various lower and higher plants. But because of the involvement of several enzymes in different metabolic processes, it is not yet possible to prove unequivocally their specific function in photosynthesis. The existence of an enzyme pattern specific for the photosynthetically active structures can only be indirectly and incompletely verified. *Phosphotriose dehydrogenase* in chloroplasts, for instance, is only active with NADP, and differs therefore from the analogous enzyme in glycolysis which is exclusively activated by NAD.

While the two enzymes generally occur together in the photosynthetically active cell, the NADP-enzyme is inactive both in chlorophyll-free seedlings which have been kept in darkness, and in bleached cells of *Euglena*. It only appears upon illumination, its activity then increasing rapidly (cf. p. 448). More recent findings show that actively photosynthesizing cells contain a specific *aldolase* with properties differing distinctly from those of the analogous enzyme occurring in most other cells. The former appears to be a permanent component of chloroplasts. Phototrophic bacteria and blue-green algae probably contain only one *aldolase* with properties very similar to those of the chloroplast enzyme.

Metabolic Transfer Between Chloroplast and Cytoplasm

Since the chloroplast membrane has turned out to be practically impermeable on the one hand for adenylates (ADP, ATP) and pyridine nucleotides, whereas on the other hand chemical energy, hydrogen and other metabolites evidently flow from the chloroplast to the cytoplasm, an indirect transfer mechanism, e.g. via "transport metabolites", must exist. This function is obviously assumed by 3-phosphoglyceric acid (3-PGA) and dihydroxyacetone phosphate (DHAP), both formed by the Calvin cycle; glyceraldehyde-3-phosphate (G-3-P) probably acts as an intermediate which, however, cannot leave the chloroplast. The following scheme (Fig. 77) indica-

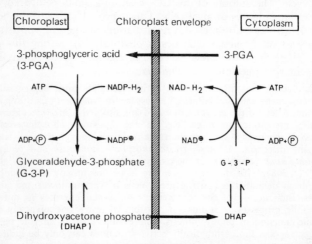

Fig. 77 Translocation of energy and hydrogen via the "phosphoglyceric acid — dihydroxyacetonephosphate shuttle" from chloroplasts to cytoplasm (after Lüttge).

tes that 3-PGA is "loaded" with chemical energy (from ATP) and hydrogen (from NADP-H + H$^{\oplus}$) yielding G-3-P, a compound of higher energy content. The latter is subsequently converted to iso-meric DHAP which is then transferred to the cytoplasm. Here, the chemically bound energy serves to phosphorylate ADP, the hydro-gen to reduce NAD$^{\oplus}$. The end product 3-PGA may re-enter the chloroplast as transport metabolite. This mechanism, by which one mole of $\sim$ P and one mole of reduction equivalent are translocated per mole of metabolite from chloroplast to cytoplasm has been termed "phosphoglyceric acid/dihydroxyacetone phosphate shuttle" (Stocking, Heber, Heldt, Santarius).

Photorespiration

With appropriate methods it can be demonstrated that, under cer-tain conditions, numerous plant species also take up O_2 and release CO_2 in light. Because of the superficial similarity to respiratory gas exchange, this process is referred to as *photorespiration* of photo-synthetically active organisms. In fact, the reactions responsible for this process are closely connected with photosynthesis and have little to do with dark respiration. This is indicated by the finding that photorespiration increases with mounting light intensity and is – in contrast to dark respiration – only saturated at relatively high oxygen concentrations: at about 20 %, compared to about 2 % (the inhibitory effect of high oxygen concentrations on the rate of photosynthesis in *Chlorella* has been known already since 1920 as the "Warburg effect"). The CO_2 compensation point of a plant (p. 81) obviously depends mainly on the rate of photorespiration: a high one indicates intensive CO_2 release in light. This behavior generally characterizes C_3 plants which fix CO_2 via the Calvin cycle (p. 145 ff). On the other hand, a low rate of photorespiration is gener-ally found in C_4 plants which incorporate CO_2 primarily by the Hatch-Slack pathway (p. 155 f). Due to their very effective CO_2 binding mechanism they probably utilize the major part of the CO_2 released in their photosynthetic apparatus. The earlier assumption that C_4 plants do not have any photorespiration does not hold any longer since the enzymes involved have been demonstrated in these plants.

The gas exchange typical of photorespiration is the result of a series of reactions closely connected with photosynthetic CO_2 reduction and pro-ceeding in various compartments of the cell (Fig. 78). The key compound is probably glycolic acid which is preferentially produced in larger quan-tities when, at high light intensities, there is a relatively high O_2 concen-tration but a low CO_2 one. Under these conditions glycolic acid is even

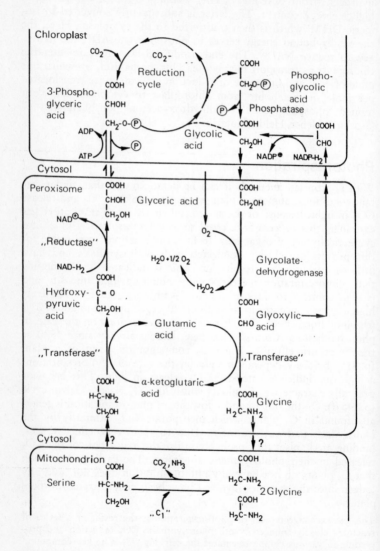

Fig. 78 Proposed reaction pathways in photorespiration involving chloro-plasts, peroxisomes and mitochondria (after Tolbert).

secreted into the culture medium by several algal species *(Chlorella, Ankistrodesmus)*. This compound is probably formed from sugar phosphate via glycolaldehyde *(transketolase* reaction! p. 151) in the Calvin cycle. The subsequent enzymatic conversion of glycolic acid to glyoxylic acid consumes oxygen. The enzyme involved, *glycolate oxidase (glycolate dehydrogenase)* is apparently a flavoprotein which transfers electrons to molecular oxygen, at least in higher plants. The H$_2$O$_2$ formed is cleaved by *catalase* (p. 418) to O$_2$ and water. This results in the net consumption of oxygen typical of photorespiration. The two enzymes are bound to "peroxisomes", a functional group of the "microbodies" (p. 227 f). It is assumed therefore that the glycolic acid leaves the chloroplasts and is further metabolized in the peroxisomes. Presumably, these also provide the site for the subsequent conversion of glyoxylic acid to glycine, mediated by glutamic acid (Kisaki and Tolbert), as the *transferase* involved is also a peroxisomal enzyme. An additional non-enzymatic cleavage of glyoxylic acid to formic acid and CO$_2$ may also occur (Zelitch). Whether the subsequent decarboxylation of glycine to serine by *serine hydroxymethyl transferase* (2 glycine → serine + CO$_2$) – the actual CO$_2$-yielding step in photorespiration! – occurs in the cytoplasm or in the mitochondria of the leaf cells is not yet clear. The major part of the serine formed is probably utilized for a light-dependent formation of carbohydrates in the chloroplast, which may proceed by a reversal of the reactions of serine biosynthesis (p. 369 f) with phosphoserine, 3-phosphohydroxypyruvic acid and 3-phosphoglyceric acid acting as intermediates.

Apart from a rapid synthesis of the amino acids glycine and serine, the biological function of photorespiration is still unknown. In view of the economy of photosynthesis and the energy balance, the release of photosynthetically fixed CO$_2$ (sometimes up to 30%!) by C$_3$ plants is difficult to comprehend.

Photosynthesis and Agricultural Production

Photosynthesis is the fundamental process by which the energy of solar radiation is transformed into chemical energy of complex organic compounds in plant organisms. These serve directly or indirectly as food and energy source for man, animals and heterotrophic microorganisms. This flow of energy within living matter and the problem of energy balance have recently aroused considerable interest, particularly the question of how efficiently solar energy is utilized biologically in various terrestrial regions. The data obtained in this question allow an estimate of maximal biomass production on the earth's surface by photosynthesis. This may permit an answer to the urgent question where the ultimate limit for the ever-growing world population lies as far as food supply is concerned.

Intensive primary plant production depends on an optimal combination of light, temperature, minerals and water. However, in the various kinds of natural environment that exist on the earth, certain restrictions of these factors are imposed. Nevertheless, many different species of plants are able to survive and to reproduce in each of these environments, because they adapt functionally to the conditions of their respective habitat. In the desert high light intensities and high temperatures but a low water supply prevail; the shady tropical rain forest offers enough humidity and heat, but too little sunshine; in the forests of the temperate zone the essential nutritional mineral salts are often lacking while the other factors are normal during the vegetation period. Thus, the natural vegetation is a rather accurate indicator of the climate that prevails in the respective habitat. The same holds true for the agricultural production which also depends strongly on climatic conditions.

The average period of solar radiation often matches the average temperature of a habitat. But there are significant exceptions. Thus the British Isles receive the same quantity of sunshine as certain regions of the United States, yet the vegetation period is significantly longer in Britain due to the influence of the Gulf Stream. In a very dry area both heat and solar radiation are generally at a maximum. The solar energy reaching the earth's surface per annum varies considerably in different areas. A maximum of about 220 kcal per annum and cm^2 is reached in deserts, a minimum of about 70 kcal per annum and cm^2 was estimated in the arctic regions. The plant production of a certain habitat, however, depends primarily on the daily radiation dose available during the vegetation period. It amounts to 500–700 cal/cm^2 in forests or farmland of the United States. Taking into consideration the photosynthetically active wavelengths (p. 85 ff) and the respiration rates of plants, the efficiency of energy transformation, i.e. the conservation of chemical energy in the plant biomass, has been calculated as being about 12%. Under ordinary circumstances, however, this value is significantly lower and ranges from 2.7 to 6.5%. Grain crops convert an even smaller proportion of the solar energy into chemical energy of plant compounds under adequate conditions. Calculations of the possible photosynthetic production of grain crops in the various habitats of the earth show that the developed industrialized countries reach an average daily production of about 27 g/cm^2 during the four months of the summer vegetation period. In the underdeveloped regions, however, this value is distinctly lower – even with improved agricultural techniques. These regions suffer from an additional disadvantage in their agricultural production due to the unfavorable climatic conditions.

In an attempt to increase crop yields by enhancement of the photosynthetic capacity in a given physical environment, recent research into the biochemistry, physiology and genetics of plants possessing the C_4-dicarboxylic acid cycle, the so-called "C_4 plants" (p. 155 f), has opened up interesting new possibilities. As described previously, these plants exhibit highly efficient photosynthesis in a hot and arid environment due to their excellent adaptation to intense radiation, high temperature, limited water supply, and low CO_2 concentration within the leaves. To explain this phenomenon the water economy of these plants must be taken into account. Let us recall that transpiration (p. 275 ff) and CO_2 uptake are linked insofar as closure of the stomata simultaneously prevents CO_2 from entering the interior of the leaf. Accordingly, the success of a desert plant largely depends on its capacity to maintain a high rate of photosynthesis even when the CO_2 concentration within the leaves falls to a low level, as is the case when the stomata are partly closed and the loss of water is low. In other words: the reduction in stomatal aperture increases the plant's ratio of CO_2 fixation with respect to water loss; thus its "photosynthetic water-consumption efficiency" (= CO_2 fixed per unit of water lost in transpiration) will increase. The metabolism of the C$_4$ plants shows some striking similarities to the *Crassulaceae acid metabolism* (p. 156 f) employed by plants which are adapted to more or less constant aridity.

Now we are coming to another important aspect of the C$_4$ mechanism. Since some C$_4$ plant species, like maize and sugar cane, belong to the most productive in the world it has been assumed that the C$_4$ mechanism is generally associated with a higher primary production rate due to its higher photosynthetic rate in comparison to C$_3$ plant species. This may be true in some cases, but in others there is no evidence for consistently higher rates of crop growth among C$_4$ plants. High efficiency at the metabolic level must not necessarily be correlated with a high rate of plant growth. Apparently the capacity of the C$_4$ mechanism for operating at very low ambient CO_2 concentrations is of major significance, maintaining growth in the water-deficient conditions of intermittent aridity.

Undoubtedly C$_4$ plant species have several potential advantages over C$_3$ plant species. They may well prove to be ideal crops in regions where water is limited. Attempts to breed the C$_4$ pathway into C$_3$ plants have been unsuccessful so far; apparently the complete coordination required between anatomical and biochemical properties of the leaf accounts for the failure to introduce the C$_4$ mechanism into plants lacking it. More promising are the efforts to increase the stock of important C$_4$ crop plants by breeding certain wild species into agriculturally useful domestic species, yielding forage or even grain for flour supplement. Generally, plants with the C$_4$ mechanism of CO_2 fixation are likely to play a key role in future programs for developing new agricultural practices and crop varieties that will help to enhance the productivity of arid land.

Synthesis of Other Compounds Associated with CO_2 Reduction

The autoradiogram of the soluble compounds produced after pulse labeling with $^{14}CO_2$ in light (Fig. 75) shows that not only the characteristic sugar phosphates of the cycle, but also a number of other compounds are ^{14}C-labeled. Consequently, synthesis of the latter must be closely linked to the reduction of CO_2. In fact, the cyclic reaction sequence involved has an additional important function: apart from the end product fructose-6-phosphate it supplies certain intermediates to synthesis reactions coupled with it. The cycle resembles an assembly line which picks up components at the beginning and releases end products, as well as supplying building blocks formed during the production process to side lines. The latter require an intermediate product from the main assembly line, but turn out a different end product. All the production pathways belonging to the system are regulated at the beginning of the main assembly line by changes in the input of components, and at its end by the rate of removal of the end products.

This important feature of cyclic reaction pathways, i.e. the availability of several outlets for intermediate products, will be encountered frequently when analyzing cellular metabolism. The regeneration of a specific acceptor compound for a starting substance is a further characteristic of these reaction cycles (e.g. ribulose 1,5-biphosphate; p. 150 ff).

Sugar phosphates, amino acids, organic acids, sucrose, and in a few cases free sugars are among the compounds which are formed relatively rapidly from intermediates or from the end product of the photosynthetic CO_2 reduction cycle. They can be demonstrated in both algae and green leaves by the radioactivity incorporated after one to several minutes of photosynthetic $^{14}CO_2$ fixation.

The synthesis of various kinds of compounds taking place simultaneously with or a little later than CO_2 reduction contradicts the notion that chloroplasts are merely the production center for carbohydrates which are further metabolized in the dark. On the contrary, the findings described above indicate that via intermediates of the CO_2 reduction cycle, photosynthesis seems directly to supply precursors for the synthesis of cellular macromolecules: amino acids for proteins, glucose for reserve starch and cellulose, pentose for nucleotides and nucleic acids, and heptulose for aromatic compounds.

Looking at it from this angle, the effect of photosynthesis on the "building metabolism" is a multiform and lasting one. On the other hand, the cell is hereby enabled to control the activity of photo-synthesis on a mostly qualitative basis.

These concepts contradict an earlier notion that the processes of "assimi-lation" and "dissimilation" in autotrophic and heterotrophic cells are rather similar and that the photosynthetic apparatus is merely an addi-tional device for production of carbohydrates by CO_2 assimilation, these carbohydrates being used up by respiration in the dark. This rigid con-cept of autotrophic metabolism is no longer valid and requires revision insofar as a greater plasticity and a far-reaching effectiveness must be conceded to photosynthesis. The profound effect of photosynthesis on cellular metabolism is due to the fact that it is the source for all the starting substances required in the various cellular "construction sites" for the synthesis of species-specific structural elements. Moreover, photo-synthesis fulfills a second important function by supplying hexose as the "fuel" for the process of "dissimilation" or "respiration" which produces the energy enabling the cell to carry out biological work. This important process will be discussed in detail later (p. 201 ff).

In a broad sense, the composite sugars – disaccharides, oligo- and polysaccharides – are also products of the photosynthetic CO_2 re-duction, resulting from the provisional end product, hexose sugar. A discussion of these sugars in connection with photosynthetic CO_2 reduction, therefore, seems logical.

Sugar Phosphates. The level of hexoses reached in the CO_2 reduc-tion cycle is represented by fructose-6-phosphate and fructose-1,6-biphosphate. Other hexose phosphates appear besides in green algae and leaves, as shown by pulse labeling with $^{14}CO_2$. These are mainly glucose-6-phosphate and glucose-1-phosphate. They are formed from fructose-6-phosphate which undergoes an intramole-cular rearrangement (ketose to aldose form!) catalyzed by the en-zyme *phosphohexose isomerase*:

Fructose-6-phosphate   Glucose-6-phosphate

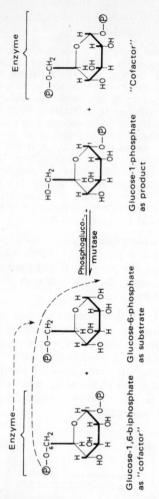

The conversion of glucose-6-phosphate to glucose-1-phosphate is catalyzed by *phosphoglucomutase*.

This reaction is formally also an intramolecular rearrangement, but a more complicated one. The phosphoryl moiety is displaced via the specific intermediate of glucose-1,6-biphosphate. This hexose biphosphate functions as "cofactor" in the reaction. The phosphoryl moiety is transferred from position 6 of glucose-1,6-biphosphate to position 1 of the substrate molecule of glucose-6-phosphate. The latter molecule now assumes the role of the "cofactor", while the glucose-1-phosphate left behind becomes the end product of the conversion. Every glucose-1-phosphate molecule has hence acted once as "cofactor". We shall encounter a similar mechanism of rearrangement in the biosynthesis of 2-phosphoglyceric acid (p. 212 f).

Amino Acids. In algae and leaves incubated in the presence of $^{14}CO_2$ in light, radiocarbon appears very soon in alanine, serine and glycine, a little later in glutamic acid and aspartic acid, and relatively late in proline, arginine and the remaining amino acids. The mechanism whereby these amino acids are synthesized must accordingly be more or less closely connected to the CO_2 reduction cycle.

Recent studies indicate that glycine and serine are formed as intermediates of a by-path of carbohydrate synthesis which is dependent on light. Glycolic acid, the starting substance of this pathway (hence the name "glycolic acid pathway"), originates from the Calvin cycle. This complex reaction sequence, accompanied by consumption of oxygen and release of CO_2, has already previously been referred to as "photorespiration" (p. 159 ff).

The biosynthesis of the individual amino acids and their properties will be discussed together with the protein biosynthesis (p. 354 ff).

Organic Acids. Some organic acids will also have been labeled after short periods of photosynthesis in the presence of $^{14}CO_2$: the monocarboxylic acid glycolic acid (see above), the dicarboxylic acids malic acid and succinic acid, and, in a few cases, the tricarboxylic acid citric acid, too. We shall become more closely aquainted with this group of compounds when discussing the "citric acid cycle" (p. 219 ff).

Disaccharides. Since sugars have free and reactive hydroxyl groups, two of their molecules may react to form a "disaccharide". This gives rise to the characteristic bond linking the carbohydrate units in composite sugars, including oligo- and polysaccharides: the "glycosidic bond". It generally results from combination of a sugar with an alcohol accompanied by loss of water:

D-Glucose   α-D-Glucoside (glycoside)

The reaction product is a "glycoside" corresponding to the acetal of the aldehyde group. The starting substances are recovered by hydrolysis since the reaction is reversible. The corresponding glucose derivatives are also termed "glucosides", those of galactose "galactosides".

The glycosides are a very heterogeneous class of compounds; besides alcohols, a great number of substances from other classes are able to form a bond via an OH group (phenolic or carboxylic) with a sugar. These are generally referred to as "aglycones". Glycosides occur in many species of higher plants.

Generally, there are two mechanisms by which a disaccharide may be formed: 1. Both the hemiacetal hydroxyl groups participate; the disaccharide formed is non-reducing and shows no mutarotation*. All disaccharides constructed in this way, e.g. sucrose (p. 169) belong to the "trehalose type", according to the smallest native compound of this kind. 2. The hemiacetal hydroxyl group of one sugar reacts

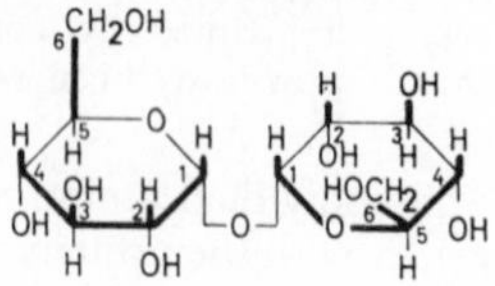

Trehalose

(1-α-D-Glucopyranosido-1-α-D-glucose)

with the free hydroxyl group of another; the latter thereby retains its reducing capacity, shows mutarotation and is able to form another glycosidic bond with an alcohol or another sugar molecule.

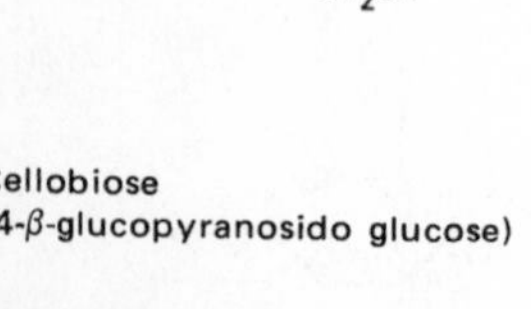

Cellobiose
(4-β-glucopyranosido glucose)

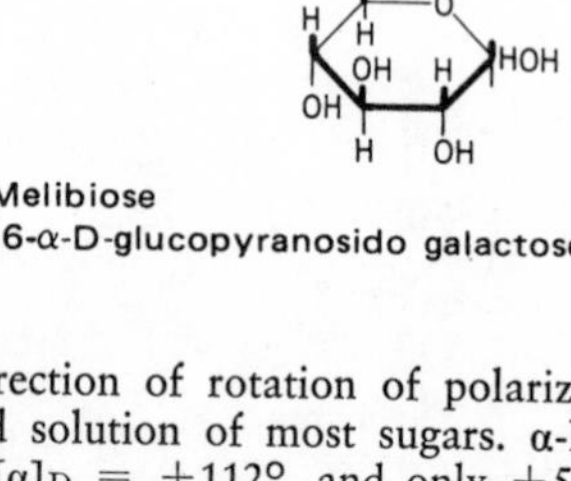

Melibiose
(6-α-D-glucopyranosido galactose)

---

* Mutarotation is the change in the direction of rotation of polarized light passing through a freshly prepared solution of most sugars. α-D-glucose in water initially has the value $[\alpha]_D = +112°$, and only $+52°$ after a few hours. This change reflects the establishment of equilibrium between the two "anomeric" forms: α-D-glucose and β-D-glucose ($[\alpha]_D = +19°$) via a common transition form.

In the latter case, an oligosaccharide is formed. The prototype for compounds of this second group is maltose (structure on p. 206). All corresponding disaccharides, e.g. cellobiose, melibiose, and the oligosaccharides derived from them belong to the "maltose type".

Trehalose is a typical disaccharide of fungi, blue-green algae and red algae. Apart from its occurrence in *Selaginella,* the compound has recently also been identified as a rapidly labeled photosynthetic product, along with sucrose and maltose, in leaves of the fern *Botrychium lunaria.* This appearance of labeled maltose in leaves after short periods of photosynthesis in the presence of $^{14}CO_2$ is inconsistent with the earlier view that maltose occurring in leaf cells is exclusively a product of starch degradation. On the contrary, its rapid synthesis classifies maltose as an intermediate product comparable with the suger phosphates. Like the latter, maltose exhibits the highest level of radioactive labeling after 5 min.

Sucrose, a disacccharide of fructose and glucose ("α-glucopyranosyl-(1→2)-β-fructofuranoside") is characteristic of plant organisms. The two sugar moieties are formed as monophosphates in close connection with the Calvin cycle which explains the relatively fast appearance of labeled sucrose in $^{14}CO_2$ incorporation experiments (cf. p. 164).

Sucrose

Sugar cane and sugar beet are rich in sucrose which is extracted industrially to produce table "sugar". We shall later be discussing the important role of sucrose as a "transport metabolite" for carbohydrates in higher plants (p. 190).

In various higher plant tissues and, in particular, in yeast cells an enzyme has been demonstrated which cleaves sucrose by hydrolysis into its two sugar components:

sucrose + H_2O → D(+)glucose + D(−)fructose ($\Delta G° = -6600$ cal/mole)

Since the rotation of polarized light (cf. p. 354) passing through the assay is changed by this reaction, i.e. an "inversion" occurs (sucrose $[\alpha]_D = +66.5°$; hydrolysis mixture: D(+)glucose + D(−)fructose $[\alpha]_D = -20.5°$), the active enzyme was previously referred to as *invertase.* It is today classified as *β-fructofuranosidase*. Sucrose can also be cleaved by an *α-glycosidase*, since the glycosyl residue has an α-configuration. Despite its occurrence in higher plants, invertase is of no great significance in the interconversion or degradation of sucrose because the equilibrium catalyzed is unfavorable.

Sucrose biosynthesis is rather complicated, since glucose must first be "activated", i.e. converted to a higher energy level. This is a rather common phenomenon in cell metabolism which is always observed when synthesis or conversion of a complex compound is attempted. Free glucose must first be converted by means of *hexokinase* and ATP into glucose-6-phosphate and then isomerized to glucose-1-phosphate by *phosphoglucomutase* (cf. p. 166). Glucose-1-phosphate reacts in an enzymatic reaction with the nucleoside triphosphate UTP yielding uridine-diphosphate glucose:

Uridine-diphosphate glucose, UDPG

The uridine diphosphate attached to glucose constitutes the actual "activating" grouping or the "group-transferring" coenzyme for active glucose. We shall encounter other representatives of this important group of coenzymes later on. The free energy of hydrolysis of the glucosyl ($\Delta G°$ = –7600 cal/mole) is substantially higher than that of the ester bond in glucose-1-phosphate ($\Delta G°$ = –4800 cal/mole).

In UDP-glucose, the glucose molecule has formally replaced the terminal phosphoric acid residue. Instead of this residue, glucose is now transferred to other compounds. These must, however, possess a reactive OH group, as is the case in fructofuranose in sucrose synthesis. The reaction is catalyzed by a specific enzyme, an *UDP-glucose: D-fructose-2-glucosyl transferase (= sucrose synthase):*

With the equilibrium of the reaction displaced far to the right, sucrose synthesis is favored.

The UDP released is regenerated enzymatically to UTP by means of ATP and is thus ready to activate another molecule of glucose-1-phosphate. The analogous compound ADP-glucose formed by direct reaction with ATP is far less active as a precursor in sucrose synthesis than UDP-glucose. ADP-glucose is probably of significance for the biosynthesis of starch and glycogen (p. 178 ff).

UDP-glucose is one of the first compounds to be labeled in photosynthetically active cells in ¹⁴C pulse experiments. In *Chlorella* and *Scenedesmus* cells, for instance, it incorporates radiocarbon before sucrose does. Several researchers believe that sucrose synthesis takes place in the companion cells, from which it passes to the sieve tubes and is translocated (cf. p. 192).

UDP-glucose is involved in two other reactions. One is catalyzed by the enzyme *4-epimerase* and results in a shifting of the OH group at carbon atom 4 (= epimerization!) of the glucose molecule which is converted to galactose. The precise reaction mechanism is unknown; it is, however, certain that the *epimerase* contains bound NAD.

UDP-Glucose   UDP-Galactose

An interesting correlation between the activity of UDP-*galactose-4-epimerase* and cell morphogenesis has been observed in the marine "umbrella alga" *Acetabularia mediterranea*. At the beginning of the gametangia formation ("hat") which, in contrast to the rest of the cell wall, requires a relatively large number of galactose units for its fibrous substance, the activity of this enzyme increases significantly. This is obviously brought about by the increased demand for galactose (Zetsche).

In the second reaction, in which UDP-glucose is the substrate, *glucose-1-phosphate-uridyl transferase* effects an exchange of glucose for galactose. Thus a second mechanism of synthesizing

$$\text{UDP-Glucose} + \text{Galactose - 1 - } \textcircled{P} \xrightarrow[\phantom{\text{Transferase}}]{\text{Transferase}} \text{UDP-Galactose} + \text{Glucose - 1 - } \textcircled{P}$$

UDP-galactose exists. Since it originates from UDP-glucose, "activated galactose" is formed in direct relation to the photosynthetic CO_2 reduction cycle. It is of importance in the synthesis of those trisaccharides and oligosaccharides which contain galactose as an additional constituent (see below).

Oligosaccharides. In terms of structure oligosaccharides form the link between disaccharides and polysaccharides. A disaccharide is, in fact, the simplest form of an oligosaccharide. Enlargement by one monosaccharide unit gives rise to a trisaccharide; enlargement by two monosaccharide units results in a tetrasaccharide etc. Compounds with up to eight units in the molecule are termed oligosaccharides. All those containing more than eight are polysaccharides.

Species of *Gentiana* contain a typical oligosaccharide, gentianose. This is a trisaccharide consisting of sucrose with an additional glucose unit. Raffinose is constructed in a similar way, with a galactose unit in place of glucose (structure, p. 173).

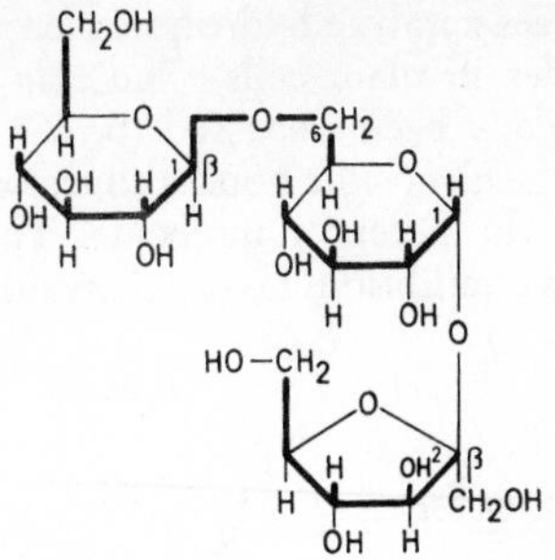

Gentianose

= β-D-Glucopyranosido-(1→6)α-D-glucopyranosido-
(1→2)-β-D-fructose

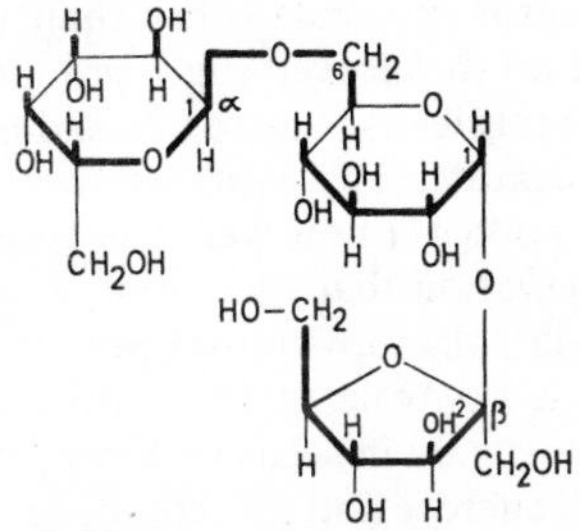

Raffinose

= α-D-Galactopyranosido-(1→6)α-D-glucopyranosido-
(1→2)-β-D-fructose

Other oligosaccharides, derived from raffinose, occur in higher plants where they function as reserve substances. Stachyose, a tetrasaccharide, contains an additional galactose unit in position 6*:

Gal-(1→6)-Gal-(1→6)-Glu-(1→2)-Fru

This sugar occurs in large quantities in *Leguminosae* seeds. In some plant species, raffinose replaces sucrose as "transport metabolite" of carbohydrate, and is hence found in the sieve tubes (cf. p. 191). Recent studies on leaves, particularly from *Labiatae*, have shown that raffinose and stachyose are relatively rapidly labeled during photosynthesis in the presence of $^{14}CO_2$.

Verbascose, a pentasaccharide, has an additional galactose unit:

Gal-(1→6)-Gal-(1→6)-Gal-(1→6)-Glu-(1→2)-Fru

---

* The arrow shows the direction of the glycosidic bond between the hemiacetal OH group and the hydroxyl or hemiacetal group of the next sugar involved.

Biosynthesis. The glycosidases catalyze hydrolytic cleavage of disaccharides and oligosaccharides in plant cells. The role of *invertase,* specific for sucrose, has already been discussed (p. 169 f). The reaction consists in cleaving of the glycosidic bond and transferring of the residue to water, acting as the acceptor molecule. The reaction is practically irreversible, as its equilibrium favors cleavage:

α-D-Glucoside   α-D-Glucose

On the other hand, an acceptor molecule other than water with a reactive OH group may take up the cleaved sugar residue. In this case the *glycosidase* involved catalyzes a "transglycosidation". If the acceptor molecule is a disaccharide, a trisaccharide would be formed by this mechanism. This hypothesis, however, has recently become questionable due to the observation that galactinol, an α-galactoside of *myo*-inositol (L-1-(O-α-D-galactopyranosyl)*myo*-inositol), with its galactosyl moiety plays a key role in the synthesis of the important members of raffinose sugars in seeds of *Vicia faba*. Raffinose synthesis apparently requires sucrose and galactinol, and is catalyzed by an enzyme purified from the same source which acts as a *galactosyl transferase* and differs from α-*galactosidase*. Biosynthesis of stachyose and verbascose also proceeds via transglycosylation of the galactosyl moiety from galactinol to raffinose and stachyose, respectively. Galactinol is apparently formed from the "activated", i.e. energy-rich form of galactose, UDP-galactose (for biosynthesis see p. 172) which then reacts with *myo*-inositol, releasing UDP. The reaction principle observed here also underlies sucrose biosynthesis: following an initial activation step the monosaccharide is first converted into the energy-rich nucleoside diphosphate sugar. We shall see later that such nucleoside diphosphate sugars also serve as precursors in biosynthesis of the polysaccharides (p. 178 ff). Obviously the galactosides of yeast and red algae are formed in the same way.

Polysaccharides. The molecules of the polysaccharides consist of chains of monosaccharide units linked by glycosidic bonds (see below). Accordingly, their molecular building blocks originate from only one class of compounds. Structural variations are hence restricted to the combination of different kinds of sugar units in the molecule, formation of different kinds of bonds between the units and formation of more or less long or more or less branched molecule chains. The polysaccharides perform functions as reserve substances, fibrous material, plant rubber and slime substances, and are therefore indispensable for the structure and the functioning of the cell.

In the polysaccharides, we encounter for the first time the macromolecular construction principle which is of particular significance in the nucleic acids and proteins (cf. p. 310 f). The macromolecules have certain properties which make them very suitable reserve or structural substances. The first one is their relative insolubility which is utilized for an economic synthesis of reserve substances without disturbing the osmotic equilibrium in the cell. A second important property is useful in accomplishing architectural functions: macromolecules can easily be formed into long-extended fibrillar structures which are particularly important for the cormus of higher plants because of the severe mechanical strain to which it is subjected.

A further possibility for classification is based on the chemical structure. All polysaccharides consisting of only one kind of sugar building block are termed "homoglycans" while those with several kinds of monosaccharide units are "heteroglycans". A third group comprises those compounds which contain a polysaccharide and another chemical constituent (protein or lipid): glycoproteids and glycolipids ("composite" or "conjugated polysaccharides"). The most important polysaccharides of the plant cell belong to the first group: starch, cellulose and glycogen, which contain glucose as the only building block.

Reserve Substances. The most important reserve polysaccharide in the plant kingdom is starch. Chloroplasts are the organelles of starch synthesis in photosynthetically active cells, amyloplasts fulfill this function in cells of storage tissues. In the first case starch is embedded between the thylakoids, as shown by electron micrographs; in the second case the typical starch granules appear. Their form, size and layering are often species-specific; thus their origin can be determined by microscopic examination. Starch is chemically a homogeneous substance, since boiling of starch granules with dilute inorganic acids always gives rise to the formation of α-D-glucose; degradation by specific enzymes, the *amylases* (p. 205 f), always produces maltose as end product. These findings, as well as the structural formula derived from elementary analysis show that starch is a polycondensate of α-D-glucose units: $[C_6H_{10}O_5]_n$.

In most cases, two molecular forms of starch occur: *amylose* and *amylopectin*. The share of the latter varies between 70 and 90 % in the various species of starch.

In amylose, chains of about 200 to 1000 glucose units are linked by 1→4-α-glycosidic bonds, forming a macromolecule.

Non-reducing end   Reducing end

Amylose

Each molecule has a glycosidic carbon atom at one end (= "reducing end") and a hydroxyl group at carbon atom 4 at the other end (= "non-reducing end"). The molecular weight obtained by physical methods (measurement of viscosity and osmotic pressure; determination of sedimentation constant) amounts to 10 000 to 100 000. In amylose there are only very few side branches from the main chain. The 1→4-α-bonds cause the amylose chain to wind up to a screw or "helix". In the specific color reaction with iodine/potassium iodide, the iodine atoms are deposited as a chain in the interior of the helix and held tightly. This brings about a strong absorption of long-wave visible radiation, giving the iodine-starch complex a bluish-violet color. Amylose is usually soluble in hot water (70–80 °C).

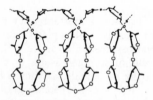

Molecular structure of amylose (section)

The amylopectin molecule consists of 2000 to 22 000 glucose units (molecular weight 50 000 to 1 million). In contrast to amylose, it is an extensively branched polysaccharide. Side chains of various lengths are attached by 1→6-α-glycosidic bonds to the main chain (1→4-α-bonds!). On average, there is one branching to about 25 glucose units. Main and branching chains may both be helical. The presence of branching is shown by the appearance of small quantities of isomaltose (formula above) during enzymatic cleavage of amylopectin. Amylopectin is insoluble in hot water. A violet to red color appears with iodine/potassium iodide.

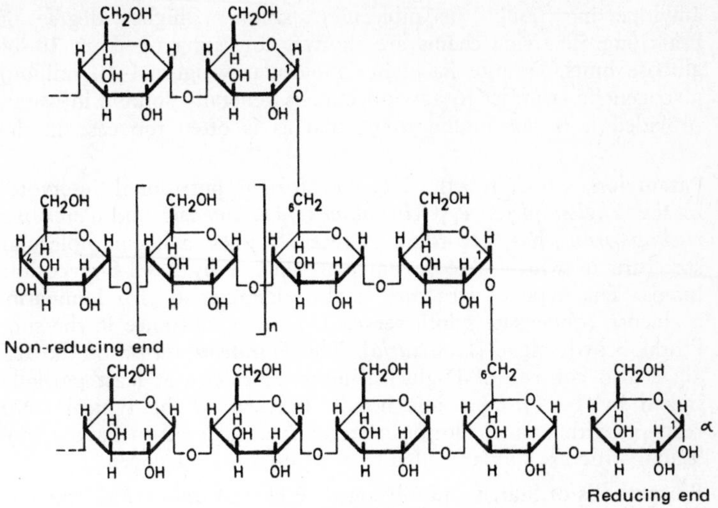

Amylopectin

Non-reducing end

Reducing end

Amylopectin

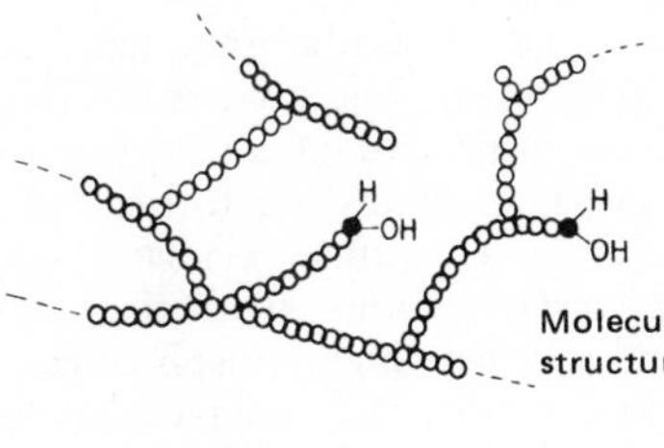

Molecular structure

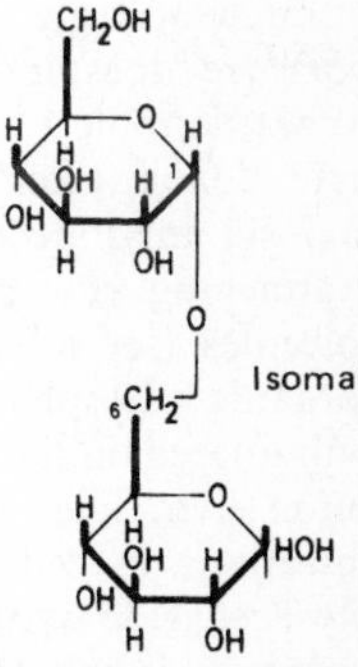

Isomaltose

By boiling with dilute inorganic acid, amylose and amylopectin are hydrolysed to D-glucopyranose via intermediate cleavage products of varying size, the "dextrins". The formation and subsequent degradation of dextrins can be followed by performing the iodine test. A purple coloration only appears down to a certain chain length, and is absent in the short-chain dextrins. The starch in the kernels of some varieties of barley, maize and rice has a waxy and glutinous quality. It consists practically only of amylopectin. On the other hand, there are breedings of maize and peas in which the reserve starch contains between 50% and 80% amylose.

Glycogen. This reserve substance, characteristic of the *Cyanophyceae*, fungi and bacteria, is constructed in a way similar to amylopectin, though its molecules exhibit a higher degree of branching. The side chains are shorter, consisting of about 10–14 glucose units. Despite its higher molecular weight (1–10 million) glycogen (in contrast to amylopectin) is generally soluble in water, provided it is not bound to protein as is often the case in the cells.

Paramylon, which functions as "protozoan nutritional reservoir" in the *Euglenophyceae, Ochromonas malhamensis* and *Perenema trichophorum,* has, according to recent studies, a linear molecular structure in which glucose units are linked by $1\to3$-β-glycosidic bonds. This type of structure is also characteristic for **laminarin,** a glucose condensate which serves as reserve substance in the sublitoral brown algae *(Laminaria)*. The *Floridean starch* of the red algae also consists of D-glucose units which are at least partially linked by $1\to4$-α-glycosidic bonds. In view of the typical color reaction with iodine, Floridean starch is considered to have a molecular structure similar to that of amylopectin or glycogen.

Biosynthesis of Starch and Glycogen. When *phosphorylase* was discovered, it was thought that the enzyme specific for the biosynthesis of starch and glycogen had been found. In a purified form *phosphorylase* does indeed catalyze the synthesis of long-chain polysaccharide molecules when an adequate quantity of glucose-1-phosphate ("Cori ester"*) and "primers", i.e. molecular fragments of amylose, amylopectin or glycogen are present. The process consists of attaching new glucose units at the chain end of added primer molecules (see scheme, Fig. 79). New $1\to4$ bonds are formed and inorganic phosphate is released. In the meantime, evidence has accumulated indicating that *phosphorylase* has no synthetic function in vivo. The high pH values and relatively high phosphate concentrations in cells of higher plants are obviously unfavorable to polysaccharide synthesis. Besides, the concentration of glucose-1-phosphate is too low to achieve measurable synthesis. Today one tends to ascribe a specific function to *phosphorylase* in the degradation of starch and glycogen (cf. p. 207 f). For the in-vivo synthesis of these polysaccharides a different reaction catalyzed by a different enzyme is probably used (see below). The existence of reaction pathways for synthesis different from those for degradation of an important compound in metabolism provides the cell with a better means of regulation. Such a separation of reaction pathways is also observed in fatty-acid metabolism (p. 409).

---

* Named in honor of the American biochemists G. T. Cori and C. F. Cori, who discovered and successfully isolated the *phosphorylase* of muscle and liver.

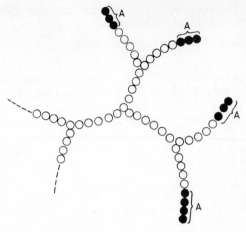

Fig. 79 Synthesis reaction catalyzed by *starch phosphorylase:* by attachment of new glucose units (A) to the free chain ends (non-reducing; OH group in position 4) of the primer molecule an elongation of the chain is achieved.

Recent studies with plant and animal storage tissues have shown that the energy-rich compounds of nucleoside diphosphate glucose, UDP-glucose or ADP-glucose, are utilized as substrates in the biosynthesis of starch and glycogen. In the biosynthesis of sucrose and oligosaccharides (p. 170) we have already encountered the principle of substrate activation effective in these processes. Probably all other polysaccharides are also synthesized from nucleoside diphosphate sugar units.

In comparison with the biosynthesis of sucrose, the participation of an appropriate primer molecule (see above) in starch synthesis is a new feature. To this molecule a D-glucose unit is attached by a 1→4-α-glycosidic bond. The primer itself must be a glucose condensate with 1→4-α-glycosidic bonds. Its chain length appears to be of minor significance; under certain conditions even the disaccharide maltose may accept glucose units from nucleoside diphosphoglucose. Oligo- and polysaccharides (dextrins, amylose, amylopectin) may equally well serve as primers. The presence of inorganic phosphate does not affect the process of synthesis which can thus be distinguished clearly from the *phosphorylase*-catalyzed reaction. The active enzyme, *UDP-glucose-starch transglycosidase* or *starch synthase* has been detected in plant cells. An enzyme with very similar properties which forms glycogen from UDP-glucose and short-chain D-glucose condensates had previously been discovered in liver cells. The active enzyme from starch granules of *Phaseolus vulgaris* catalyzes the following reaction:

$$\text{UDP-glucose} + \text{acceptor molecule} \rightarrow \text{UDP} + (\text{glucose})_{n+1}$$
$$(\text{glucose})_n$$

The reaction is exergonic ($\Delta G^\circ = -3300$ cal/mole); the equilibrium favors starch synthesis. The reaction mechanism may be regarded as a special case of synthesis by "transglycosidation", which has already been mentioned in connection with oligosaccharide synthesis (p. 174).

Starch synthase is firmly bound to the starch granules, probably to their membrane, and is difficult to solubilize. Active complexes of amylose and *starch synthase* are also formed in vitro when the solubilized enzyme is precipitated from an amylose solution by acetone. These complexes may serve as models for the close association of amylose and amylopectin molecules with the enzyme protein in starch granules. In the meantime the enzyme has also been found in starch preparations from maize and potatoes.

Of the other nucleoside diphosphate sugars, ADP-glucose (structure below) showed an activity which was, surprisingly, many times higher than that of UDP-glucose. Since ADP-glucose has also been shown to be a native component of the storage cells, it is tempting to postulate that this nucleoside diphosphate sugar is the natural substrate of starch synthesis. The hypothesis is supported by the discovery of a specific enzyme in wheat and rice grains, *ADP-glucose pyrophosphorylase,* which catalyzes the following reaction:

$$\text{glucose-1-phosphate} + \text{ATP} \rightarrow \text{ADP-glucose} + \text{pyrophosphate}$$

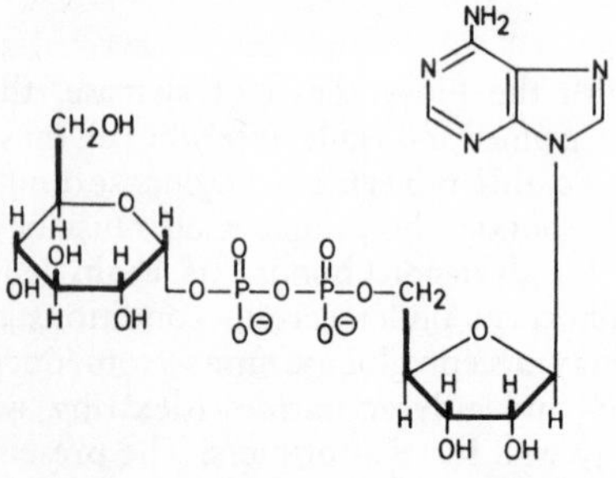

Adenosine diphosphoglucose, ADPG

It is worth-while mentioning that the enzyme acting in the biosynthesis of paramylon in *Euglena gracilis* is specific to UDP-glucose. *UDP-glucose-1→3-β-glucan transglycosidase,* with similar properties, has been found in *Phaseolus aureus* seedlings and in several other plant tissues. The enzyme appears to be generally responsible for the synthesis of 1→3-β-glucans from UDP-glucose.

So far, only starch synthesis in cells of storage tissues has been discussed. How does this process proceed in chloroplasts, the other important site of starch synthesis?

The preferential incorporation of labeled glucose from ADP-^{14}C-glucose in isolated starch particles from *Soja* leaves, and the demonstration of ADP-glucose synthesis in photosynthetically active cells of *Chlorella pyrenoidosa* suggest a similarity of the synthesis mechanisms for assimilation starch and for reserve starch. ATP derived from photophosphorylation (p. 125 ff) probably brings about a direct synthesis of ADP-glucose in chloroplasts.

The enzymic mechanism by which side chains are established in the amylopectin molecule is still obscure. An enzyme *(Q enzyme)* had, in fact, been isolated from potatoes which, in co-operation with synthetically active and purified *phosphorylase* produces 1→6-α-bonds. Its participation, however, in amylopectin synthesis catalyzed by *ADP-glucose-starch transglycosidase* has not yet been demonstrated.

Nucleoside diphosphate glucose probably also plays a key role in the conversion of sucrose to reserve starch (cf. p. 190 ff). In a combination of the two reactions catalyzed by *sucrose synthase* and *starch synthase*, respectively, there is rapid incorporation of the glucose unit from sucrose into starch, particularly in the presence of ADP.

Fructosans. Apart from the glucose homopolymers starch and glycogen, there are corresponding compounds of fructose serving as typical reserve substances, particularly in *Compositae* and *Gramineae*. They are termed "fructosans" or "polyfructosans", i.e. polymers of D-fructofuranose of relatively low molecular weight, usually containing a small percentage of glucose. The most well-known and most studied one is *inulin*, which occurs in large quantities in tubers of *Dahlia* species and *Helianthus tuberosus* (Jerusalem artichoke) in colloidal form. It is also found in the stem tissues of these plants, though not in the leaves, where sucrose and starch occur. In *Gramineae*, particularly in cereal plants, the relationship is reversed: starch is replaced by fructosan in the leaves; sucrose is also present.

The fructosans of low molecular weight ought to be classified as oligosaccharides, since they have less than eight sugar units.

With its 30–40 fructose units *inulin* reaches the highest level of polycondensation. The units form a molecular chain the members of which are linked by 2→1-β-glycosidic bonds (see formula). There

Inulin   Fructosan of "Phlein Type"

is one glucose molecule at the reducing end and possibly another one in the middle of the chain (1→3 bonding).

All fructosans possessing 2→1 bonds belong to the *inulin type*. They differ from those of the *phlein type* (named after the characteristic fructosan from *Phleum pratense*!) in which the 2→6 bond predominates (see formula). The molecules of the latter consist of 3–30 hexose units and end in a non-reducing sucrose unit; these compounds are typical for *Gramineae* and some microorganisms.

In some fructosans of the two groups side chains occur more or less frequently in which fructosyl groups are attached to the molecular chain by 2→6 or 2→1 bonds.

Apart from the relatively long-chain fructosans, a number of short-chain forms of the oligosaccharide type are found in cells. They are probably intermediates in the process of conversion to larger molecules.

The biosynthesis of fructosans is thought to originate from sucrose as the starter molecule; new fructosyl units are successively attached to the fructose moiety of the sucrose molecule. This would explain the terminal position of the glucose unit in the completed fructosan molecule. The "transglycosidation" involved is probably catalyzed by a *transfructosidase*. Plant fructosans are probably synthesized by a mechanism similar to that found in *Aerobacter levanicum*. In this organism a fructosan of relatively high molecular weight, "levan", is formed from sucrose, mediated by a *sucrase (saccharase)*. Single

D-fructose units are attached via $2 \rightarrow 6$-β-bonds to the fructose of the sucrose molecule:

$$n \text{ Sucrose} \xrightarrow{\text{\textit{levan sucrase}}} n \text{ D-glucose} + (\text{D-fructose})_{n+1}$$

The recent isolation of UDP-fructose (see formula) from *Dahlia* tubers indicates that fructose units may be transferred via this nucleoside diphosphate in fructosan biosynthesis. Hence the reaction principle applied

Uridine diphosphofructose, UDPF

would be the same as that in the biosynthesis of other reserve polysaccharides.

Mannans. The mannans constitute another group of reserve polysaccharides; they are polycondensates of the hexose sugar mannose. They usually contain an additional hexose component, glucose or galactose, so they are to be classified as "heteroglycans". The chemical structure of the mannans has not yet been completely elucidated. They probably consist of chains of mannopyranose units which are linked by $1 \rightarrow 4$-β(?)-glycosidic bonds; $1 \rightarrow 6$ bonds may also occur. The molecule chain is terminated in various mannans by galactopyranose.

The "reserve cellulose" in the endosperm of date seeds and stone fruits consists largely of mannans which are degraded during seed germination.

Mannan-type polymers which function equally well as reserve polysaccharides and as structural substances, lead up to the true structural substances.

Structural Substances. The portion of polysaccharides acting as true structural substances in the plant cell wall may amount to up to 90 % of dry weight of the primary wall. Although their chemical constitution has been intensively and successfully studied in recent years, the biosynthesis of cell-wall polysaccharides remains obscure. With a few exceptions, attempts to synthesize individual components in vitro by means of isolated enzymes have been unsuccessful.

The in-vitro formation of callose from UDP-glucose by an enzyme from bean seedlings *(Phaseolus aureus)* may serve as model for cellulose biosynthesis. The reaction product consists of $1 \to 3$-β-glycosidically linked glucose units, like the native compound. Callose is not a common cell wall component. Its deposition blocks the sieve pores in aging sieve tubes (cf. p. 190).

Cellulose. The most important constituent of the cell wall is cellulose, a linear polysaccharide of glucopyranose units which are linked by $1 \to 4$-β-glycosidic bonds:

Cellulose molecule (section)

Additional hydrogen bonding (broken line!), contributing to the stability of the chain occurs in the unbranched macromolecule. The length of the chain varies; recent studies indicate that native cellulose contains up to 14 000 glucose units which would correspond to a chain length of about 7.0 μ* and a molecular weight of 2.3 million. The threadlike molecule is probably repeatedly folded back in itself and forms an "elementary fibril" which is detectable in electron micrographs.

The secondarily modified cell wall generally contains lignins in addition to cellulose (= "lignification"). These substances are polymers of phenylpropane compounds (coniferyl, cimaryl and sinapyl alcohol) which are able to penetrate the interfibrillar spaces because of their three-dimensional molecule structure. In technical extraction of cellulose from wood, lignins must first be removed by boiling with calcium bisulfite ("sulfite leach"). There are also small amounts of hemicelluloses ("cellulosans") which probably form the matrix of the secondary cell wall. The elementary constituents of these complex macromolecules are pentoses such as D-xylose, L-arabinose ("pentosans") or hexoses such as D-mannose, D-galactose ("hexosans"). In exceptional cases, e.g. in the seed hairs of cotton, the cellulose is largely free from lignins.

Several of the most important features of cellulose derive from its high resistance to chemical and enzymatic degradation. Prolonged heating in the presence of concentrated acids (H_2SO_4!) is required

---

* 1 μ = 1/1000 mm

to hydrolyze all glycosidic bonds and convert the polymer to glucose ("saccharification of wood"). Numerous bacteria and several seedlings have specific enzymes, *cellulases,* which degrade cellulose to the disaccharide cellobiose (structure, p. 168) by hydrolytic cleavage of β-glycosidic bonds. Cellobiose may be regarded as the building unit of cellulose, though it is probably not utilized in its biosynthesis. The attack of *cellulase* is often coupled with a second enzyme, *cellobiase,* which completes the degradation of cellulose to free hexose by hydrolytic cleavage of the dimer. Both enzymes are *β-D-glycosidases.*

A few facts will help to illustrate the importance of cellulose. Of approximately 30 billion tons of carbon bound annually in organic substances by higher plants, about one third ends up in cellulose. Our clothes consist mainly of cellulose (80–90 %). Paper contains about 50 % cellulose. Cellulose constitutes about one third of the food of animals which contribute to our meat supply.

Cellulose synthesis in vitro was first accomplished with an enzyme preparation from the bacterium *Acetobacter xylinum* which catalyzes transfer of D-glucopyranose residues from UDP-glucose to cellulose. The particle-bound enzyme is only active when cellulose or short-chain cellodextrins are present in the assay. The yield, measured in terms of the incorporation of ^{14}C-labeled glucose from UDP-^{14}C-glucose, was very small (1–2 %). In contrast, an enzyme bound to subcellular particles isolated from *Phaseolus aureus* (mung bean) preferred GDP-glucose as substrate. 25 % of the ^{14}C-radioactivity were recovered in cellulose when GDP-^{14}C-glucose was given. It has recently been established that an intermediate identified as glycolipid (p. 417) is effective in the polycondensation process. It probably contains a long-chain (C_{30}) polyhydroxy alcohol and an organic base which binds the glucose. An enzyme which catalyzes the transfer of lipid-bound glucose into polymer has only been found in the culture medium of *Acetobacter xylinum.* The active glycolipid may act as transport vehicle which carries the molecules of nucleoside diphosphate sugar across the lipophilic cytoplasmic membrane. This is in contrast to another concept according to which the glucose unit bound to glycolipid is merely transported to the cell wall where synthesis of GDP-glucose or possibly UDP-glucose takes place.

On the basis of its chemical constitution, cellulose is classified as a "homoglycan". Some recent findings indicate that other sugar molecules may be linked to the molecule chains in the cell wall; hence cellulose would be a heteropolymer, at least in vivo.

Pectic Compounds ("Polyuronides"). In young cells of higher plants the matrix of the primary cell wall consists of complex polysac-

7 Richter, Physiology

charides which are not identical with cellulose. Apart from hemi-celluloses (p. 184), the most important constituents are the pectic substances. Three groups of these are known at present: pectic acids, pectinic acids, and protopectins.

Pectic acids probably consist of unbranched molecule chains of galacturonic acid, a sugar acid (structure, see below) the units of which (5–500) are linked by $1\rightarrow4$-α-glycosidic bonds ("polygalac-turonic acid"!).

α-D-Galacturonic acid   Polygalacturonic acid molecule (section)

Pectic acids of high molecular weight have the properties of colloids.

Pectinic acids are formed from pure polygalacturonic acids by esterification of several of the free carboxyl groups with methanol. They contain 100–200 galacturonic acid units and are constructed in a way very similar to the pectic acids. The extent of methylation is difficult to determine since numerous methyl ester bonds are hydrolytically cleaved during acid extraction of these compounds from the cell wall. As far as we know, there are no completely methylated derivatives in plant cell walls. The number of the methyl groups present appears to be species-specific. In general, the water-solubility of pectinic acids increases with the extent of methylation and decreases with increasing molecular size. Usually colloidal solutions result.

All the insoluble pectic compounds are termed **protopectins**. They are relatively unstable. Their structure and composition are still largely unknown, since they are destroyed by the standard extraction methods.

Pectin compounds are not only major components of the cell wall matrix and the middle lamella of higher plants and algae, they also occur in large quantities in fruit (apple, apricot, citrus fruit), from which they are technically extracted. Because of their ability to form gels in low concentrations they are used preferentially in the food industry as a gelling agent.

Due to their free unesterified COOH groups, pectic compounds are able to form additional bonds with OH groups of other sugars which may be free or part of a chain. Recent investigations of noncellulose polysaccharides from sisal, lucerne and sugar beet indicate a more complex structure of pectic substances. According to these findings, they may form the main chain or a part of it in some polysaccharides, whereas in others they form merely the side chains. Moreover, other sugar components such as galactose, arabinose, methyl xylose, rhamnose, and methyl fucose participate in the composition of these pectic substances, causing their homoglycan nature to become questionable.

Rhamnose and fucose are deoxy sugars. Rhamnose is 6-deoxy-L-mannose, while fucose is 6-deoxy-L-galactose:

L-Rhamnose   L-Fucose

= 6-Deoxy-L-mannose   = 6-Deoxy-L-galactose

An interesting and promising possibility for studying the structure and composition of native pectic compounds is provided by cell suspension cultures of *Acer pseudoplatanus,* which secrete water-soluble polysaccharides containing galacturonic acid into the culture medium. Using them as test material has the great advantage that no methodically difficult extraction from cell walls is required; thus a creation of artifacts is avoided. The overall composition of these "external" polysaccharides is identical with that in the walls of their cells of origin.

Among the enzymes which degrade pectic substances, the *pectin esterases* occur in phytopathogenic organisms as well as in higher plants. They are probably located in the cell wall and catalyze the hydrolytic release of methyl groups from ester bonds.

The 1→4-α-bonds of polygalacturonic acid, however, are hydrolyzed by *pectinase* (also called *pectin depolymerase* or *polygalacturonase).* Preparations have been obtained from tomatoes and carrots. They are suitable for the isolation of single cells from plant tissues since they gently dissolve the middle lamella.

Biosynthesis of Pectin Compounds. A synthesis of pectin compounds in a cell-free system has not yet been achieved. We are fairly well informed about the formation of the probable starting substance, UDP-galacturonic acid. This nucleotide-bound sugar acid is an "uronic acid" which is generally formed by oxidation of the CH_2OH group of a sugar (oxidation at carbon atom 1 gives rise to an "aldonic acid", e.g. 6-phosphogluconic acid, p. 256). Oxidation of glucose at carbon atom 6 produces glucuronic acid. In the cell this reaction does not involve the free sugar but its "activated" form,

UDP-glucose. The UDP-glucuronic acid formed is the key substance in polyuronide biosynthesis. The reaction is catalyzed by an *UDP-glucose dehydrogenase:*

―――――― UDP-Glucose dehydrogenase ――――――▶

Uridine diphosphoglucose   UDP-glucuronic acid

The enzyme has been isolated from pea seedlings; it catalyzes the double dehydrogenation step.

UDP-glucuronic acid is converted to UDP-galacturonic acid by a *4-epimerase:*

UDP-glucuronic acid   UDP-galacturonic acid

Since up to now no *UDP-galactose dehydrogenase* has been found, the formation of UDP-galacturonic acid seems to be possible only via this epimerization of UDP-glucuronic acid.

An *UDP-galacturonyl transferase* is thought to be involved in the biosynthesis of polygalacturonic acid from UDP-galacturonic acid. A similar enzyme occurs in the microsomal fraction from mammalian liver. The methyl groups which are attached to pectic acids, converting them into pectinic acids, derive from methionine, a sulfur-containing amino acid (p. 358) which readily supplies a methyl group.

Chitin. This structural substance occurs as a characteristic component in the cell wall of most fungi, and is also major component of the shells of crabs and insects. It resembles cellulose insofar as the building block N-acetylglucosamine is a derivative of glucose. Formally, this is an amino sugar since the OH group at carbon atom 2 is replaced by the modified amino group ($-NH-CO-NH_3$):

β-D-Glucosamine   N-Acetyl-   Chitobiose
glucosamine

There is a further similarity in respect of the shape of the macromolecule which is also folded back in itself because of the β-glycosidic bonding (probably 1→4).

The molecular weight of chitin is unknown, but it is probably very high. Chitin is an example showing that not only pure sugars but also their derivatives may be used for the formation of polysaccharides. Other compounds of this type are found in the animal kingdom, in man (hyaluronic acid, chondroitin sulfates) and in the cell walls of bacteria and *Cyanophyceae* as "mureins" (= "peptidoglycans").

The enzyme *chitinase* which occurs in a few species of bacteria and in the digestive secretions of the edible snail, degrades the polysaccharide to its N-acetylglucosamine units. The disaccharide "chitobiose" (structure, see above) is formed after partial hydrolysis of chitin. Chitobiose may be regarded as the building unit of chitin and is thus analogous to cellobiose in cellulose.

According to first results obtained with an in-vitro system derived from *Neurospora crassa*, nucleoside sugars are probably the starting substance in chitin biosynthesis, too.

References

Arnon, D. I.: Photosynthetic activity of isolated chloroplasts. Physiol. Rev. 47: 317, 1967

Björkmann, O., J. Berry: High-efficiency photosynthesis. Sci. Amer. 229: 80, 1973

Calvin, M., J. A. Bassham: The photosynthesis of carbon compounds. Benjamin, New York 1962

Gibbs, M.: Photosynthesis. Ann. Rev. Biochem. 36: 757, 1967

Ginsburg, V.: Nucleotides and the synthesis of carbohydrates. Advanc. Enzymol. 26: 35, 1964

Hatch, M. D., C. R. Slack: Photosynthetic CO₂ fixation pathways. Ann. Rev. Plant Physiol. 21: 141, 1970

Heber, U.: Metabolic exchange between chloroplasts and cytoplasm. Ann. Rev. Plant Physiol. 25: 393, 1974

Jackson, W. A., R. J. Volk: Photorespiration. Ann. Rev. Plant Physiol. 21: 385, 1970

Laetsch, W. M.: The C₄ syndrome: a structural analysis. Ann. Rev. Plant Physiol. 25: 27, 1974

Marx-Figini, M., G. V. Schulz: Zur Biosynthese der Cellulose. Naturwissenschaften 53: 466, 1966

Preston, R. D.: Plants without cellulose. Sci. Amer. 218: 102, 1968

Transport of the Assimilates

In higher plants, the photosynthetic production of substances largely takes place in the assimilation parenchyma of the leaves. It supplies substrate not only for all tissues and cells with a heterotrophic metabolism, but also for those cells whose specific function it is to build macromolecular reserve substances, such as polysaccharides, fats or proteins. The assimilates must hence be transported and distributed in an appropriate chemical form to the various parts of the plant's cormus (growing points, storage organs and tissues, fruits and seeds). The plant accomplishes this task by means of a transport system in the shoot, a continuous stream of dissolved assimilate molecules flowing mainly from the leaves to the root. The transport elements are the sieve tubes or sieve cells (gymnosperms and vascular cryptogams) in the phloem of the vascular bundles or in the bast if there is secondary growth.

Direct evidence for the participation of sieve elements in phloem transport is obtained by the following rather elegant method: an aqueous solution of ^{3}H-labeled sucrose or glucose is applied to leaves of an intact plant (Cucurbita, Cucumis); in longitudinal sections of the conducting tissues below the site of application, radioactivity coinciding with the location of sieve tubes can be demonstrated after 3–4 h by means of autoradiography (p. 144).

The sieve tubes are particularly adapted for the mechanism of translocation. In the functionally active state the nucleus and plastids have disappeared from the cytoplasm; the mitochondria are partly degraded; the endoplasmic reticulum, however, remains intact. The occurrence of numerous enzymes of carbohydrate metabolism is striking; they are closely connected with the function of sieve tubes (see below). Since the tonoplast is partly eliminated, the distinction between cytoplasm and vacuole disappears in places, particularly near the transverse walls. The term "lumen" is therefore more suitable than "vacuole". The plasmalemma remains intact. The functional activity of a sieve tube is generally restricted to one vegetation period; the sieve pores are subsequently blocked by callose formation (p. 184). There is apparently a close physiological contact with the companion cells as indicated by numerous plasmodesmata between them and the sieve tubes. The role of the companion cells in the transport function of the sieve tubes in still unknown.

Analysis of the sap of the sieve tubes shows that the major part of the dissolved and translocated substances consists of sucrose. It represents the most important transport form for carbohydrates, and possibly also for fats which must first be converted to carbohydrates (cf. p. 226 f). Thus the assimilation starch (p. 181) formed in the chloroplasts in the day-time is degraded during the dark period, i.e. at night, and conducted as sucrose to the sites where it is required. In addition to sucrose, small amounts of amino acids, amides, organic acids, nucleotides, as well as some vitamins and hormones are transported. Virus particles are also believed to be distributed via the sieve tubes. In exceptional cases, raffinose (p. 173) assumes the role of sucrose. The speed of phloem transport in the sieve tubes differs greatly among the various plant species (from a few cm to several meters per hour).

Aphids are successfully used to remove selectively the contents of sieve tubes in intact plants. With the tip of their stylet they tap the interior of a single sieve tube element: in this state the insect body is removed by a cut through the stylet. This remains attached as before and now acts as a minute cannula in the sieve tube wall through which the sap from the sieve element exudes, driven by the turgor release. For hours or even days the sap is collected in this way and analyzed.

Direction of Phloem Transport. The "downward" movement of assimilates and nutrients described above is not quite exact insofar as organic compounds may also move "upward" depending on the location of the producing organ or tissue (= "source") and the receiving organ or tissue (= "sink"). Apparently this occurs in young, rapidly growing plants where young leaves near the growing tips of branches utilize assimilates from older leaves below. Shifts in the direction of phloem movement may also be induced experimentally, e.g. by removal of the leaves from a certain region of the stem thus depriving it of its original source of nutrients; supply of assimilates will then be taken over by the leaves of the region below acting as a new source. Thus, the direction of movement depends entirely on the position of the source. Often, a source may be located between two sinks which would give rise to a *bidirectional movement*. The mechanism of the latter is still a matter of controversy.

Mechanisms of Phloem Transport. The exact mechanism of assimilate flow, i.e. the "long-distance transport" (cf. p. 278) of sugars and other organic compounds, is still unknown. In contrast to the water transport (p. 276), purely physical processes cannot explain this type of long-distance transport. Among the various hypotheses and models proposed, the *pressure flow hypothesis* of Münch (1926–1930) has remained the principal concept till today, despite a great deal controversy. He assumed that a gradient in sugar concentration extending through the sieve tubes causes a pressure flow which brings about the transport movement: in leaves, the synthesized carbohydrates are transferred into the sieve tubes resulting in a fairly concentrated solution of sugar (= "source"!). Consequently, water moves into the sieve tubes causing a rise of turgor pressure. In the region of consumption, on the other hand, assimilates are continuously removed from the sieve-tube sap leading to a decrease of the osmotic value (= "sink"!). Because of the resulting difference in osmotic potential or hydraulic pressure the solution in the sieve tubes is thought to move in a "mass flow" from the regions of high concentration (source) to those of low concentration (sink). The flow is maintained as long as the assimilates accumulate in the sieve tubes of the source end and are removed at the sink end of the transport line. The velocity of movement would thus depend on the osmotic potential gradient and on the resistance encountered within the sieve tubes. If the hypothesis of a directed movement in the phloem, as proposed by Münch, is correct, it follows that the major and minor solutes in the sieve tubes should follow the same path from source to sink; there is, indeed, some evidence that this is the case (even virus particles ride in the same direction as sugar molecules!). It is rather difficult, however, to demonstrate that all materials move at the same rate, due to methodical difficulties. Though these and other findings are compatible with Münch's hypothesis, several objections have been raised against it; one major problem is the narrowness of the sieve plate pores with the cytoplasmic strands passing through them, which militates against a mass flow of solution through the sieve tubes as postulated by the hypothesis.

In the *electroosmotic hypothesis* (Fensom, Spanner) the sieve plates are regarded as "pumping stations" transferring solutes from one sieve tube into the next one. An electrical potential difference exists across the sieve plate resulting either from a differential permeability to $H^\oplus$ and $HCO_3^\ominus$ ions or from an active transport of $K^\oplus$ ions across the sieve plate. This difference in electropotential is assumed to bring about an electroosmotic flow of water. So far, however, there are no convincing experiments supporting this hypothesis.

All mechanisms of phloem transport proposed so far have one feature in common: they depend on a continuous energy supply, i.e. on active metabolism, since the transfer of assimilates into the sieve tubes at the source region and their removal at the sink region are "active", energy-requiring processes (p. 283). Probably a close physiological contact exists with the metabolically rather active companion cells which may supply energy equivalents and essential metabolites to the adjacent sieve elements.

References

Eschrich, W.: Biochemistry and fine structure of phloem in relation to transport. Ann. Rev. Plant Physiol. 21: 193, 1970

Lüttge, U.: Stofftransport der Pflanzen. Heidelberger Taschenb., Vol. 125. Springer, Berlin–Heidelberg–New York 1973

Preston, R. D.: Phloem transport in plants. In: Progress in biophysics, Vol. 13, Pergamon Press, Oxford 1963

Wardlaw, I. F.: Phloem transport: physical chemical or impossible. Ann. Rev. Plant Physiol. 25: 515, 1974

Ziegler, H.: Symposium Stofftransport. Ber. dtsch. botan. Ges. Neue Folge 2 (1968)

Zimmermann, M. H.: Transport in the phloem. Ann. Rev. Plant Physiol. 11: 167, 1960

Chemosynthesis

Only the photosynthetically active cell is able to separate hydrogen from oxygen in the photolysis of water by utilizing absorbed radiant energy. Subsequently the hydrogen is raised to the higher energy level of carbohydrate by reacting with CO_2. Thus a potential between oxygen and hydrogen is created. This is also formed by certain species of bacteria though they lack pigments and hence cannot absorb and utilize radiation energy. In these organisms, the photochemical reaction is replaced by a chemical reaction which provides the "energy equivalents" and probably also the "reduction equivalents" required for the reduction of CO_2 to carbohydrate. Pfeffer has termed this process *chemosynthesis*. Organisms exhibiting this type of metabolism are accordingly termed "chemoautotrophic" since they build their carbon compounds from CO_2 like the "photo-autotrophic" organisms.

Based upon the fundamental investigations of Winogradsky on "sulfur bacteria" and "nitrifying bacteria" towards the end of the last century, the obligatory chemoautotrophic type of metabolism has been demonstrated in a number of bacterial species and intensively studied. Most of these species are cultured in the laboratory (though this is often a difficult task), an essential premise for a detailed analysis of their metabolism.

The energy-yielding chemical reactions involved are oxidations of inorganic compounds which exhibit a low (negative!) redox potential compared to oxygen and, accordingly, an "electron pressure". Hence they donate electrons to an adequate acceptor rather readily in an exergonic reaction during which they are oxidized.

Organisms and Substrates

Sulfur-Oxidizing Bacteria

The energy-yielding reaction in the pigmentless sulfur bacteria *Beggiatoa*, *Thiotrix* and *Thiobacillus* which must be clearly distinguished from the phototrophic *Thiorhodaceae* (= *purple sulfur bacteria*) consists in oxidation of hydrogen sulfide or thiosulfate to elementary sulfur or sulfate:

$$2\ H_2S + O_2 \longrightarrow 2\ H_2O + 2\ S \quad (\Delta G° = -118\ kcal\ )$$

$$2\ S + 2\ H_2O + 3\ O_2 \longrightarrow 2\ H_2SO_4 \quad (\Delta G° = -280\ kcal\ )$$

$$H_2S_2O_3 + 2\ O_2 + H_2O \longrightarrow 2\ H_2SO_4 \quad (\Delta G° = -100\ kcal\ )$$

The sulfur is thus oxidized from the negative divalent to the positive hexavalent state. There is a considerable energy yield. The organisms exhibit a surprisingly high resistance to the sulfuric acid produced; even $1/10$ normal H_2SO_4 is tolerated without harm.

Nitrifying Bacteria

The genera *Nitrosomonas* and *Nitrobacter* are responsible for biological nitrification, i.e. the oxidation of ammonia to nitrate (cf.

p. 288). *Nitrosomonas* carries out the first step, the conversion of ammonia to nitrite:

$$2\ NH_3 + 3\ O_2 \longrightarrow 2\ HNO_2 + 2\ H_2O\ (\triangle G° = -158\ kcal)$$

Nitrobacter converts nitrite to nitrate:

$$2\ HNO_2 + O_2 \longrightarrow 2\ HNO_3 \qquad (\triangle G° = -36\ kcal)$$

Knallgas Bacteria

These organisms (example: *Hydrogenomonas*) oxidize molecular hydrogen and utilize the energy released to synthesize cellular carbon compounds from CO_2. We have already met this biological "Knallgas reaction" as dark reaction in the hydrogen-adapted green alga *Scenedesmus* (p. 141):

$$2\ H_2 + O_2 \longrightarrow 2\ H_2O \qquad (\triangle G° = -114\ kcal)$$

In contrast to most sulfur bacteria and nitrifying bacteria, the knallgas bacteria are only facultatively chemoautotrophic: in the presence of organic substances they grow heterotrophically like most bacteria and animal organisms.

In adaptation to the prevailing external conditions they switch to the appropriate kind of metabolism.

Iron Bacteria

Members of this group also obtain energy for the assimilation of CO_2 anoxybiontically by conversion of ferrous to ferric salts, e.g. the genus *Ferrobacillus*:

$$Fe^{2\oplus} \xrightarrow{\ -\ e\ } Fe^{3\oplus} \quad (\triangle G° = -16\ kcal)$$

Because of the small energy yield of this reaction relatively large amounts of substrate must be converted to meet the energy requirements of the organism. It is not surprising that a few bacteria accumulate considerable amounts of ferric salts in their environment.

Formation of "Energy Equivalents" and "Reduction Equivalents"

The most striking feature of the chemoautotrophic metabolism is not the oxidation of inorganic compounds – other bacteria also metabolize sulfur and nitrogen compounds in this way – but the ability of these organisms to utilize the energy released in these exergonic reactions for the assimilation of CO_2; this is a feature exhibited only by chemoautotrophic organisms. They are equipped with specific enzymes which set free hydrogen or electrons from the inorganic substrate. Such enzymes are, for instance, those referred to as *ammonia dehydrogenase, nitrite dehydrogenase* and *hyponitrite dehydrogenase*. They have been identified in some instances.

The electrons are transferred enzymatically to a transport chain which is probably very similar to the "respiratory chain" (p. 235 f) in the oxidative degradation of glucose. Several cytochromes of the c type and *cytochrome oxidase* (p. 237 ff) are among the redox systems involved; they have been detected in a number of chemoautotrophic bacteria. The terminal electron acceptor is oxygen or, in the anoxybiontic organisms, an oxidized inorganic compound ($SO_4^{2\ominus}$, $NO_3^{\ominus}$). The actual transfer of energy to chemical bonds occurs during the "downhill" movement of electrons following the natural energy gradient via the redox systems of the electron transport chain. Here, we encounter again the mechanisms of biological energy production which we discussed first in photosynthesis (p. 125 ff). It is hence plausible that here too, ATP formation is coupled to the oxidation of inorganic substrate; this has actually been demonstrated in cell-free extracts of several chemoautotrophic organisms. The adenylic acid system also acts as the central collection site for free energy in the "anorgoxidation" process. The precise mechanism of this process is still obscure. Adenosine-5'-phosphorylsulfate ("APS") may participate as an intermediate:

Adenosine-phosphorylsulfate, APS

This compound corresponds to an "activated", i.e. energy-rich sulfate which is thought to be formed in the following way:

$$4\ H^{\oplus} + 4\ e + 2\ S_2O_3^{2\ominus} \longrightarrow 2\ SO_3^{2\ominus} + 2\ H_2S \qquad ①$$

$$\left[2\ H_2S + O_2 \longrightarrow 2\ S + 2\ H_2O \right] \qquad ②$$

$$2\ SO_3^{2\ominus} + 2\ AMP \rightleftharpoons 2\ APS + 4\ e \qquad ③$$

$$2\ APS + 2\ PO_4^{3\ominus} \rightleftharpoons 2\ ADP + 2\ SO_4^{2\ominus} \qquad ④$$

$$2\ ADP \rightleftharpoons AMP + ATP \qquad ⑤$$

Thiosulfate disproportionates to sulfite and sulfur (equations 1 and 2); sulfite reacts with adenosine monophosphate (AMP) forming adenosine-5'-phosphorylsulfate ("APS") while being simultaneously oxidized to sulfate (equation 3). Sulfate and ADP arise from APS and $PO_4^{3\ominus}$ (equation 4). Two molecules of ADP react to form AMP and ATP (equation 5). The reactions 3–5 are reversible. All the enzymes involved are known.

Among the processes not yet elucidated in chemoautotrophy is the formation of "reduction equivalents" required for assimilation of CO_2. As regards the latter process, a series of experimental results obtained with *Thiobacillus, Nitrobacter* and *Hydrogenomonas,* particularly the demonstration of *carboxydismutase* and other enzymes of the "Calvin cycle", have confirmed the concept that CO_2 reduction in chemoautotrophic organisms proceeds via the same or a very similar reaction cycle as the one in photoautotrophic organisms. Ribulose biphosphate acts as primary CO_2 acceptor and gives rise to 3-phosphoglyceric acid as the first stable product of fixation. With the formation of triose phosphate, hydrogen reaches a higher energy level. Thus the potential difference to oxygen, which has been mentioned previously, is established.

Chemosynthesis probably preceded photosynthesis in phylogeny. As long as organisms lacked the photochemical apparatus, fixation of carbon was only possible with simultaneous oxidation of inorganic or organic substrate. Only after acquisition of a reaction mechanism for the transformation of radiant energy into chemical energy, water could be utilized as an electron donor. Reduction of CO_2 and hence conversion of carbon into organic compounds became possible without any restrictions. A terminal acceptor for the oxygen originating from water, as in the case of the photosynthetically active bacteria, was no longer required since it could now be eliminated as molecular oxygen.

Since species of purple bacteria and green algae are also able to activate H_2 in the light, they may form the link (from the point of view of comparative biochemistry) between photosynthesis and chemosynthesis. The demonstration of the knallgas reaction in hydrogen-adapted green algae of the genera *Scenedesmus* and *Ankistrodesmus* in the dark emphasizes the role of their metabolism as connecting link between the two forms of autotrophy. The uniformity of photosynthesis, photoreduction and chemosynthesis is particularly apparent in the mechanism of CO_2 fixation. Formally, photosynthesis may also be considered as a special case of

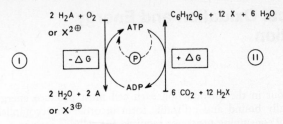

I: Energy-yielding process   =   oxidation
II: Synthetic process   =   energy-consuming reduction

General Scheme of Chemosynthesis

chemoautotrophic metabolism in which radiant energy, via photolysis of water, provides the electron donor and removes the oxidized cleavage product as molecular oxygen. "Photosynthesis without oxygen" results from the absence of the latter process.

References

Elsden, S. R.: Photosynthesis and litho-trophic carbon dioxide fixation. In: The bacteria, Vol. 3/1. Ed. by I. C. Gunsalus, R. Y. Stanier. Academic Press, New York 1962

Kelly, D. P.: Autotrophy: Concepts of lithotrophic bacteria and their organic metabolism. Ann. Rev. Microbiol. 25: 177, 1971

Lees, H.: Biochemistry of autotrophic bacteria. Butterworths, London 1955

Schlegel, H. G.: Physiology and biochemistry of Knallgasbacteria. Advanc. comp. Physiol. Biochem. 2: 185, 1966

Biological Oxidation and Energy Production

As pointed out in the introduction each cell must produce energy in a chemically bound and utilizable form in order to accomplish all the energy consuming processes required for cellular activity and survival. We are already familiar with these tasks, collectively designated as "biological work", i.e. osmotic, mechanical and chemical work (p. 3). Chemical work is particularly important for the biosyntheses of the cell. This part of the metabolism (= "building metabolism") will be treated in separate parts.

If the process of photosynthesis is interpreted as an energy-dependent cleavage of water and subsequent elevation of hydrogen to the higher energy level of the metastable carbon compound glucose, the process of cellular energy production, "dissimilation", may be regarded as the reversal of photosynthesis. The establishment of the energy potential between oxygen and hydrogen has formerly been compared with the tension of a spring (p. 114). Dissimilation then corresponds to the release of the spring tension. The energy set free in this process is partially conserved as chemical energy. Figuratively speaking, hydrogen drops "downhill" to oxygen and is finally again bound to water. The amount of energy released during this transition exactly equals that which had to be spent in establishing the energy potential, i.e. in "loading the spring".

Formally, the dissimilatory "back-reaction" of hydrogen with oxygen corresponds to the strongly exergonic "knallgas reaction" mentioned previously (p. 194). In the cell, however, the relatively large amount of energy is not released spontaneously with production of heat, but in a controlled manner and small portions so that it can be conserved as chemically bound energy in ATP. The carbon which served originally as acceptor for the activated hydrogen is of no further use to the cell once it has lost the hydrogen; it is therefore released as CO_2.

Thus the overall process of dissimilation or "respiration" corresponds to a reversal of photosynthesis. Glucose is degraded to CO_2 and H_2O, accompanied by consumption of O_2 and concomitant release of energy:

$$C_6H_{12}O_6 + 6\ O_2 \longrightarrow 6\ CO_2 + 6\ H_2O$$

$$(\Delta G° = -675 \text{ kcal/mole glucose})$$

The energy potential established between oxygen and hydrogen and stabilized in the glucose molecule is not only exploited by the autotrophic producer cell, but by all those plant cells which lack chloroplasts and carry on a heterotrophic metabolism, e.g. root tissues. Glucose is supplied to them mainly as sucrose, the transport form of carbohydrates (p. 190 f). Moreover, all heterotrophic organisms such as animals, fungi and most bacteria utilize the energy potential between oxygen and hydrogen created by autotrophy to meet their energy requirements. They oxidize mostly glucose which is mainly derived from macromolecular organic compounds such as starch or glycogen.

For the sake of completeness, it must be mentioned that apart from carbohydrates also fat and, in a few cases, proteins after decomposition into smaller fragments may enter the dissimilatory reaction mechanism, and are then oxidized yielding energy (p. 398 f, p. 412 f). This is the case particularly among those plant organisms in which the reserve material is fat or protein. A few saprophytic organisms also depend on complex organic compounds to meet their energy requirements.

In the cells of most autotrophic and heterotrophic organisms glucose degradation proceeds "aerobically", i.e. with consumption of oxygen; these are hence termed "aerobic organisms" or "aerobes". On the other hand, certain tissues of higher plants (at least for a short period of time), and a number of microorganisms dissimilate glucose also in the absence of oxygen. Such cells or organisms exhibit the "anaerobic" type of dissimilation. Oxygen has even a toxic effect on the metabolism of another group of organisms; these are accordingly referred to as "obligate anaerobes" to distinguish them from the "facultative anaerobes" of the former group. In anaerobic dissimilation, another hydrogen acceptor replaces oxygen. The difference to oxygen-dependent respiration is, however, only a relative one if oxidation is defined generally as a removal of electrons. In the aerobes, the electrons are finally taken up by molecular oxygen, in the anaerobes by some other oxidant (= electron acceptor).

Historical Aspects. Until recently, biological oxidation has been regarded as a combustion process. This view has its origins in the concepts of Lavoisier (1780). As we know today, the conformity between the two processes is restricted to the overall equation and the end products; the fundamental difference lies mainly in the form of energy release. In the combustion process, it is radiated as heat; in biological oxidation, however, energy is partially conserved in chemical "energy equivalents", i.e. in ATP.

Wieland and Thunberg observed that oxygen can be replaced by a series of hydrogen acceptors (e.g. methylene blue) in oxidation of various organic substrates by tissue homogenates. In 1912, these results were used for formulating a hypothesis proposing that the crucial process of biological oxidation (= "respiration") is an "activation of hydrogen" which may then react with oxygen or another acceptor. Warburg put forward the view that respiration is characterized by an "activation of oxygen". When Keilin postulated (1925) that biological oxidation requires activation of both oxygen and hydrogen, with cytochromes assuming a mediatory role, the concepts of Wieland and Warburg – at first sight incompatible – were united in a new overall concept of respiration. Its accuracy was later confirmed in almost all aspects by experiments.

Methods of Measuring Respiration

Gas Exchange. In accordance with the overall equation of respiration (p. 198), an estimate of the occurrence and extent of respiration is possible if one measures the gaseous starting substances and end products: oxygen will be removed from the reaction space, while the amount of CO_2 will increase. Possible intermediate reactions and compounds are of minor significance for the measurements. Green, i.e. photosynthetically active plant tissues or organs are better studied in the dark in order to avoid interference with the gas exchange of photosynthesis. The manometric, chemical and physical procedures discussed previously in connection with photosynthesis (p. 28 ff) are also suitable for the quantitative determination of oxygen and CO_2. The fresh or dry weight of the tissues or cells under study, or their total nitrogen or protein content may serve as reference quantity for the gas exchange measurements. A commonly employed quantity for characterizing the O_2 consumption of an organism or a tissue is the "respiratory rate":

$$Q_{O_2} = \frac{mm^3 \ O_2}{mg \ dry \ weight \times h}$$

If one relates the values of CO_2 release and O_2 uptake obtained in a test object, one obtains the "quotient of gas exchange" or the "respiratory quotient" ("RQ")

$$RQ = \frac{volume \ of \ CO_2 \ evolved}{volume \ of \ O_2 \ consumed}$$

Within certain limits, the value of the respiratory quotient indicates the substrate of the respiratory process measured. Complete degradation of carbohydrate according to the equation given above results in a RQ of 1.0. Deviations from this value may be due to various reasons: 1. Substances of different chemical composition are being dissimilated, either along with carbohydrates or as the only substrate. 2. Besides CO_2 and H_2O, the degradation of carbohydrate is producing intermediates which are either accumulating in the cells or entering synthesis pathways as precursors. Thus the RQ is below 1.0 when organic acids accumulate which bind CO_2 (p. 230 f) or when fat is converted to carbohydrate. RQ values below 1.0 also indicate that fat (0.7) or protein (0.8) is being dissimilated; both substrates have a relatively low oxygen content. When organic acids form the substrate of respiration the RQ value is greater than 1.0. Definitive conclusions regarding the actual conditions giving rise to a RQ value are however, only possible when the results of other experimental procedures are taken into account as well.

Isotope Incorporation Studies. Deeper insights into the respiratory processes have been gained by tracing the pathway of carbon atoms in individual compounds with radiocarbon (^{14}C). The methods developed correspond in principle to those employed successfully in elucidating the photosynthetic CO_2 reduction cycle (p. 143 ff). Two types of experiments are mainly performed: 1) the release of $^{14}CO_2$ from the test object is monitored; 2) the appearance and the distribution of ^{14}C in intermediates and end products is observed after previous feeding with ^{14}C-labeled precursors. Misinterpretations of the results obtained are possible, resulting from cyclic processes, renewed binding of $^{14}CO_2$, degradation of cellular substances and isotope exchange in the cells. The adequacy and the limitations of the method must hence be critically reviewed in each case.

Specific Inhibitors. Certain inhibitors proved to be valuable tools in analyzing the complex processes of respiration. Inhibitors (p. 10) are substances which block either a particular reaction step or a reaction sequence by inactivating the enzymes involved. The type of inhibition obtained often gives significant hints as to the properties of the enzyme or enzymes affected, e.g. presence of a heavy metal (Fe, Cu) or SH groups. Other inhibitors such as dinitrophenol or some types of antibiotics uncouple oxidation and phosphorylation; respiration then continues without yielding energy (cf. p. 130). This "inhibitory effect" often gives rise to an increase in the respiratory rate. Complete reaction complexes of respiration such as glycolysis or the citric acid cycle can be blocked by means of specific inhibitors.

Aerobic Degradation of Carbohydrate ("Respiration")

According to our present knowledge the production of chemical energy is achieved in most organisms by biochemical reactions, which are the same in a green or pigmentless plant cell as in an animal or bacterial cell. The fuel hexose, mainly glucose, is aerobically degraded in a complex process involving a large number of individual reactions. When we distinguish between several reaction complexes for the sake of better understanding, it must not be forgotten that these are interconnected in the cell and form a functional entity. The whole process was divided into various reaction complexes originally for the sake of clearness, but this is justified to a certain extent by the fact that several intermediate processes have been found to take place in different compartments of the cell. In vivo these are spatially but not functionally separated from each other.

In *glycolysis** the glucose molecule, after having been phosphorylat-

---

* Also termed the "Embden-Meyerhof scheme" or "Embden-Meyerhof-Parnas scheme" after the most prominent contributors.

ed, is cleaved to two molecules of triose phosphate. These are dehydrogenated and converted to pyruvic acid. The negative change in free energy connected with these reactions is utilized to form ATP by coupling of two strongly exergonic steps with the adenylate system (= "substrate chain phosphorylation"). The reactions of glycolysis take place in the cytosol. Some of the enzymes involved may be bound to membranes of the endoplasmic reticulum. The close connection between "operating metabolism" and "building metabolism" is evident in glycolysis: several intermediates (both C_3 and C_6 compounds) serve as precursors for synthesis of cellular substances.

Up to the stage of pyruvic acid formation the reactions of aerobic and anaerobic glucose degradation are identical; from here they differ. In the absence of oxygen, decomposition of the sugar ends up with a C_3 or C_2 compound, i.e. a relatively energy-rich end product. Oxygen as the terminal acceptor for hydrogen is replaced by an organic compound. Glycolysis is thus the makeshift procedure for energy production during failing or poor oxygen supply. A separate section is devoted to these processes of "fermentation" (p. 246 ff).

In the presence of oxygen, the carbon chain is further degraded: pyruvic acid is converted by "oxidative decarboxylation" to a reactive C_2 compound, the "activated acetic acid".

The C_2 fragment of the "activated acetic acid" plays a key role in metabolism, as will be discussed later (p. 218). Not only are fragments of large carbon chains (fat, protein) aerobically degraded to this compound, but it also serves as a precursor in the synthesis of numerous cellular compounds (fatty acids, carotinoids, terpenes).

The "activated acetic acid" molecule serves as immediate fuel of the *citric acid cycle** where it is completely degraded to CO_2 and hydrogen. The latter is bound to specific "transport metabolites" and thus remains reactive.

The name given to this reaction complex indicates that citric acid holds a key position and that the reaction sequence is cyclic by nature. It starts with the donation of "activated acetic acid" to a specific acceptor which is regenerated by the cycle at each turn while the C_2 fragment fed into the cycle is degraded as described above. We have already encountered a similar cyclic process in photosynthetic CO_2 reduction (p. 145 ff).

We wish to point out beforehand that the citric acid cycle not only plays an important role in catabolism, but also supplies important precursors

---

* Also frequently termed "Krebs cycle". H. A. Krebs succeeded (simultaneously with Knoop and Martius) in deducing the cyclic mechanism which gained him the Nobel Prize in 1953.

for the synthesis of cellular material. In cell metabolism it functions as a kind of "emporium" for important intermediate compounds of synthesis and catabolism. The interlocking of the two processes and the resulting interdependence are analyzed in a separate section (p. 231 f).

The citric acid cycle, including the formation of activated acetic acid, is located within a specific cellular organelle, the mitochondrion (see below). Some of the enzymes involved, however, are also present in the cytoplasm.

In the *endoxidation* the hydrogen transferred by the transport metabolites finally reacts with molecular oxygen in a series of reactions catalyzed by the enzymes of the *respiratory chain*. The energy of these highly exergonic reactions is largely retained as chemical energy in the form of ATP due to a stepwise decline in free energy in this reaction sequence. The respiratory chain is thus the main supplier of chemical energy in the cell; it is the actual oxygen-consuming process.

The respiratory chain and hence also the reaction sequence of "endoxidation" is located in the mitochondria as well. Its presence in the same cellular organelle with the citric acid cycle is obviously a consequence of the close functional relationship of the two processes.

Mitochondria as Organelles of Dissimilation. As mentioned above, the reactions of the citric acid cycle and the respiratory chain, including ATP formation, take place in the mitochondria. These contain all the enzymes involved – more or less tightly bound to the mitochondrial structures. Being selfsufficient units, they are set off from the surrounding cytoplasm. Small quantities of DNA and ribosomes have recently been found in mitochondria (cf. p. 329). The number of mitochondria per cell varies (10–20 in microorganisms and up to 200000 in large cells) depending on the function of the cell and its metabolism, etc. There are only a few mitochondria (15–20) in cells of *Euglena* which grow autotrophically. However, the mitochondria increase considerably in number when the cells switch to heterotrophic metabolism. Mitochondria are present in practically all aerobic organisms, whereas they are absent in strict anaerobes.

The mitochondrion has an outer membrane which is rather permeable, and an inner membrane which acts as actual permeability barrier and also carries the components of the respiratory chain (p. 235). The surface area of the inner membrane is considerably enlarged by numerous invaginations into the inner space (formation of "cristae", "tubuli" or "sacculi"). The inner surface of this membrane is occupied by numerous knoblike structures consisting of a headpiece and a stalk, which can be identified only with optimal resolving power in the electron microscope; they are probably connected with oxidative phosphorylation (p. 241). Between the two membranes there is the "perimitochondrial" or "intermembrane space", which may contain a fluid. The semi-fluid material of the inner compartment is called "matrix"; it also contains several enzymes, e.g. those of the citric acid cycle (p. 219 ff).

Due to the great diversity of their metabolic activities, mitochondria exchange chemical energy, reduction equivalents and metabolites with the surrounding cytoplasm. The inner mitochondrial membrane plays a key role in controlling these processes (see above). Details of the various uptake and release mechanisms involved will be discussed later (p. 244 f).

Active mitochondria are isolated from plant cells by fractional centrifugation of tissue homogenates. The methodical difficulties are far greater than those experienced in isolation of pure mitochondria from animal cells. Isolated mitochondria have been successfully used to study the reaction of the citric acid cycle since they metabolize several intermediates also in vitro. There are some intermediates, however, that cannot be utilized; they may even act as inhibitors. All such intermediates are organic acids or their salts.

The Fuel of Glycolysis and Its Supply

Phosphorylation. The photosynthetic CO_2 reduction cycle has made us familiar with the fact that sugars are mostly active as phosphoric acid esters in metabolism. In the reactions of glycolysis too, they are involved exclusively in this form*.

Sugar molecules hence undergo a specific phosphorylation before entering glycolysis if they have not been released from a previous reaction as a phosphoric acid ester. Phosphorylation occurs at the hexose level and concerns mainly glucose, in some cases also fructose and mannose. Accordingly, the cells contain specific enzymes, *kinases,* which accelerate the phosphorylation reaction.

The transfer of a phosphoric acid residue from ATP to the 6-position of glucose is catalyzed by the enzyme *hexokinase.* The special ability of ATP to transfer phosphoryl residues, due to its high group-transfer potential, has already been discussed (p. 20). Glucose is raised to a higher energy level by phosphorylation. The free energy of hydrolysis of the phosphoric acid ester formed is about −3300 cal/mole. This ester represents the reactive form of glucose.

Of the hexose phosphates present in photosynthetically active cells, glucose-6-phosphate and fructose-6-phosphate may flow directly into the process of glycolysis while glucose-1-phosphate must first be isomerized to glucose-6-phosphate. We are already acquainted with this reversible conversion which proceeds with the mediation of *phosphoglucomutase* (p. 166 f); each converted molecule functions as "cofactor" (= glucose-1,6-biphosphate) in the reaction.

---

* Some are named after their discoverers or after researchers who contributed to the elucidation of glycolysis: Harden and Young, Neuberg, Fischer, Negelein, Robison, Nilsson.

Phosphorylation of free fructose in the plant cell is carried out by a specific *hexokinase*. The fructose-6-phosphate formed enters glycolysis directly since it is one of the intermediate compounds. However, fructose may be degraded by another reaction mechanism too.

Enzymatic Cleavage of Reserve Carbohydrates. The polysaccharides represent the most important fuel source for glycolysis particularly starch and, in some cases, also glycogen (bacteria, fungi and *Cyanophyceae*). In plants these macromolecules are degraded enzymatically via two pathways:

1. Hydrolytic cleavage by *amylases* to maltose which is split to glucose by the enzyme *maltase*. This is typical for the mobilization of reserve starch in cells of storage organs.

2. Phosphorolytic degradation to glucose-1-phosphate by the enzyme *phosphorylase*. This is the mechanism which generally appears to bring about the cleavage of reserve polysaccharides in plant and animal cells.

Hydrolytic Starch Degradation by Amylases. The *amylases*, responsible for the degradation of starch, are *glycosidases* belonging to the major class of the *hydrolases*. According to their different mode of action, we distinguish between α-*amylase* and β-*amylase*.

α-*amylase* is found in many lower and higher plants, in fungi, bacteria, but also in human saliva and pancreatic juice. It has been isolated in pure crystalline form from barley endosperm as well as from various microorganisms. The molecular weight is about 60 000. $Ca^{2\oplus}$ is essential as cofactor. The distribution of β-*amylase* is probably restricted to the plant kingdom. The enzyme has been isolated and crystallized from sweet potatoes and germinating seeds of barley and wheat.

α-*amylase* initially cleaves the macromolecules of amylose and amylopectin of which starch is generally composed (p. 175 f) into smaller fragments of 6–7 glucose units (see reaction scheme). It acts as an "endoenzyme". The formation of these oligosaccharides presumably results from a separation of the windings in the helical macromolecule (cf. p. 176). The additional side chains present in amylopectin do not prevent the attack by α-amylase, though the enzyme cannot cleave the characteristic 1→6 bonds. In case of longer exposure α-amylase eventually also degrades the oligosaccharides to maltose (below). The 1→6 bonds appear in the disaccharide isomaltose (p. 177) which is also among the degradation products.

The effect of α-amylase on colloidal starch solutions can be observed directly during solubilization which is accompanied by disappearance of the characteristic iodine reaction (p. 176). Reducing sugars are not detectable; they only appear after longer incubation periods.

β-amylase can only degrade amylose and amylopectin from the non-reducing end of their molecules (see scheme). Two glucose units in the form of maltose are split off each time. β-amylase is a typical "exoenzyme". While amylose molecules are completely degraded to maltose in this way, those of amylopection give rise to residues of relatively high molecular weight ("limit dextrins"). The points of branching with their 1→6 bonds represent insurmountable barriers to β-amylase; its hydrolytic action ends here. This enzyme can hence hydrolyze only about 50% of amylopectin.

As they are high molecular weight fragments, the limit dextrins exhibit a positive reaction with iodine. Reducing sugars can be demonstrated immediately after initiation of the cleavage reaction by β-amylase.

Maltose

The complete degradation of amylopectin is only brought about when α-amylase is present too. The enzyme can also be replaced by isoamylase which specifically cleaves 1→6 bonds. Starch hydrolysis by a cooperation of the two enzymes (e.g. in germinating cereal seeds) yields mainly maltose. Besides isomaltose there are also small amounts of glucose, formed by cleavage of maltotriose to maltose.

Cleavage of maltose to two molecules of glucose is catalyzed by another enzyme, *maltase,* which is usually associated with the *amylases.* It is specified as *α-glycosidase* because of its group specificity (p. 10). Enzymatic hydrolysis of maltose corresponds to a "transglycosidation": a glucose residue is transferred to a molecule with an OH group, in this case to water. This type of reaction is of some significance in synthesis of oligo- and polysaccharides (p. 174).

Maltose   α-D-Glucose

The glucose molecules formed during hydrolytic degradation of starch must be phosphorylated before entering the process of glycolysis. We are already familiar with this reaction: A phosphoric acid residue is transferred from ATP to position 6 in glucose, catalyzed by *hexokinase* (p. 20).

Phosphorolytic Cleavage of Polymeric Carbohydrates. Most plant cells and some animal cells as well (muscle, liver) tap their carbohydrate reserves by means of another reaction which is more economical in terms of energy than hydrolytic cleavage. In contrast to the latter phosphoric acid now assumes the role as glucose acceptor in place of water.

Glucose-1-phosphate

As shown in the scheme above, the polysaccharide chain is shortened at the non-reducing end by one glucose unit per reaction step. The phosphoric acid residue binds to the glucose molecule released while hydrogen is transferred to the new terminal glucose unit of the chain. The free end group formed enables the enzyme to split off another glucose unit as glucose-1-phosphate. This transglycosidation (see above) is catalyzed by a specific glycosidase, *phosphorylase* (= *"α-1,4-glucan phosphorylase"*; cf. p. 178 f). Its name indicates that phosphorolysis has replaced hydrolysis in cleaving the glycosidic bond in the polycondensate.

Amylose molecules are completely degraded to glucose-1-phosphate by *phosphorylase*, those of amylopectin and glycogen only partially. Neither *β-amylase* nor *phosphorylase* can cleave the 1→6 bonds of the branched-chain molecules, since they are specific for 1→4 bonds only. Their activity is blocked by the branch points if these are not eliminated by an enzyme specific for 1→6 bonds *(R-enzyme)*.

The energy of the glycosidic bond ($\Delta G^\circ = -4300$ cal/mole) which is lost as heat in hydrolytic cleavage is largely preserved in phosphorolysis in the phosphate ester bond of glucose-1-phosphate. It is the great advantage of the latter mechanism that it yields glucose as a phosphorylated compound. The initial phosphorylation step is unnecessary; one molecule of ATP which would otherwise be required, is spared. Glucose-1-phosphate merely has to be converted by *phosphoglucomutase* to glucose-6-phosphate before entering glycolysis.

Phosphorylase has been detected in cells of green and pigmentless plant tissues and in algae. It has been isolated in pure form from potatoes. The homogeneity of the enzyme was shown by means of electrophoresis and ultracentrifugation (p. 335 f). Its molecular weight is about 200 000. Pyridoxal-5-phosphate is thought to function as prosthetic group. This compound also plays an important role in amino acid metabolism (p. 363 f). The mammalian enzyme which has been studied intensively has a molecular weight of 360 000 and is composed of four identical subunits. It differs from the potato enzyme in that it occurs in a relatively inactive (dephosphorylated) form "b", and in an active (phosphorylated) one called "a". In muscle, the enzyme is subject to a complex regulation mechanism (p. 441 f).

Two *phosphorylases* have been found in the blue-green alga *Oscillatoria princeps*, but this is probably an exception. One of these enzymes is activated by ATP exactly like the mammalian enzyme. The ability of *phosphorylase* to catalyze also the synthesis of long chain polysaccharide molecules in vitro has already been discussed (p. 178).

The Reactions of Glycolysis

Conversion of the C_6 Molecule. The initial reactions of glycolysis are characterized by interconversions of the phosphorylated hexose molecule which is thereby fitted with the chemical structure required for the subsequent cleavage and oxidation reactions.

Glucose-6-phosphate is converted to fructose-6-phosphate by iso-merization. The enzyme involved is accordingly termed *phospho-hexose isomerase* (p. 165). The equilibrium of the catalyzed reaction favors glucose-6-phosphate. Phosphorylation of fructose-6-phos-phate in position 1 by *phosphofructokinase* with ATP as phosphate donor yields fructose-1,6-biphosphate (see scheme). By analogy with the first phosphorylation of free glucose in position 6, energy must also be invested in binding of the second phosphoric acid residue. Of the 2×7000 cal/mole in the two energy-rich bonds in ATP,

only about 2×3500 cal/mole are retained as chemical energy in the two ester bonds of fructose biphosphate. This consumption of two molecules of ATP at the beginning of glycolysis and the creation of an energy deficit appear paradoxical for a process which, on the contrary, ought to generate energy in the form of ATP. We shall soon see, however, that this is a thoroughly reasonable investment of energy which eventually – as in national economy – results in a profit in the final balance. The input of two ATP molecules acts as a "priming" which sets the whole process in motion.

Formation of the C_3 Molecule and the 1st Energy-Conserving Reaction. In fructose biphosphate the hexose molecule has attained a

structure which permits enzymatic cleavage to two largely symmetrical triose phosphates. The carbon atoms 1–3 are found in dihydroxyacetone phosphate, the carbon atoms 4–6 in 3-phosphoglyceraldehyde (see reaction scheme p. 209). The establishment of equilibrium in this reaction (89 % fructose biphosphate and 11 % triose phosphate) is brought about by *aldolase*. The reversion of this reaction is an important step in the photosynthetic CO_2 reduction cycle (p. 150). The next glycolytic reaction is also a part of this cycle, i.e. the enzymatic equilibration of ketotriose phosphate and aldotriose phosphate by *triose phosphate isomerase*. The following reaction can only use 3-phosphoglyceraldehyde as substrate. Since only 4 % of this compound is present in the equilibrium established by the action of *triose phosphate isomerase*, it is delivered at a high rate by continuous re-establishment of the equilibrium, each time the aldolase reaction removes 3-phosphoglyceraldehyde.

Triose phosphate isomerase is able to convert several hundred thousand molecules per minute. Formally, the splitting of fructose biphosphate thus supplies two molecules of 3-phosphoglyceraldehyde which are subject to identical conversions. Therefore, we shall follow the path of one molecule only.

The reactions of glycolysis discussed up to now proceed with relatively small changes in free energy, i.e. they proceed on about the same energy level. However, this does not apply to the subsequent highly exergonic oxidation of 3-phosphoglyceraldehyde (ΔG° = $-16\,000$ cal/mole). Chemically, this is the oxidation of an aldehyde to a carboxylic acid. Since biological oxidation is tantamount to the removal of hydrogen or electrons, a suitable acceptor for these equivalents is necessary. This is available to the cell in the form of the transport metabolites $NAD^\oplus$ or $NADP^\oplus$, as explained previously. In glycolysis $NAD^\oplus$ accepts hydrogen or electrons (we recall that in normal photosynthesis $NADP^\oplus$ serves exclusively as acceptor!). Thus hydrogen is preserved in a reactive form in cell metabolism, not, however, the considerable amount of energy liberated by the oxidation of triose phosphate. In the cell loss of energy in the form of heat, which is useless for cell metabolism, is avoided by the uptake of phosphoric acid (more precisely: of $H^\oplus$ ions and $HPO_4^{2\ominus}$ ions) and by the formation of an energy-rich intermediate, 1,3-biphosphoglyceric acid. This acyl phosphate compound has a high free energy of hydrolysis, namely about $-13\,000$ cal/mole ($\rightarrow$ 3-phosphoglyceric acid + orthophosphate).

By means of this "artifice", most of the energy normally released in oxidation of an aldehyde is conserved as chemical energy in the phosphorylated derivative of the oxidation product. The reaction scheme below illustrates this complex process*. The aldehyde group of 3-phosphoglyceraldehyde first reacts with a sulfhydryl group of the enzyme *phosphotriose dehydrogenase (glyceraldehyde phosphate dehydrogenase)*. The substrate is thereby made accessible to dehydrogenation by means of NAD⊕. The oxidation product, an acyl-sulfur compound, is energy-rich since hydrolysis of the thioester bond (between carboxyl and sulfhydryl group) proceeds with a marked decline in free energy. In the cell, however, the reaction product and the ("apo"-) enzyme are not separated by hydrolysis but by phosphorolysis (p. 207 f). The chemical energy of the thioester bond is thus retained in the energy-rich anhydride bond of the

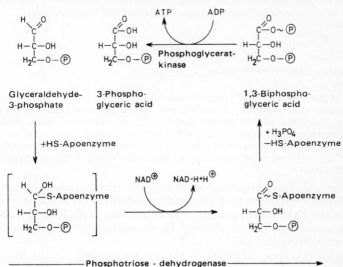

1,3-biphosphoglyceric acid formed. Because of its very high group transfer potential, the carboxyl phosphate group is transferred completely to ADP in an exergonic reaction catalyzed by *phosphoglycerate kinase;* 3-phosphoglyceric acid remains behind. The formation of ATP by this reaction mechanism is termed "substrate chain phosphorylation". Thus about 7000 cal/mole of the energy of aldehyde oxidation are finally conserved as chemical energy in ATP.

---

* It was elucidated by Warburg in 1939 as the first reaction in which ATP is generated from an enzymatic oxidation process.

The enzymatic oxidation of 3-phosphoglyceraldehyde showed for the first time that chemical energy can be recovered from a biological oxidation process in the form of ATP. The reaction mechanism represents a model generally applicable to all energy-conserving processes in the cell. These are characterized by the coupling of an energy-yielding oxidation process with the energy-requiring synthesis of ATP, during which an energy-rich intermediate is formed. The free energy change in the energy-yielding process must amount to at least -7000 cal/mole in order to meet the energy requirements of ATP synthesis ($\Delta G^\circ = +7000$ cal/mole).

The presence of small amounts of arsenate prevents the formation of 1,3-biphosphoglyceric acid, thus also preventing energy conservation in the transfer of the carboxyl phosphate group to ADP. Because of its close chemical affinity arsenate assumes the role of phosphate in releasing the apoenzyme and forming an anhydride bond at carbon atom 1, but this is so unstable that it is immediately hydrolyzed and loses its energy:

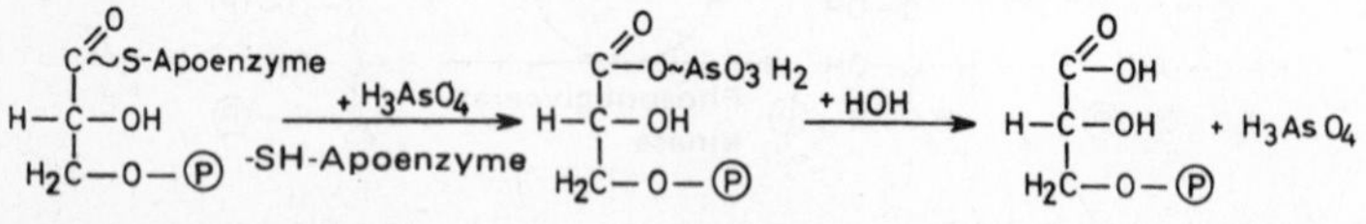

The generation of ATP normally coupled to the formation of 3-phosphoglyceric acid is thereby abolished, and with it the actual net efficiency of glycolysis, namely energy conservation by substrate chain phosphorylation. The cells affected suffer from a gradual energy depletion, although, in the presence of arsenate, they react with a strong activation of glycolysis or fermentation, as Harden and Young (1932) discovered in yeast.

After this first energy-conserving step the energy balance is even. The two molecules of ATP initially invested in each molecule of glucose have been recovered. The desired net profit in chemically bound energy will be obtained by the subsequent conversion of 3-phosphoglyceric acid.

Formation of Pyruvic Acid and the 2nd Energy-Conserving Reaction. By an intramolecular displacement of the phosphoric acid residue catalyzed by *phosphoglyceromutase*, 3-phosphoglyceric acid is converted to 2-phosphoglyceric acid. This reaction follows the

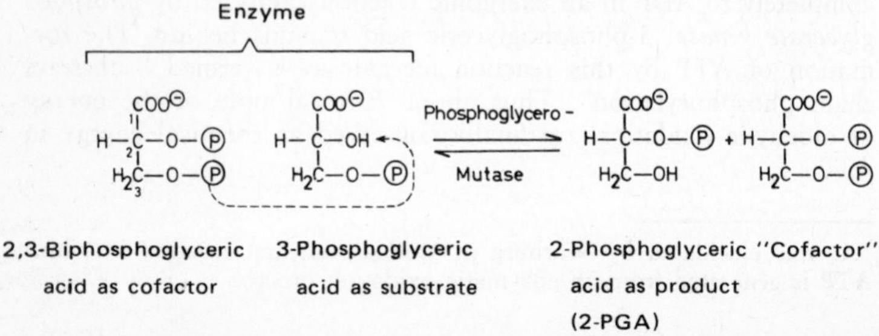

2,3-Biphosphoglyceric   3-Phosphoglyceric   2-Phosphoglyceric "Cofactor"
acid as cofactor   acid as substrate   acid as product
(2-PGA)

same mechanism as that in enzymatic transformation of glucose-6-phosphate to glucose-1-phosphate (p. 166 f): each molecule of 2-phosphoglyceric acid formed has previously acted as "cofactor" of the reaction, i.e. as 2,3-biphosphoglyceric acid.

In the next step the enol compound phosphoenol pyruvic acid is formed from 2-phosphoglyceric acid by elimination of a molecule of water. The enzyme active in this reaction is accordingly termed *enolase* (= *2-phosphoglycerate hydrolyase*). The most important consequence of this reaction is the transfer of the phosphoric acid residue from the ester bond to the energy-rich bond of an "enol phosphate". The free energy of hydrolysis of this high-energy compound ($\rightarrow$ pyruvic acid + orthophosphate; $\Delta G° = -13\ 000$ cal/mole) is so high that a high group-transfer potential for the phosphoric acid residue can be ascribed to it, comparable to that of

2-PGA · Phosphoenol pyruvic acid · Pyruvic acid

1,3-biphosphoglyceric acid. Consequently, phosphoenol pyruvic acid then donates its phosphoryl residue to ADP in a reaction catalyzed by the enzyme *pyruvate kinase*. This exergonic reaction goes far toward completion yielding ATP and pyruvic acid ($\Delta G° = -5000$ cal/mole).

In the 1st energy-conserving reaction of glycolysis ATP formation was associated with removal of hydrogen and its transfer to $NAD^{\oplus}$. In the second one an intramolecular oxidation-reduction reaction gives rise to ATP formation. These are both oxidation reactions, since biological oxidation is, by definition, a removal of hydrogen or electrons. In the first case, the electrons are exchanged between two different molecules; in the second case, electrons shift within the same molecule. Thus the model for an energy-conserving reaction described above is valid for the second ATP-producing step in glycolysis, too. Here the postulated energy-rich intermediate arises from an intramolecular oxidation-reduction reaction ("oxidoreduction").

After feeding with ^{14}C-labeled glucose the characteristic intermediates of glycolysis were identified in tissue sections and in algal cells by the radiocarbon incorporated. Another important proof of the occurrence of glycolytic reactions in pigmentless and green plant cells may be seen in the successful demonstration of all enzymes involved in extracts from these plants. Moreover, several enzymes such as *phosphoglucomutase*, *phosphohexose isomerase, aldolase,* and *phosphotriose dehydrogenase* have been partially purified. Finally, the specific action of inhibitors should be mentioned. Iodacetate or p-chloromercuribenzoate block respiration in cells of *Chlorella pyrenoidosa* because these substances specifically inactivate the key enzyme of glycolysis, *phosphotriose dehydrogenase.* Fluoride brings about a similar inhibition of respiration in plant organisms, caused by a marked reduction in the activity of *enolase.*

Balance of Glycolysis

Let us first consider energy. As a "reaction priming", the glucose molecule is first charged with two phosphoric acid residues, a process consuming two molecules of ATP. This investment, however, is a profitable one: four ATP (two for each triose phosphate) are gained by coupling the adenylate system with two strongly exergonic reactions. The net profit is hence two ATP per molecule of glucose. At first sight this gain appears modest, but we must not overlook that the preliminary end product pyruvic acid represents a relatively energy-rich substance which may be further energetically exploited by degradation. The cleaved hydrogen in the form of two molecules of $NAD-H + H^{\oplus}$ is also a potential source of energy since it reacts in a strongly exergonic reaction with oxygen via the "respiratory chain" (p. 235 ff).

We notice that in glycolysis the problem of glucose degradation in the absence of oxygen with extensive preservation of chemical energy is solved with amazing efficiency.

Activation of Pyruvic Acid

Pyruvic acid, formed in the 2nd energy-conserving step of glycolysis, does not only mark the end of glycolysis, but holds a key position in the further progress of dissimilatory degradation as well. Pyruvic acid may enter two different reaction pathways – apart from being an important precursor for various synthesis processes (see below):

1. Oxidative degradation leading to decomposition of the pyruvic acid molecule to H_2O and CO_2;

2. Fermentative metabolism, as carried out in facultative and strict (= "obligatory") anaerobes.

Since the first pathway is typical for most organisms, it will be discussed first.

Like the glucose molecule, which is "activated" by phosphorylation before entering glycolysis, pyruvic acid is converted by a corresponding chemical reaction enabling it to be utilized in the citric acid cycle. This "activation" differs from that of glucose in that no phosphorylation by ATP occurs. Instead, pyruvic acid is converted by "oxidative decarboxylation", i.e. by cleavage of CO_2 and hydrogen, to acetic acid which is immediately transferred to a specific "carrier" or metabolite, *coenzyme A ("CoA")*, giving rise to the high-energy derivative "acetyl-coenzyme A" ("acetyl-CoA") also termed "activated acetic acid" or "activated acetate".

Formation of Acetyl-Coenzyme A. This complex process is mediated by several enzymes organized in a "multienzyme complex" (p. 406 ff), the *pyruvate dehydrogenase complex*. Besides CoA several other coenzymes participate in this complex; we shall become acquainted with them while discussing the various reaction steps.

With the mediation of the coenzyme thiamine pyrophosphate ("ThPP"), the complex enzyme system first liberates CO_2 in such a way that the C_2 compound formed is linked to the coenzyme

(= "active acetaldehyde"), first to carbon atom 2 of the thiazole ring, then to the amino group of the pyrimidine ring which donates it to another coenzyme (see reaction scheme). ThPP assumes two functions here: as the coenzyme of a *lyase* which releases CO_2 *(co-decarboxylase)*, and as the group-transferring coenzyme for "active acetaldehyde". We have already encountered ThPP in the latter function in the reaction catalyzed by *transketolase*, i.e. transfer of another C_2 fragment, "active glycolaldehyde". The chemical structure of ThPP was described at the time (p. 151). The compound must be present in the diet of man and of many higher animals, as it represents the vital vitamin B_1. The green plant cell, however, possesses the mechanism required to synthesize ThPP and is thus self-sufficient in this respect.

ThPP was one of the first coenzymes to be purified and to have its structure elucidated. It was the first example of a vitamin acting as a coenzyme.

In the next reaction step "active acetaldehyde" is transferred to another coenzyme, lipoamide, the amide of lipoic acid. ThPP is released and can react with another molecule of pyruvic acid.

Lipoic acid (= "thioctic acid") originally discovered as a nutritional factor for microorganisms contains eight carbon atoms in a chain which bears a disulfide ring at one end, and at the other one a carboxyl group. The latter gives rise to an amide linkage, particularly when bound to the enzyme complex. Lipoic acid is classified as a hydrogen-transferring coenzyme, although it has a group-transferring function in the case of "active acetaldehyde".

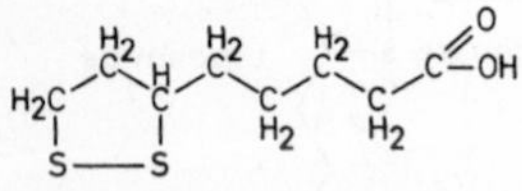

Lipoic acid (Thioctic acid)

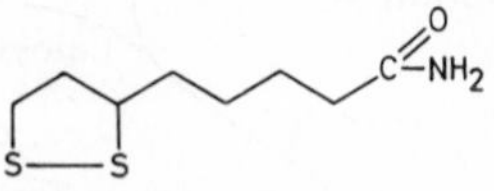

Lipoamide
(Abbreviated formula)

The characteristic disulfide ring of lipoamide is opened in such a way that a hydrogen atom of the aldehyde group is added to the one sulfur atom, its residue to the other. Hydrogen is thus separated from acetaldehyde, which is thus oxidized to acetic acid. As an acetyl residue it is then transferred to the coenzyme with the formation of a thioester bond. This is energy-rich, as explained previously in connection with the structure of 1,3-biphosphoglyceric acid (p. 210 f). Like there, the energy of the aldehyde oxidation is partially conserved in an energy-rich bond; in this case, however, the hydrogen is not donated to a specific transport metabolite, but to another site within the same molecule. The lipoamide is accordingly converted to a derivative of dihydrolipoic acid.

Since the acetyl residue is linked by an energy-rich bond, it can easily be transferred to coenzyme A leaving behind dihydrolipoamide with two sulfhydryl groups. These are oxidized by a specific *dihydrolipoyl dehydrogenase** transferring the hydrogen to NAD$^\oplus$. Having been thus regenerated to the oxidized form, lipoamide again participates in the reaction. Finally, we shall discuss the transfer of the acetyl residue to coenzyme A.

Coenzyme A ("CoA") is a typical representative of the group-transferring coenzymes, since it provides acyl groups for enzymatic reactions or accepts them from such reactions. Since the acyl groups are linked to CoA by a high-energy bond the complex formed has a high group-transfer potential (see below). The CoA molecule consists of three building blocks: adenosine-3′,5′-diphosphate, pantothenic acid phosphate and the sulfur-containing cysteamine (mercaptoethylamine), a decarboxylation product of cysteine. They are linked together in the following structure:

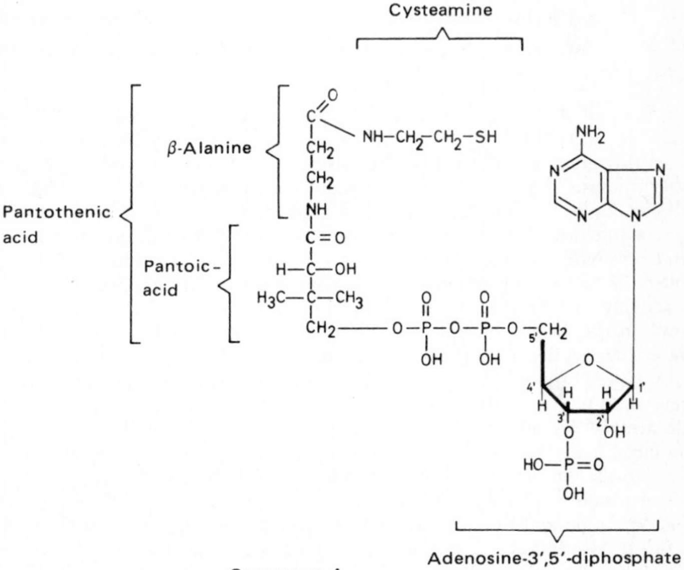

Coenzyme A

_____

* Recent studies indicate that it is a flavoprotein with a highly negative redox potential ($E_0' = -0.34$ volt).

Pantothenic acid* is composed of pantoic acid (= α,γ-dihydroxy-β, β-dimethylbutyric acid) and β-alanine. It is an important B-group vitamin and has to be supplied to man, to a number of higher animals and to certain microorganisms. The compound is used mainly for the synthesis of coenzyme A. Several microorganisms merely require pantoic acid or β-alanine, others pantetheine (pantothenic acid + cysteamine), e.g. *Lactobacillus bulgaricus*.

The main producers of pantothenic acid are higher and lower plants, and a number of microorganisms. The latter have been studied to elucidate the synthetic pathway. The starting substance is ketoisovaleric acid (p. 369).

The reactive site of the CoA molecule is the free SH-group of cysteamine. This reacts with the acetyl residue linked to lipoamide. The transfer occurs practically on the same energy level, since the energy-rich thioester bond is retained in the acetyl-CoA formed (free energy of hydrolysis: $\Delta G° = -8000$ cal/mole).

"Activation" of the C_2 fragment is complete when the acetyl residue has been incorporated into the high-energy compound of acetyl-CoA. Here, the initial investment of energy is made to the debit of the substrate which is energetically devaluated by an oxidation step.

The Role of Acetyl-CoA in Metabolism. It must be pointed out beforehand that acetyl-CoA or "activated acetic acid" is not only an intermediate in aerobic glucose dissimilation, but also acts as a precursor for important biosyntheses. The activated C_2 fragment thus assumes a key position in cell metabolism where reactions of degradation and of synthesis meet as if at a junction. The terms "reservoir" or "pool" have been introduced for compounds which, like acetyl-CoA, hold a central position in metabolism. A "reservoir" or "pool" comprises the amount of substance which is actively metabolized. Such a reservoir is usually fed by various reactions. In the case of acetyl-CoA the influx does not only originate from the oxidative decarboxylation of pyruvic acid, but also from fat catabolism (p. 411 ff) and from various reactions converting amino acids. There are several pathways by which incoming molecules may leave the reservoir. For acetyl-CoA an important pathway leads to oxidative degradation in the citric acid cycle and the respiratory chain (see below). Other outlets run into the building metabolism: the biosynthesis of fatty acids (p. 405 ff), of carotinoids (p. 431 ff), of terpenes and steroids. The side chains of plastoquinone and ubiquinone (p. 119 f and p. 135) also start with acetyl-CoA. It also plays an important role in the formation of esters and amides in the living cell.

---

* The name derives from the wide distribution of this compound in animal and plant cells and in microorganisms. Pantothenic acid is probably present in each cell, the greater part being bound as coenzyme A.

We shall encounter the terms "reservoir" or "pool" quite often in this treatise, connected with the close interrelation of catabolic and anabolic reaction mechanisms. Both phenomena emphasize the dynamic nature of metabolism and the intertwining of various metabolic pathways.

The Citric Acid Cycle

The acetyl residue bound to coenzyme A constitutes the actual fuel of the citric acid cycle. It is metabolized in a cyclic reaction sequence similar to that which we have observed in the photosynthetic CO_2 reduction cycle. There is also a specific acceptor at the beginning: oxaloacetic acid. The C_2 fragment is donated to this C_4 compound giving rise to the C_6 tricarboxylic acid, i.e. citric acid. Free co-enzyme A is regenerated. The citric acid is subjected to a series of reactions which results in regeneration of the acceptor oxaloacetic acid while the C_2 fragment introduced is formally degraded to CO_2 and coenzyme-bound hydrogen. The circle is thereby closed, and a new acetyl residue can be fed into the cycle. The balance is given by the following overall equation:

$$CH_3-C{\overset{O}{\diagup}}\!\!\sim S-CoA + 3\ H_2O \longrightarrow 2\ CO_2 + 8\ [H] + HS-CoA$$

([H] = coenzyme-bound hydrogen)

The degradation of the C_2 fragment is characterized by decarboxylations leading to a release of CO_2, and by dehydrogenations in which hydrogen is transferred to specific transport metabolites. The kind of decarboxylation taking place depends on the structure of the keto acid involved. β-keto acids are decomposed directly; α-keto acids are subject to "oxidative decarboxylation" (p. 215 ff), i.e. simultaneous removal of CO_2 and hydrogen. This process is the same as the one during formation of acetyl-CoA from pyruvic acid. By analogy with glycolysis, oxygen is apparently not involved in reactions of the citric acid cycle. The oxidations occurring are restricted to removal of hydrogen or of electrons.

The Individual Reactions. The scheme on p. 220 shows the whole sequence of reactions. It is obvious that all compounds concerned are organic acids. In contrast to glycolysis and the photosynthetic CO_2 reduction cycle, phosphorylated intermediates do not occur. Depending on the pH value in the cell the acids are present in the form of salts, the names of which are therefore often used. For the sake of better distinction the individual reactions have been numbered consecutively, and will be discussed in this order.

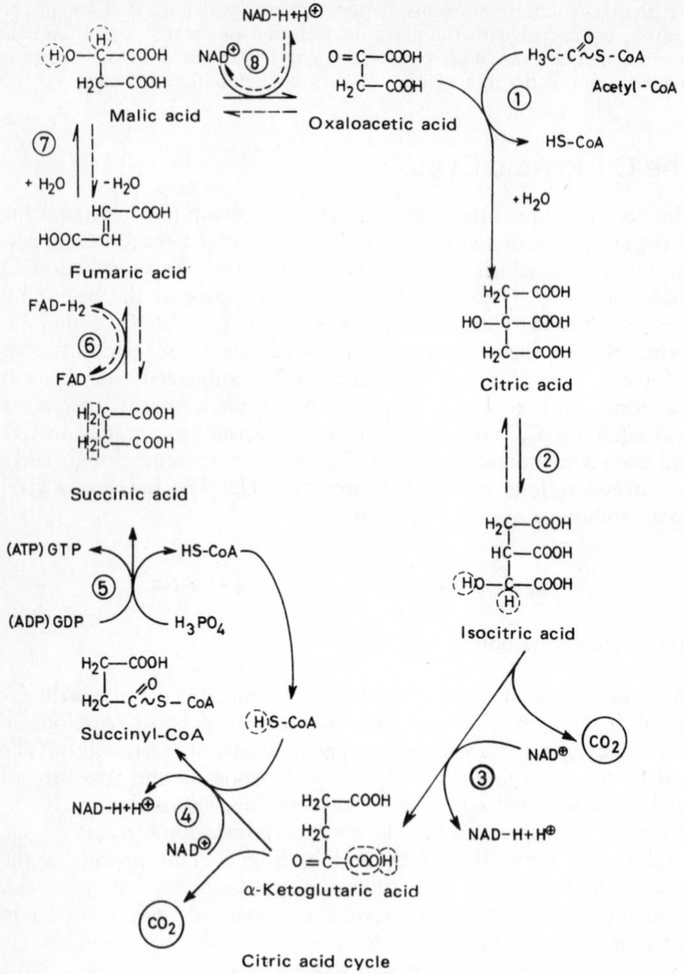

Citric acid cycle

① The monohydroxytricarboxylic acid* citric acid and free CoA are formed from acetyl-CoA and oxaloacetic acid. In this condensa-

---

* The citric acid cycle or "Krebs cycle" is occasionally also termed "tri-carboxylic acid cycle".

tion process the acetyl residue reacts via its methyl group with the keto group of oxaloacetic acid:

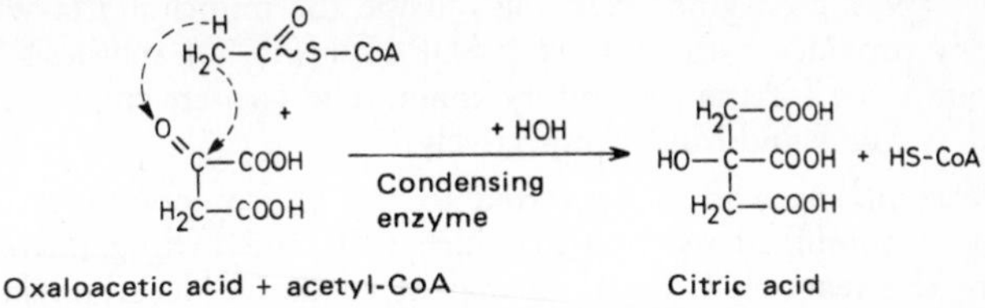

Oxaloacetic acid + acetyl-CoA   Citric acid

The energy required is derived from the energy-rich thioester bond in acetyl-CoA (p. 218). The equilibrium of the reaction strongly favors citric acid. The *condensing enzyme* or *citrate synthase* involved has been isolated and purified from pigs' liver. It has also been partially purified from leaves of *Xanthochymus guttiferae*.

② The enzyme *aconitase* is responsible for the next step, the isomerization of citric acid to isocitric acid. The enzyme is specifically inhibited (see below) by monofluorocitric acid which is formed in cells on addition of fluoroacetic acid (CH_2F-COOH). The equilibrium mixture of products from the *aconitase*-catalyzed reaction consists of 89 % citric acid, 8 % isocitric acid and 3 % *cis*-aconitic acid which is the common anhydride. According to recent concepts a "carbenium cation" is the intermediate for all three compounds:

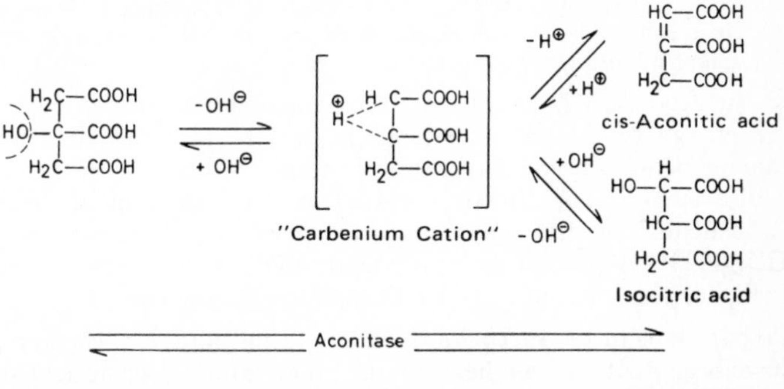

③ Hydrogen is removed from the secondary hydroxyl group of isocitric acid by *isocitrate dehydrogenase* and donated to $NAD^{\oplus}$. This oxidation gives rise to the formation of a keto group: from isocitric acid, oxalosuccinic acid is formed which is converted by the same enzyme to α-ketoglutaric acid via decarboxylation. Since CO_2 is split off from a β-keto-acid, the reaction is reversible. The $NAD^{\oplus}$-specific *isocitrate dehydrogenase* has been isolated from yeast, pea seedlings and heart muscle. The enzyme is always bound to mitochondria and is activated by ADP (p. 444). There is also a NADP⊕-specific enzyme occurring outside the mitochondria which probably provides hydrogen via NADP-H + H⊕ for synthesis, but apparently not for the respiratory chain. The two enzymes are activated by $Mg^{2\oplus}$ and $Mn^{2\oplus}$, respectively.

④ α-ketoglutaric acid is converted to the C_4 compound succinic acid by a combined reaction in which CO_2 and hydrogen are released. The reaction is equal to that of oxidative decarboxylation of pyruvic acid (p. 214 ff) and is characterized by the participation of ThPP, lipoamide, CoA and NAD⊕. As shown in the reaction scheme below, the succinic acid residue is transferred as decarboxylated product from ThPP to lipoamide. In this step, the aldehyde is oxidized to an acid (cf. p. 216 f); the free energy of the reaction is conserved as chemical energy in the thioester bond. Dihydrolipoamide is formed upon transfer of the succinic acid residue to CoA. It is dehydrogenated by *dihydrolipoyl dehydrogenase* (formerly referred to as *diaphorase)* which transfers the hydrogen to NAD⊕ (p. 217).

The active "multienzyme complex" composed of *α-ketoglutarate decarboxylase, lipoyl-reductase transsuccinylase* and *dihydrolipoyl dehydrogenase* has been demonstrated in preparations of plant mitochondria. However, it has not yet been isolated in a free form.

As in the case of acetyl-CoA, succinyl-CoA is metabolized in various ways: 1) as a starting substance in biosyntheses; 2) by further cleavage in the citric acid cycle. As will be discussed later (p. 231 f), the cycle has at this stage an important "outlet valve" for succinyl-CoA.

⑤ Succinyl-CoA is converted to succinic acid in such a way that the energy of the thioester bond is used to phosphorylate GDP, thereby being conserved as chemical energy. In this case, the adenylic acid system is not directly involved in the formation of "energy equivalents"; it acts only indirectly via the equilibrium with the GDP/GTP system. On the other hand, ATP may be formed directly in higher plants as indicated by recent findings (see below).

Preparations of the *succinate thiokinase* catalyzing this reaction step have been isolated from heart muscle; they exhibit specific activation by GDP/GTP. However, the corresponding enzyme isolated from spinach leaves *(succinyl-CoA synthase)* is highly specific for ADP/ATP as co-substrate.

⑥ Dehydrogenation of succinic acid by *succinodehydrogenase (succinic acid dehydrogenase* or *succinate dehydrogenase)* gives rise to the formation of a double bond in the molecule. The *trans-*compound, fumaric acid, is formed. Hydrogen is removed without participation of NAD⊕. In oxidation processes of this kind, leading to formation of a carbon double bond in the molecule, other active groups are apparently competent for taking up hydrogen. In this case, it is flavin adenine dinucleotide ("FAD") since *succinodehydrogenase* is a flavoprotein, a "yellow ferment". It is closely associated with other enzymes and cofactors in the mitochondria or their subunits which catalyze the hydrogen or electron transfer

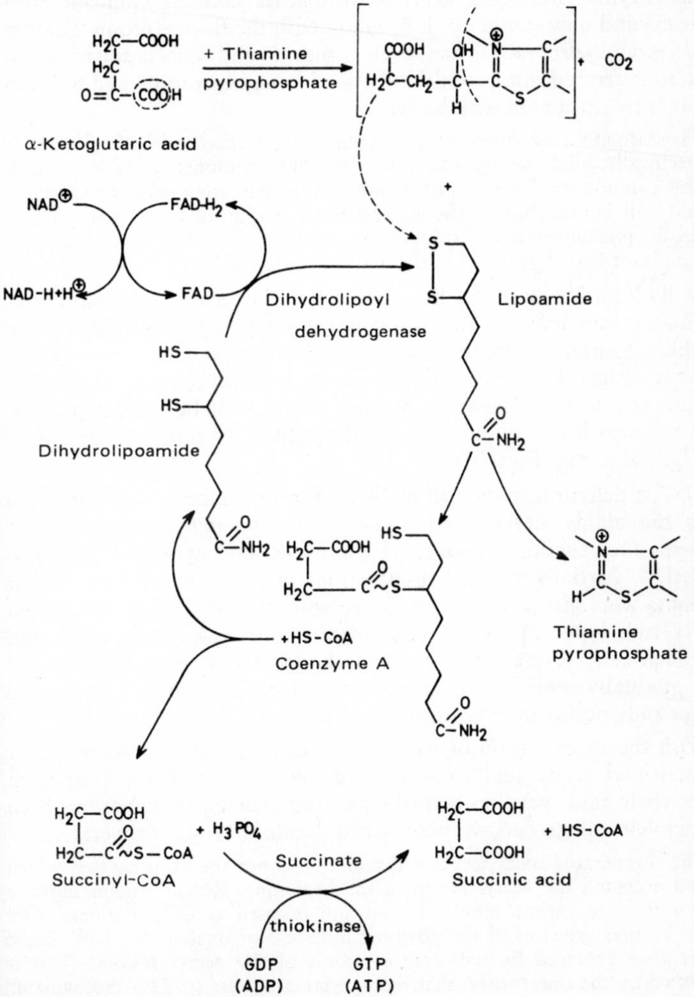

α-Ketoglutaric acid

NAD⊕

NAD-H+H⊕

FAD-H₂

FAD

Dihydrolipoyl dehydrogenase

Lipoamide

Dihydrolipoamide

Coenzyme A

+HS-CoA

Thiamine pyrophosphate

Succinyl-CoA

+ H₃PO₄

Succinate

thiokinase

GDP (ADP)

GTP (ATP)

Succinic acid

+ HS-CoA

directly to oxygen or to an artificial acceptor (methylene blue!), i.e. without mediation of the nicotinamide coenzymes. The system is hence referred to as *succinoxidase complex*.

This complex is part of the "respiratory chain" and, therefore, tightly bound to the mitochondrial structure (see below). *Succinodehydrogenase* is hence difficult to solubilize and usually contains other active components of the respiratory chain. Preparations of the enzyme from beef heart mitochondria contain nonheme iron atoms and ubiquinone (p. 135) along with the flavine group. Soluble *succinodehydrogenase* has been isolated and concentrated from acetone-treated mitochondria obtained from bean roots and tobacco leaves by extraction with buffer.

The formation of fumaric acid, catalyzed by *succinodehydrogenase,* is specifically inhibited by addition of suitable amounts of malonic acid. This compound which is in its chemical structure rather similar to succinic acid will be attached to the enzyme but not cleaved by it. This characteristic phenomenon of "competitive inhibition" (p. 10) has contributed significantly to elucidation of the mode of action of *succinodehydrogenase*.

⑦ By addition of water fumaric acid is converted to L-malic acid. The enzyme active in this reversible reaction is *fumarate hydratase* which ensures a strictly stereospecific configuration of the ligands: the hydrogen is in *trans*-position to the hydroxyl group. At equilibrium about 80 % L-malate (malate = salt of malic acid) is present. In tobacco leaves about 90 % of the activity of *fumarate hydratase* is bound to mitochondria.

⑧ The dehydrogenation of malic acid to oxaloacetic acid, catalyzed by the highly stereoselective enzyme *malate dehydrogenase* completes the reaction sequence of the cycle. The hydrogen passes to $NAD^{\oplus}$. Partially purified preparations of this enzyme from higher plants were also found to be active with $NADP^{\oplus}$. There are generally two types of *malate dehydrogenase* in eukaryotic plant and animal cells. One of these is firmly bound to the mitochondria and is gradually inhibited by the oxaloacetic acid formed. The second type only occurs in the cytoplasm.

With the regeneration of oxaloacetic acid, the specific acceptor for "activated acetic acid", one turn of the cycle has been completed; the citric acid cycle has turned over once. The C_2 fragment has been completely degraded, the acceptor molecule has been regenerated.

The regenerated oxaloacetic acid molecule is not the same as that which had accepted the acetyl group in the beginning. Recent studies indicate that the two carbon atoms cleaved and released as CO_2 originate from the carbon skeleton of the acceptor molecule of oxaloacetic acid. These are then replaced by two carbon atoms of the acetyl residue. This is shown by the observation that the enzyme *aconitase* (p. 221) "recognizes" citric acid as a biologically unsymmetrical compound, and converts it accordingly: water is removed from the molecule only in one specific direction; therefore, the enzyme is able to distinguish between the terminal COOH groups which are chemically identical.

Balance of the Citric Acid Cycle. Two molecules of CO_2 and coenzyme-bound hydrogen (3 NAD-H + H$^\oplus$ and 1 FAD-H$_2$) are formed from one molecule of "activated acetic acid" (= acetyl-coenzyme A) with uptake of water. In one turn of the cycle, there is a relatively small decline in free energy (ΔG° = −25 000 cal/mole) for the whole reaction sequence. It results mainly from cleavage of the energy-rich thioester bond in acetyl-CoA and from the two decarboxylations. Only the formation of 1 GTP which for once replaces ATP as energy carrier contributes to the conservation of chemical energy.

Thus, the achievement of the citric acid cycle in dissimilatory degradation consists in supplying reactive hydrogen bound to coenzymes, and in liberating CO_2. The direct yield of chemically utilizable energy is relatively small.

Occurrence of the Citric Acid Cycle in Plant Cells. Today, the participation of the citric acid cycle in the oxidative degradation of carbohydrates in plant cells is a well established fact. Most of the cells contain the enzymes involved; in some cases these have been isolated and partially purified, as pointed out while discussing the individual reactions. All the characteristic intermediates of the citric acid cycle are natural components of many plant cells. By feeding these intermediates the respiratory activity of plant tissue slices is markedly enhanced. It has been possible to analyze the reactions involved, particularly by means of ^{14}C-labeled acetate or pyruvate. In the course of incubation ^{14}C radioactivity appears in the typical intermediates of the cycle. Release of $^{14}CO_2$ also occurs in accordance with the postulated reaction scheme.

A further factor demonstrating the activity of the citric acid cycle in plant cells is the inhibition of respiration by specific inhibitors. The competitive inhibition by malonic acid which is abolished by addition of succinic acid, has already been described (p. 224). Monofluoroacetic acid is another effective inhibitor (p. 221). As was first shown in animal tissues, this compound is initially converted to fluoroacetyl-CoA and, after condensation with oxaloacetic acid, fed into the citric acid cycle as fluorocitric acid. This compound competitively inhibits the enzyme *aconitase* thus preventing further metabolism of citric acid and blocking the cycle. The unused citric acid accumulates. In this way the cells produce a highly potent inhibitor from the rather weak toxic precursor fluoroacetic acid.

In animal cells the citric acid cycle is located in the mitochondria. The presence of the citric acid cycle has also been confirmed in mitochondrial preparations from seedlings and other plant tissues using the following criteria: presence of the characteristic enzymes, uptake and degradation of intermediates of the citric acid cycle, and inhibitory action of malonic acid and fluoroacetic acid.

The Glyoxylic Acid Cycle

In germinating plant seeds in which fat in the cotyledons or in the endosperm is the sole reserve substance, an important modification of the citric acid cycle has been found, the "glyoxylic acid cycle", named after its key compound. This reaction pathway has also been demonstrated in microorganisms which use acetate, ethanol or fatty acids as carbon source for synthesizing their cellular substances, and in tissues of higher plants which convert fat to carbohydrate.

Reactions of the glyoxylic acid cycle

The glyoxylic acid cycle was originally discovered in microorganisms which use exclusively acetate as carbon source for growth. Degradation of this precursor by the citric acid cycle provides energy, but no precursors for the synthesis of cellular substances since the cycle does not bring about a net synthesis of intermediates. If one of these were to be removed as a precursor for synthesis the citric acid cycle would come to a stop because the essential regeneration of oxaloacetic acid would be prevented. It was hence inferred that acetate-metabolizing microorganisms must be equipped with an additional set of reactions to produce oxalo-acetate from acetate. This task turned out to be accomplished by the glyoxylic acid cycle, discovered by Kornberg and Krebs (1957).

The reaction sequence (see scheme) starts with the formation of citric acid from oxaloacetic acid and acetyl-CoA, as is characteristic for the citric acid cycle (p. 220 f).

The isocitric acid formed is decomposed to glyoxylic acid and succinic acid by *isocitratase (isocitrate lyase)*. Then *malate synthase* catalyzes the condensation of glyoxylic acid and acetyl-CoA to malic acid. This process corresponds to the formation of citric acid in-sofar as here too the methyl group of the acid residue reacts with a carbonyl group:

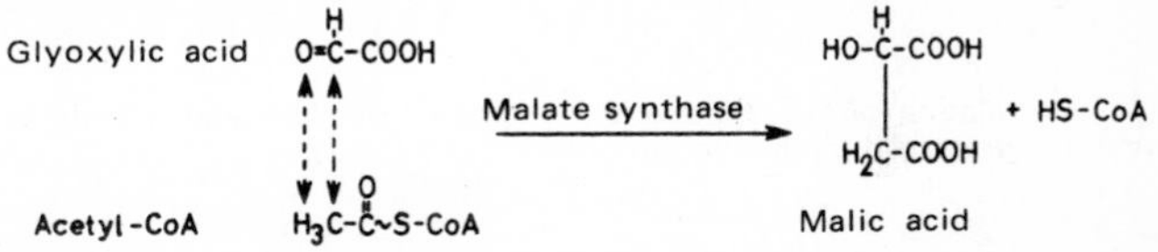

Malic acid is then converted to oxaloacetic acid by dehydrogenation mediated by *malate dehydrogenase* and $NAD^{\oplus}$ (cf. p. 224). The succinic acid initially formed may be converted to oxaloacetic acid, too, via the reactions of the citric acid cycle, and may thus contribute to the regeneration of this important key compound.

In seeds rich in fat (peanut, sunflower, *Ricinus*), the maximum of *isocitratase* activity coincides with the period of maximal fat degradation during germination (Fig. 80). The occurrence of this enzyme seems to be restricted to fat-containing plant tissues and acetate-metabolizing micro-organisms.

Studies with fat-containing cotyledons have shown that the typical enzymes of the glyoxylic acid cycle are obviously not bound to mito-chondria but to other cytoplasmic particles, the "glyoxysomes" (Beevers). The glyoxylic acid cycle obviously operates as an independent reaction sequence in a compartment different from that of the citric acid cycle, although a few of the reactions are identical. It is unknown to what extent glyoxysomes possess the enzymes required to convert succinic acid to oxaloacetic acid. It is possible that this reaction product must be taken up by mitochondria and metabolized to oxaloacetic acid there. On the other hand, it seems certain that glyoxysomes are equipped with all enzymes necessary for β-oxidation of fatty acids (p. 412 ff), and with catalase, too.

Glyoxysomes and the "peroxisomes" (p. 161) are relatively well-definable subcellular particles due to their biochemical functions. They are classified

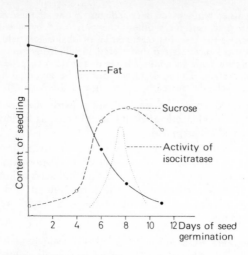

Fig. 80 Correlation of fat degradation, sucrose formation and activity of the enzyme *isocitratase* during seed germination of *Ricinus*.

as "microbodies" (de Duve), and were first discovered in mammalian cells, later in cells of lower and higher plants as well. They differ morphologically from mitochondria by reason of their roundish shape and their size (ϕ 0.2–1.5 μ), the single outer membrane without invaginations, the fine granular matrix without ribosomes, and the higher density in sucrose-containing media.

With the introduction of glyoxylic acid a reaction pathway is installed which, by adaptation to the particular metabolic situation (see above), permits an increased formation of oxaloacetic acid since an additional C_4 compound arises from two C_2 fragments. Here, the acetyl residue in acetyl-CoA is not degraded by the citric acid cycle in the usual way but is utilized to synthesize C_4 compounds, particularly oxaloacetic acid. On the one hand the large supply of acetyl-CoA, the typical degradation product of fatty acids (p. 412 ff), is skimmed off, while on the other hand the amount of oxaloacetic acid is increased. This is due to the function of the latter compound as precursor for several important synthetic pathways. In this connection the formation of hexose phosphate from oxaloacetic acid is of great interest since the process is characteristic for the mobilization of fat reserves and their conversion to carbohydrate.

In an initial reaction oxaloacetic acid is phosphorylated by GTP with simultaneous decarboxylation; the reaction is catalyzed by *phosphoenolpyruvate carboxykinase* (cf. p. 157):

$$
\begin{array}{c}
\text{COOH} \\
| \\
\text{C=O} \\
| \\
\text{CH}_2 \\
| \\
\text{(COOH)}
\end{array}
\quad
\xrightarrow[\text{Phosphoenolpyruvate carboxykinase}]{\text{GTP} \quad \text{GDP}}
\quad
\begin{array}{c}
\text{COOH} \\
| \\
\text{C-O-}\circledP \\
\| \\
\text{CH}_2
\end{array}
\;+\; CO_2
$$

Oxaloacetic acid   Phosphoenolpyruvic acid

In an endergonic reaction phosphoenolpyruvic acid is formed which is subsequently converted to fructose biphosphate and finally to hexose monophosphate via 3-phosphoglyceric acid and triose phosphates, a reversion of the reaction sequence of glycolysis (see scheme below). Fatty acid degradation (p. 412 ff) supplies reduced coenzyme while ATP is provided by the respiratory chain. Up to 3-phosphoglyceric acid, the reactions proceed on about the same energy level and are hence reversible. ATP is only required in the reduction of 3-phosphoglyceric acid in order to overcome the energy barrier which this strongly endergonic reaction poses for the reversion of glycolysis. Formally, the reduction corresponds to the formation of triose phosphate from 3-phosphoglyceric acid in the photosynthetic CO_2 reduction cycle (p. 145 ff), except that the *phosphotriose dehydrogenase* involved in the latter case is activated by NADP specifically.

This reversal of glycolysis has been termed "gluconeogenesis". Acetyl-CoA, the specific degradation product of fatty acids, serves as starting

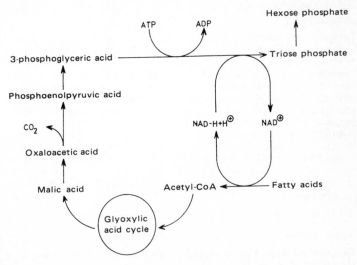

Scheme of gluconeogenesis

substance. In mammals it is lactic acid, the end product of anaerobic glycolysis (= "fermentation", p. 246 ff). Several amino acids may also assume this function, e.g. in the conversion of reserve protein to carbohydrate which occurs in germinating seeds or in storage tissues of higher plants. The proteins are first degraded to peptides and amino acids by specific enzymes, the *proteases* (for details, see p. 397). Of the latter compounds only those convertible to one of the dicarboxylic acids of the citric acid cycle are suitable for synthesis of hexose phosphate. They are converted to oxaloacetic acid either directly or indirectly via the reactions of the cycle. By phosphorylation and simultaneous decarboxylation in the endergonic reaction described above, phosphoenolpyruvic acid is formed from oxaloacetic acid. The compound is transformed to hexose phosphate via the reverse reaction sequence of glycolysis (see above).

Acetyl-CoA and pyruvic acid are also formed by conversion of amino acids. While acetyl-CoA enters the citric acid cycle directly and is converted to oxaloacetic acid, pyruvic acid must first undergo oxidative decarboxylation (p. 215 ff). From the formation of oxaloacetic acid onwards, the reactions are identical with those mentioned above.

All amino acids which contribute in this way to the formation of carbohydrate are termed "glucoplastic amino acids". Gluconeogenesis from amino acids or proteins is of great physiological importance, particularly in animal and human organisms.

Integration of the glyoxylic acid cycle and the citric acid cycle in synthetic processes indicates the close cooperation of "operating metabolism" and "building metabolism". This second important function of the citric acid cycle as a mediator between catabolic and anabolic processes in the cell will be discussed in a separate section (p. 231 f).

Formation of Oxaloacetic Acid. Since the occurrence of the glyoxylic acid cycle is limited, it has to be replaced in most organisms by other reactions in order to ensure an adequate supply of oxaloacetic acid in the cell (= "anaplerotic reactions" according to Kornberg). This provision is necessary because normal function of the citric acid cycle would soon be seriously impaired by intense tapping of oxaloacetic acid or of other cycle intermediates for syntheses: the specific acceptor for acetyl-CoA would be depleted! Therefore, a reaction is required in which oxaloacetic acid is formed by carboxylation of pyruvic acid. It has been named the "Wood-Werkman reaction" after its discoverers (see also p. 142) and was first demonstrated in microorganisms. The CO_2 is initially converted to its "activated form" with the help of ATP and attached to the biotin enzyme (p. 405 f). This complex reacts with pyruvic acid according to the following equation:

$$CH_3\text{-}CO\text{-}COOH + HOOC{\sim}Biotin\ enzyme \rightleftharpoons HOOC\text{-}CH_2\text{-}CO\text{-}COOH + Biotin\ enzyme$$

Pyruvic acid   Oxaloacetic acid

One mole of ATP must be expended per mole of oxaloacetic acid since the process is endergonic.

Another mechanism for formation of oxaloacetic acid again starts with pyruvic acid which is reductively decarboxylated. Malic acid is the reaction product:

$$CH_3\text{-}CO\text{-}COOH + CO_2 + NADP\text{-}H + H^\oplus \underset{\text{Malic enzyme}}{\xrightarrow{\hspace{2cm}}} HOOC\text{-}CH_2\text{-}CHOH\text{-}COOH + NADP^\oplus$$

Pyruvic acid   Malic acid

The enzyme specific for this reaction is called *malic enzyme*. It is activated by NADP$\oplus$ as coenzyme. Malic acid is then converted to oxaloacetic acid by *malate dehydrogenase* and NAD$\oplus$. *Malic enzyme* has been found in animals as well as in higher plants, particularly in species with "*Crassulaceae* acid metabolism" (p. 157). As discussed previously, a further reaction produces oxaloacetic acid in these plants by carboxylation of phosphoenolpyruvic acid. Carboxylation of the same compound also produces oxaloacetic acid and ATP when catalyzed by *carboxykinase* (cf. p. 157):

phosphoenolpyruvic acid $+$ CO_2 $+$ ADP $\rightleftharpoons$ oxaloacetic acid $+$ ATP
                                   (GDP)   (GTP)

Since the reaction equilibrium favors formation of phosphoenolpyruvic acid, the reaction is more important for conversion of fat to carbohydrate (p. 228 f) than for synthesis of oxaloacetic acid.

The Citric Acid Cycle as Metabolic Pool

So far, we have only considered the dissimilatory function of the citric acid cycle. Its second important role is to act as "emporium" for numerous intermediates. A series of precursors leading to various synthetic pathways are produced by reactions of the citric acid cycle. The C_2 fragments flowing in from the metabolism of carbohydrates, fats and proteins are not only fuel for dissimilation, but also starting compounds for assimilation. Their conversion enables intermediates to leave the cycle and enter processes of synthesis, provided that oxaloacetic acid is additionally formed; the feeding of acetyl-CoA into the cycle depends on this key compound. The citric acid cycle must be seen as an important "metabolic pool" (cf. p. 218) in intermediary metabolism into which flow molecules from quite different sources; they are mixed and converted in this pool before entering various reaction pathways.

The citric acid cycle is one of the enzyme systems with self-adjusting and self-regulating features which exist in a dynamic steady state (p. 440).

In some ways, the citric acid cycle may be compared to a reaction vessel in which due to a mechanical construction a cyclic production process is taking place. Apart from an inlet for starting substances and an outlet for products, there are additional adjustable taps. These permit removal or feeding of intermediates at various stages of the production process if required.

The citric acid cycle emphasizes how difficult it would be to draw a clear line between catabolic and anabolic metabolism, i.e. between "operating metabolism" and "building metabolism"; the two are closely interlocked.

The processes of synthesis will be discussed in detail later. Therefore, we shall summarize here only the most important correlations with the citric acid cycle (see scheme). α-Ketoglutaric acid is the precursor of the amino acid glutamic acid, and of all the substances derived from it that are members of the same class of compounds (p. 364 ff). Aspartic acid, another amino acid, is formed from oxaloacetic acid. Various other amino acids are derivatives of aspartic

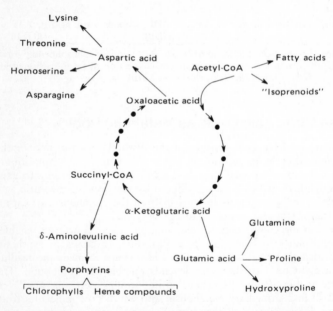

acid (p. 366 ff). Succinic acid or succinyl-CoA is a precursor of δ-aminolevulinic acid, the key substance in porphyrin biosynthesis which produces hemoproteins (cytochromes!) and chlorophylls (p. 418 ff). To complete the picture a number of other pathways depending on the "pool" of acetyl-CoA (p. 218) have been added to the scheme.

Endoxidation

As pointed out before, the major gain in chemical energy during dissimilatory degradation derives from the combination with molecular oxygen of the hydrogen cleaved from the substrate and bound to coenzyme. This fourth reaction complex comprises the "endoxidation" performed by the reactions of the "respiratory chain".

The overall process is characterized by a large decline in free energy ($\Delta G^\circ = -52\,000$ cal/mole) which the cell utilizes to gain chemical energy. The hydrogen bound as NAD-H $+$ H$^\oplus$ does not combine with oxygen in a direct reaction, but via a series of intermediate reactions and intermediates which make up the respiratory chain. The strongly exergonic "knallgas reaction" (here, indeed, hydrogen and oxygen react to form water!) thus assumes the character of a moderated and well controlled reaction chain. In consequence, the relatively large amount of energy is not released instantaneously, but in a series of small portions. They are not lost as heat but are partially conserved by the cell as chemical energy. This is brought about by lowering the energy of the electrons gradually in several intermediate reactions which are coupled with the ADP/ATP system. Hence the free energy changes of these reactions can be used for synthesis of ATP. This process, consisting in synthesis of ATP from ADP and inorganic phosphate, is called "oxidative phosphorylation"; it is the most important function of the respiratory chain.

The two light reactions of photosynthesis have made us familiar with the chemical machinery of this complex reaction sequence, the "electron transport chain". It consists of a sequential arrangement of redox catalysts along which hydrogen or electrons travel from a compound with high "electron pressure" (= reductant!) to another with a high "electron affinity" (= oxidant!) (p. 24).

The principle of lowering the electron pressure step-wise and taking advantage of the decrease in free energy to form ATP also underlies "endoxidation". The term "respiratory chain" is based on this principle as well: a sequence of redox catalysts along which hydrogen or its electrons migrate to oxygen. The coupling of the intermediate reactions taking place with the ATP-producing system has been mentioned previously.

The respiratory chain constitutes the final common pathway along which hydrogen or electrons derived from the various "fuels" travel to oxygen. It acts as the final electron acceptor or end-oxidant in the aerobic cell.

Methods. Detection and characterization of enzymes and coenzymes belonging to the respiratory chain often proved to be extremely difficult since most of them are tightly bound to the mitochondrial structure (p. 203 f). When the latter is destroyed the enzymes are released, but their native properties are then greatly changed and their biological activity is lost. Consequently, it is rather difficult to simulate individual reactions of the respiratory chain with enzymes separated from the native structure. A combination of different methods is required to obtain information on the affiliation and arrangement of the individual components of the respiratory chain. The most important methods will be briefly discussed here:

1. Measurement of uptake and metabolism of oxidized and reduced substrates by intact cells or isolated mitochondria. Sensitive optical methods had to be developed for this purpose. They monitor the specific absorption changes brought about by the participating redox enzymes.

Two modifications of the normal spectrophotometer are employed which ensure that interference of non-specific absorption by the experimental material is largely eliminated. In a "double-beam" device the change in optical density at the test wavelength is measured and related to an appropriate reference wavelength. In measuring NAD-H + H$^\oplus$, for example, radiation of wavelength 340 nm is passed through the sample, followed by radiation of 374 nm. The concentration of reduced coenzyme is calculated from Δ optical density$_{340}$ minus Δ optical density$_{374}$.

In the "split-beam photometer" a light beam of definite wavelength is split and passes alternately through two separate samples of the same assay: in the one cuvette the components of the respiratory chain are placed in an oxidized state due to lack of substrate, while the other one contains the same components in a reduced state under anaerobic conditions with substrate present. The difference in optical density of the two samples, determined by passing through light of different wavelengths, gives rise to a "difference spectrum" which provides important hints as to the involvement of individual compounds. This refined procedure was also successfully used to identify redox catalysts in photosynthetic electron transport (p. 105 ff).

2. Desintegration of mitochondria into fragments by chemical procedures (treatment with detergents, digitonin, deoxycholate) or by mechanical disruption (homogenization, ultra-sound) and determination of their active enzymes.

3. Reconstitution of the whole respiratory chain or of segments with definite reaction sequences from isolated subunits or highly purified mitochondrial components. In a few cases an artificial respiratory chain was, in fact, obtained the properties of which were partly identical with those of active mitochondria. However, since certain elementary structural requirements are lacking, only limited conclusions can be drawn with respect to the in-vivo features. Nevertheless, it is a suitable method to demonstrate the lack of structural factors and other functional components or to decide whether individual enzymes found by isolation are in fact active in the respiratory chain.

Redox Systems of the Respiratory Chain. The enzymes active in the respiratory chain may be arranged theoretically on the basis of the redox potentials (p. 21) of their coenzymes or prosthetic groups. The sequence shown in the scheme (p. 236) is characteristic for the respiratory chain of animal mitochondria and has been largely confirmed experimentally. Its composition in plant mitochondria is still uncertain because of the methodical difficulties mentioned above. Apparently differences exist with respect to the chemical nature of several components; we shall comment on this later.

The first redox system is $NAD-H + H^{\oplus}/NAD^{\oplus}$ which due to its strongly negative potential of -0.32 volt exhibits the highest electron pressure and hence the highest reduction potential. Consequently, hydrogen is easily transferred to flavoproteins which are lower in the redox scale ($E_0' = 0.00$ volt) and are therefore appropriate hydrogen acceptors. This decrease in electron pressure of 0.32 volt equals a change in free energy of $\Delta G^{\circ} = -12\ 000$ cal/mole (cf. energy scale in the scheme on p. 236). Hydrogen has moved to a lower, i.e. more positive carrier, by following the energy gradient.

It is generally accepted today that only $NAD-H + H^{\oplus}$ arising within the mitochondria can donate its hydrogen directly to the respiratory chain. The reduction equivalents formed outside the mitochondria do not normally travel across the inner mitochondrial membrane (cf. p. 204). The hydrogen of $NAD-H + H^{\oplus}$ produced in glycolysis in the cytoplasm probably enters mitochondria only after having become attached again to appropriate metabolites which subsequently deliver it to the respiratory chain (p. 244 f). In a few instances uptake of $NAD-H + H^{\oplus}$ has been observed in plant mitochondria, but it may have been caused by damage of the membranes. The NAD system seems to be present to a relatively large extent in mitochondria. The proportion of NAD to cytochrome c is about 40 : 1 in plant mitochondria. In contrast to animal mitochondria the NADP system appears to be absent. The question of attachment of nicotinamide nucleotides to the mitochondrial structure is under discussion.

Flavoproteins comprise a relatively large group of enzymes in plant and animal cells. As indicated by the name they contain flavin bound as flavin mononucleotide ("FMN") or flavin adenine dinucleotide ("FAD"; for structures see p. 17). These form the prosthetic group which enables the enzymes to transfer hydrogen (cf. p. 18). Hydrogen may be removed directly from the substrate, e.g. by *succinodehydrogenase* (p. 224) or accepted from $NAD-H + H^{\oplus}$, as is characteristic for the respiratory chain.

Accordingly, the latter is provided with a *NAD-H – cytochrome c reduc-. tase*. Recent studies indicate that this enzyme is not a homogeneous flavoprotein, but a complex enzyme system (see below). Its firm attachment to the mitochondrial substructures makes it very difficult to isolate and

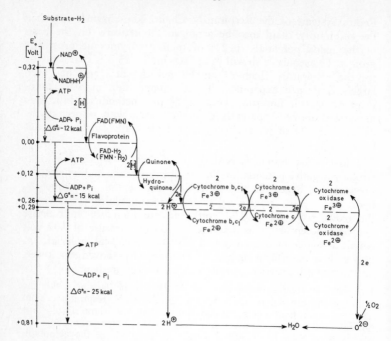

Arrangement of redox systems in the respiratory chain and formation of energy equivalents (ATP).

characterize its active components. This explains the contradicting data on flavoproteins active in the respiratory chain. *Succinodehydrogenase* (cf. p. 224) also belongs to these flavoproteins. There are probably several *NAD-H reductases* which differ from each other in their prosthetic group (FMN or FAD), in the number of active SH groups and in the number of iron atoms. The latter are bound to proteins but not via the porphyrin system; for this reason these compounds are termed "nonheme-iron proteins". They are believed to fulfill a certain function in the ionization of hydrogen $(2 [H] \rightarrow 2 H^{\oplus} + 2 e^{\ominus})$ in the respiratory chain (see below). The ferredoxins (p. 112 ff) apparently belong to the same class of proteins.

The reduced flavoproteins are probably oxidized by a quinone which is thereby converted to hydroquinone. In the respiratory chain an ubiquinone (also termed "coenzyme Q") is probably active; it occurs in plant and animal mitochondria.

Quinone systems are also active as carriers of hydrogen or electrons in photosynthesis: plastoquinone in the 2nd light reaction (p. 119 f), and ubiquinone in bacterial photosynthesis (p. 135). Their chemical constitution and mode of action have already been discussed. As lipophilic compounds, i.e. fat-soluble substances, quinones are associated with lipoproteids in the structures of chloroplasts and mitochondria. Several researchers hence consider coenzyme Q to be an appropriate metabolite for hydrogen or electron transport across the lipid regions of the mitochondrial membrane. The exact position of coenzyme Q in the respiratory chain is still unclear; possibly, it is located in a side path: enzymes have been detected which catalyze a direct hydrogen transfer from flavoprotein to cytochrome.

Specific cytochromes are active in the oxidation of reduced quinone or reduced flavoproteins. They are termed "c", "b" and "a" and are apparently not homogeneous substances but groups of closely related components. One of these electron-transferring enzymes is *ubihydroquinone-cytochrome c reductase (ubiquinone-cytochrome oxidoreductase)* which has recently been identified and isolated; it contains the cytochromes b and c_1, iron-sulfur protein ("nonheme-iron protein"; see above) and ubiquinone.

At this stage of the respiratory chain the transfer of hydrogen atoms (= "2-electron transport") changes to pure electron transport because hydrogen is ionized ($2 [H] \rightarrow 2 H^{\oplus} + 2 e^{\ominus}$). Since one molecule of cytochrome can carry one electron only at a time ($Fe^{3\oplus} + e^{\ominus} \xrightarrow{} Fe^{2\oplus}$), a "1-electron transport" results. Accordingly, two cytochrome equivalents are required per transfer step (see scheme p. 236).

Uncertainty prevails as to the role of cytochrome b; there are many indications that it has a specific function in the respiratory chain of higher plants. Cytochrome a or *cytochrome oxidase* finally catalyzes the transfer of electrons to oxygen. Cytochrome a is the sole enzyme of the respiratory chain which can react directly with oxygen; it acts as *endoxidase*. The oxygen ion ($O^{2\ominus}$) formed reacts immediately with two hydrogen ions ($H^{\oplus}$) to form water. This event terminates the process which was referred to previously as a controlled "knallgas reaction".

In photosynthesis (p. 120 ff) we have already observed the way in which the cytochromes act as oxidoreductases, transferring electrons by means of a valence change between $Fe^{2\oplus}$ and $Fe^{3\oplus}$. The iron is a constituent of the prosthetic groups formed in cytochromes by the porphyrin system of the heme. We are well informed about cytochrome c, which occurs in plant and animal mitochondria since this enzyme is readily soluble and rather easy to purify. Its molecular weight is 12 000; each molecule contains a heme group. Between this and the protein, covalent bonds exist via sulfur bridges ("thioether groupings") formed by reaction of the two vinyl side chains (positions 2 and 4) with the SH groups of two cysteine residues in the protein. In addition, histidine and methionine occupy the two free coordination sites of the iron atom. The amino acid sequence of the protein has also been elucidated. A very similar cell hemin, cytochrome c_1, has recently been found in yeast and in animal cells; in contrast to cytochrome c, this is firmly bound to the mitochondrial structure. Cytochrome c_1 is probably substituted by a very similar form ("c") in plant mitochondria. The molecular weight of c_1 is about 37 000. In the reduced state, the enzyme exhibits characteristic absorption maxima in the "Soret region" (p. 120) at 553–554 nm, 523–524 nm and 418 nm. Cytochrome b, a typical component of yeast mitochondria and animals cells,

Cytochrome c, reduced form

has a molecular weight of 30 000. Its prosthetic group is identical with that of hemoglobin. Plant mitochondria apparently lack cytochrome b of this type; three other b-type cytochromes seem to be present of which only cytochrome b_7 has been examined closely up to now. Compounds of the b-type cytochromes are also constituents of microsomes ("b_3" in plants, "b_5" in animals). The reader is reminded of the role of the chloroplast cytochromes f and b_6 in photosynthetic electron transport (p. 114 ff).

The prosthetic group of cytochrome a ("cytohemin" or "hemin a") closely resembles the "green" hemins. It was possible to elucidate its structure only after gentle separation of this cytochrome from the mitochondrial structure. A long unsaturated side chain is attached to ring I, a formyl group to ring IV:

Cytohemin ("Hemin a")

Cytochromes a and a_3 were originally distinguished from each other by their characteristic light absorption which probably results from different states of the same enzyme. It has been shown to be identical with Warburg's *Atmungsferment* and with *cytochrome-c oxidase*. The molecular weight is about 500 000, and the molecule contains six cytohemins as prosthetic groups. It is probably an oligomeric product of six subunits. A striking feature is the presence of copper in *cytochrome-c oxidase* (1–3 atoms per cytohemin). This enzyme is also present in plant mitochondria as part of the respiratory chain; the latter possibly contains a second *endoxidase*. A corresponding enzyme found in bacteria and blue-green algae has been termed "cytochrome o".

Composition of the Respiratory Chain in Plant Mitochondria. Our knowledge of the respiratory chain in plant mitochondria is far too limited to allow a final concept of its structure. Although the effective reaction mechanism is consistent with the respiratory chain of animal mitochondria, there are differences with respect to the redox enzymes involved, particularly the cytochrome components (see above). These differences are shown in the scheme below. Although all current data available have been used it is quite possible that corrections will have to be made in the near future since important details remain to be elucidated. The sequence of individual components given here is hence to be regarded as preliminary.

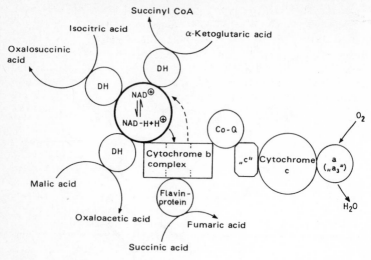

(DH = Specific dehydrogenase)

The respiratory chain of plant mitochondria

"Electron Transport Particles" as Constituents of the Respiratory Chain.

It has already been mentioned that, with the exception of cytochrome c, the enzymes of the respiratory chain are tightly bound to the inner mitochondrial membrane, particularly to the abundant lipid components (p. 203 f). Ultrasonic treatment of this membrane gives rise to formation of vesicles comparable with the "chromatophores" of phototrophic bacteria (p. 69 f); these vesicles, after enrichment and purification, contain the active enzymes of the respiratory chain. These "electron transport particles" are able to oxidize NAD-H + H$^\oplus$ or succinic acid with concomitant formation of ATP and consumption of oxygen. They lack, however, the enzymes of the citric acid cycle. Substrate is rapidly and efficiently converted via many intermediates since the enzymes involved are located next to each other in a functional sequence. These functional particles are obviously not identical with those substructures of the inner mitochondrial membrane, the "elementary particles", which appear in the electron micrograph as knoblike projections with a headpiece and a stalk (ϕ 90 Å). They contain proteins which probably assist in oxidative phosphorylation (see below). It is still uncertain whether "electron transport particles" in this form also exist in plant mitochondria.

Dynamic Steady State in the Respiratory Chain. The respiratory chain is a biochemical reaction sequence in which a balanced dynamic steady state (p. 7 ff) is maintained. Details are given in Fig. 81.

Respiratory Chain Phosphorylation. Up to now, we have left an important matter dangling: the processes of ATP synthesis coupled to the respiratory chain by which free energy of the intermediary reactions is conserved as chemical energy. Since these reactions are oxidative – hydrogen or electrons are donated! – the mechanism of ATP formation is termed *oxidative phosphorylation.*

As can be seen from the redox scale of our scheme (p. 236) the drop in electron pressure is 1.13 volt ($-0.32 \rightarrow 0.00 \leftarrow +0.81$ volt) as 2 H or a pair of electrons travel along the entire chain. This corresponds to a decrease in free energy of $\Delta G° = -52\,000$ cal/mole NAD-H + $H^{\oplus}$ which is sufficient to bring about the formation of 3 moles of ATP per mole of reduced NAD. The scheme shows that the sites of ATP formation coincide with the steps in hydrogen or electron transport in the respiratory chain: NAD-H + $H^{\oplus} \rightarrow$ flavoproteins; flavoproteins $\rightarrow$ cytochrome c and cytochrome c $\rightarrow$ oxygen. Since succinic acid is dehydrogenated directly by flavoprotein, the first

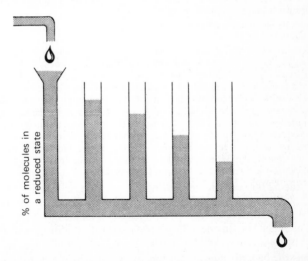

Fig. 81 Model of dynamic steady state of the respiratory chain. With the water inflow carefully adjusted, the levels in the open tubes will remain nearly constant. The water corresponds to electrons, each tube to a redox catalyst. The nearer the latter is located towards the outlet, the more positive it is, i.e. with increasing distance from the point of inflow the tendency to hold electrons — and hence the electron pressure — decreases steadily (after Lehninger).

coupling reaction is not needed for this substrate. Consequently, only 2 moles of ATP are formed per mole of water. Little in known so far about the reactions, the enzymes and intermediates involved in the phosphorylation process.

From animal mitochondria proteins have been successfully isolated from the knoblike, stalked substructures of the inner membrane (see above) which probably play a role in the coupling of respiration and oxidative phosphorylation ("coupling factors" according to Racker). If they are removed from the "electron transport particles" the ability to form ATP is lost; it can be restored by addition of these proteins. An energy-rich intermediate may cooperate with these coupling factors; its chemical nature and mode of action are unknown.

By oxidative phosphorylation ATP is formed in the inner space, i.e. in the mitochondrial matrix. Since it is mainly required in the cytoplasm outside the mitochondria, ATP must be translocated from the matrix to the outside. Recent findings indicate that an enzyme system (= *translocase*) is active in this transport. As a constituent of the inner mitochondrial membrane this enzyme catalyzes phosphorylation of external ADP with the help of ATP formed within the mitochondrion. Thereby the concentration of ADP molecules within the mitochondrion is maintained which is very important for their regulatory function in respiration (p. 438). The *translocase* system can be specifically inhibited with atractyloside, a plant glycoside (p. 168).

Apart from the "coupling hypothesis" (see above) a "chemiosmotic hypothesis" was formulated to account for phosphorylation in mitochondria; its mechanism has been discussed in connection with photophosphorylation (p. 125 ff). The conditions are quite similar in mitochondria: both cell organelles have membranes which are relatively impermeable to ions; they contain products of metabolic reactions, mostly as anions, and synthesize ATP coupled to membrane-bound electron transport. However, the effective proton gradient is established in different ways: in the chloroplast it originates from the light reactions; in the mitochondrion it arises from redox reactions. Another difference is that in the mitochondrion the protons are transported to the outer surface of the membrane (intermembrane space), the hydroxyl ions to its inner surface (matrix space); in the chloroplast a reverse translocation takes place.

Of the 52 000 cal released during the reaction of 1 mole NAD-H + $H^{\oplus}$ with $1/2\ O_2$, $3 \times 7000 = 21\ 000$ cal are utilized for the synthesis of 3 moles of ATP. This corresponds to an amazingly high efficiency of about 40 %. In the cell, where standard conditions certainly do not prevail, it may well be higher, i.e. up to about 60 %.

Thus the overall equation for the respiratory chain is as follows:

$$\text{NAD-H} + \text{H}^{\oplus} + 3\ \text{ADP} + 3\ \text{H}_3\text{PO}_4 + 1/2\text{O}_2 \longrightarrow \text{NAD}^{\oplus} + 3\ \text{ATP} + 4\ \text{H}_2\text{O}^{*}$$

$$(\Delta G^{\circ} = -31\ \text{kcal/mole})$$

---

* 3 additional moles of water arise from binding of phosphoric acid to ADP.

In various test objects synthesis of 3 moles of ATP correlated with formation of 1 mole of water by the respiratory chain has been found, provided the hydrogen of the substrate being degraded had been transferred to $NAD\oplus$ (see above). Accordingly, the "P/O ratio" (ratio of ATP formed to oxygen consumed) has a value of 3. This parameter is crucial in studies of the respiratory chain reactions in intact cells, in isolated mitochondria and in sub-mitochondrial particles. On addition of various intermediates of the citric acid cycle its value changes in a characteristic way: with α-ketoglutaric acid as substrate the P/O ratio amounts to 4, whereas it only amounts to 2 with succinic acid. The P/O ratio also changes significantly when ATP formation is reduced or abolished during the normal transfer of hydrogen or electrons to oxygen via the respiratory chain. This state of "uncoupling" (p. 130) is readily produced in intact cells and mitochondria by addition of certain cell poisons. The most wellknown uncoupler of oxidative phosphorylation is 2,4-dinitrophenol:

Dicumarol (or dicoumarol) and the antibiotics oligomycin and antimycin act in a similar way. These inhibitors are extremely useful in the elucidation of oxidative phosphorylation. In the uncoupled state the transfer of hydrogen or electrons in the respiratory chain continues, but the energy of oxidation is dissipated completely as heat. Today, it is assumed that uncoupling enables the cell to regulate ATP formation in the respiratory chain. However, specific uncoupling substances have not yet been found in living cells.

The relatively small quantity of ADP available has a self-regulating effect on the dissimilation of substrate. It acts as limiting factor in respiratory chain phosphorylation, usually preventing it from reaching its full capacity. Only an increase in consumption of ATP makes ADP available in larger amounts for oxidative phosphorylation. By this regulatory mechanism the cells are primarily prevented from literally "respiring to death"!

"Reversed Electron Transport". It has recently been found that electron transport along the respiratory chain in animal mitochondria may also proceed in the opposite direction (Klingenberg). Since energetically this entails an "uphill" reaction, ATP or an energy-rich intermediate of respiratory chain phosphorylation must be consumed. The biological significance of this process is still a matter of controversy. In plant cells electrons may, in principle, move against the natural energy gradient with consumption of ATP. The role of this "reversed electron transport" in the photosynthesis of purple bacteria has been explained (p. 137 f); it is a model for this type of reaction. Chemoautotrophic organisms such as *Nitrobacter* or *Ferrobacillus* (p. 193 f) appear to form the "reduction equivalents" required for CO_2 fixation by an analogous ATP-dependent "reversed flow" of electrons from cytochrome c to $NAD\oplus$. In plant mitochondria an energy-dependent reduction of $NAD\oplus$ by succinic acid was found during which, in contrast to the same reaction in animal mitochondria, ATP is consumed. Since the succinic acid/fumaric acid system ($E_0' = 0.00$ volt) will not donate hydrogen to the NAD system ($E_0' = -0.32$ volt) for energetic reasons, energy must be supplied to raise the hydrogen or the electrons to a higher energy level, i.e. "uphill".

Interaction between Photosynthesis and Respiration. Since photosynthesis and respiration have the same overall equation (though, of course, they proceed in opposite directions), it is difficult to analyze a possible interaction. It was necessary to use ^{18}O (cf. p. 25) together with a combination of manometric and mass-spectrographic techniques to measure photosynthetic O_2 production and respiratory O_2 uptake independently.

The results show that respiration is more or less depressed in low or medium light intensities, with values significantly lower than those of respiration in the dark which was simultaneously determined. This light-induced inhibition of respiration is apparently caused by the long-wave components of radiation (> 680 nm) which are relatively ineffective in photosynthetic O_2 evolution. Since CO_2 release is depressed at the same time, it seems likely that photophosphorylation, enhanced by long-wave radiation, continuously deprives respiration of ADP, thereby suppressing oxidative phosphorylation (cf. p. 243). To some extent, the opposite effect is observed on illumination with high light intensities: in *Anacystis*, O_2 uptake is markedly increased, particularly by the short-wave regions of visible radiation; perhaps O_2 replaces CO_2 as end acceptor of hydrogen or electrons. However, a direct connection between photosynthesis and this light-dependent increase of O_2 uptake does not necessarily have to exist. As shown by recent studies with *Chlorella pyrenoidosa* (Kowallik), blue radiation (455 nm) significantly increases the O_2 uptake of endogenous respiration even when photosynthesis has been blocked by DCMU (p. 123).

Metabolic Exchange Between Mitochondria and Cytoplasm

We have already seen how a controlled uptake and release of intermediary products, ATP and hydrogen is brought about in chloroplasts (p. 158 f). Rather similar processes occur in mitochondria. Apparently, these are accomplished by an "exchange" mechanism: for each molecule leaving the mitochondrion another one must enter the organelle, and vice versa. The principle described for chloroplasts also governs the transfer of hydrogen across the mitochondrial membrane: it must first be bound to an appropriate metabolite, i.e. malic acid, before entering the matrix space of a mitochondrion. Here malic acid is cleaved to NAD-H $+ H^\oplus$ and oxaloacetic acid. The latter is either degraded via the citric acid cycle or converted to aspartic acid with glutamic acid as a partner, α-ketoglutaric acid is formed as second product of this transamination (p. 363 f). Both compounds may be converted again to glutamic acid and oxaloacetic acid in the cytoplasm after leaving the mitochondrion. Oxaloacetic acid is then available for another hydrogen transfer into the mitochondrion where hydrogen is oxidized in the respiratory chain giving rise to $\sim$ P; the latter is translocated to the cytoplasm by the *translocase* system (p. 242). Pyruvic acid or pyruvate easily pass through the inner mitochondrial membrane and are converted to acetyl-CoA within the mitochondrion. Apart from being the precursor of citric acid, this key compound acts as substrate in fatty acid synthesis (p. 405 ff) and must hence be transferred to the cytoplasm. Citric acid serves as specific transport metabolite: it is formed by the reaction of acetyl-CoA with oxaloacetic acid, and then travels across the membrane to the cytoplasm. Here, citric acid is split again into the two precursors by an *ATP-citrate lyase (= citrate cleavage enzyme)*, the process consuming 1 ATP; acetyl-CoA is thus provided for processes of biosynthesis in the cytoplasm. The exchange partner for citric acid is oxaloacetic acid; after having been reduced to malic acid (see above) it reenters the mitochondrion.

It is obvious that these mechanisms of metabolic exchange permit an effective regulatory control (p. 439).

Energy Balance of Aerobic Glucose Dissimilation

Finally, the question arises as to how much chemical energy bound in the form of ATP is produced by aerobic degradation of hexose to CO_2 and H_2O? To answer this question, we must sum up all the ATP molecules formed during the entire process.

For the process of glycolysis, i.e. the cleavage of one molecule of glucose to two molecules of pyruvic acid, the following overall equation results:

$$C_6H_{12}O_6 + 2\ NAD^{\oplus} + 2\ ADP + 2\ H_3PO_4 \longrightarrow 2\ C_3H_4O_3 + 2\ NAD\text{-}H\text{+}H^{\oplus} + \boxed{2\ ATP}$$

Glucose   Pyruvic acid

The two molecules of NAD-H + H$^{\oplus}$ formed in glycolysis yield additional chemical energy when oxidized in the respiratory chain via phosphorylation of ADP:

$$2\ NAD\text{-}H\text{+}H^{\oplus} + 6\ ADP + 6\ H_3PO_4 + O_2 \longrightarrow 2\ NAD^{\oplus} + 8\ H_2O + \boxed{6\ ATP}$$

Representing a relatively energy-rich substrate, the two molecules of pyruvic acid are initially converted to acetyl-CoA by oxidative decarboxylation giving rise to two molecules of reduced coenzyme:

$$2\ C_3H_4O_3 + 2\ NAD^{\oplus} \xrightarrow{\ +\ CoA\ } 2\ \text{Acetyl-CoA} + 2\ NAD\text{-}H\text{+}H^{\oplus} + 2\ CO_2$$

Pyruvic acid

Transfer of this coenzyme-bound hydrogen to oxygen via the respiratory chain results in a further gain in chemical energy:

$$2\ NAD\text{-}H\text{+}H^{\oplus} + 6\ ADP + 6\ H_3PO_4 + O_2 \longrightarrow 2\ NAD^{\oplus} + 8\ H_2O + \boxed{6\ ATP}$$

Two revolutions of the citric acid cycle are required to oxidize the two acetyl residues completely to CO_2 and H_2O; in each turn eight atoms of hydrogen or four electron pairs are supplied; each pair yields an average of 3 molecules of ATP in travelling along the respiratory chain to oxygen:

$$2\ acetate + 24\ ADP + 24\ H_3PO_4 + 4\ O_2 \longrightarrow 4\ CO_2 + 28\ H_2O + \boxed{24\ ATP}$$

The formation of 3 molecules of ATP per molecule of H_2O formed results from the different yields of ATP on oxidation of the various substrates. As mentioned previously (p. 241 ff), α-ketoglutaric acid provides 4 ATP, succinic acid only 2 ATP. Oxidation of isocitric acid or of malic acid, however, yields 3 ATP. For the whole citric acid cycle therefore a theoretical (and experimentally confirmed) average of three phosphorylations of ADP per hydrogen or electron pair results.

The following overall equation is obtained for the complete aerobic oxidation of 1 molecule of glucose:

$$C_6H_{12}O_6 + 38\ ADP + 38\ H_3PO_4 + 6\ O_2 \longrightarrow 6\ CO_2 + 44\ H_2O + \boxed{38\ ATP}$$

This shows a yield of 38 ATP corresponding to $7000 \times 38 = 266\,000$ cal/mole under standard conditions. Since the free energy of the complete oxidation of glucose is $\Delta G^o = -675\,000$ cal/mole, the approximate efficiency amounts to 40 %. In vivo, it may reach a still higher value, since standard conditions very likely never occur. Accordingly, in the cell the free energy of ATP hydrolysis would also be higher than 7000 cal/mole (cf. p. 19).

Anaerobic Dissimilation of Carbohydrate ("Fermentations")

A number of microorganisms, as well as tissues of higher plants, of animals and of man, are able to dissimilate organic substrates, particularly glucose, in the absence of oxygen concomitant with conservation of energy. They make use of a reaction mechanism termed "fermentation" which is, strictly speaking, characteristic of obligate anaerobic organisms. In contrast to these, however, most organisms or tissues, being "facultative anaerobes", immediately switch to aerobic glucose dissimilation when oxygen is again available.

The anaerobic form of energy conservation may be older in phylogeny. It was only possible for organisms to utilize oxygen for this purpose when they had reached a higher level of specialization. The original ability may, however, have been retained in facultative anaerobes. This would explain why fermentation and aerobic glucose dissimilation proceed initially via the same glycolytic reactions (see below).

Since glucose is the most important "fuel" in fermentation, we shall restrict our comments to the various ways in which this compound may be dissimilated anaerobically. For the sake of completeness, it should be mentioned that a number of microorganisms may also ferment amino acids, fatty acids, and sugars other than glucose.

As indicated while discussing glycolysis, relatively energy-rich cleavage products – mostly C_2 and C_3 compounds – are left behind during anaerobic degradation of glucose. This is because the glucose molecule is cleaved into two fragments, one of which is oxidized by the other. Part of the energy released in this "intramolecular redox reaction" is conserved as chemical energy in ATP by coupling of the reaction to the adenylic acid system. The NAD-H + $H^\oplus$ formed during the oxidation step cannot donate its hydrogen to the respiratory chain because of the absence of oxygen. The functioning state of the coenzyme, i.e. its oxidized state, is only restored if an appropriate substrate accepts the hydrogen. This acceptor function is fulfilled by a compound which has been formed from the oxidized cleavage product of glucose during the process. Since it is not the same in all organisms exhibiting fermentation, different end products result. They are used to mark the various types of fermentation: alcoholic fermentation, lactic acid fermentation, formic acid fermentation, propionic acid fermentation or butyric acid fermentation.

Alcoholic Fermentation

Ethyl alcohol is the specific end product of anaerobic glucose degradation by various yeasts, fungi and tissues of higher plants. CO_2 is released during the process:

$$C_6H_{12}O_6 \longrightarrow 2\ C_2H_5OH + 2\ CO_2$$

Historical Aspects. Since earliest times man has known how to use alcoholic fermentation for the production of intoxicating beverages. Details of the chemical reactions involved only became known about the middle of the last century through the fundamental investigations of L. Pasteur. In 1815 Gay-Lussac had made an important contribution by formulating an overall equation for alcoholic fermentation. Pasteur's view that fermentation, as an expression of life, only proceeds in living cells was upset by Buchner, who demonstrated in 1897 that in a cell-free press juice from yeast alcoholic fermentation takes place, too. He called the active principle of this process "zymase". In the course of subsequent studies this was shown to comprise a complex of numerous enzymes. In 1905, Harden and Young discovered that phosphoryl esters of sugars are formed in yeast press juice during alcoholic fermentation. These research workers also succeeded in separating a heat-stable component of low molecular weight from the "zymase" which they termed "coferment of fermentation"*. Intensive studies eventually led to a precise knowledge of the reaction sequence and the enzymes involved. Alcoholic fermentation was the first complex biochemical process successfully elucidated. It was later shown that its reactions are largely identical with those of biological oxidation in muscle (glycolysis). The first important indication for this similarity was the observation that a boiled press juice from muscle is able to restore fermentation in an inactive yeast press juice because of its "coferment" content. Another significant discovery was made by Embden (1925) who found sugar-phosphoric acid esters in press juice from muscle after addition of fluoride. A great number of researchers have contributed to the present state of knowledge on fermentation and glycolysis. Only a few can be mentioned here: C. Neuberg, G. Embden, O. Meyerhof, J. K. Parnas, O. Warburg, K. Lohmann.

Enzymatic Formation of Alcohol. As already described in detail, glucose is split into two molecules of pyruvic acid via the reactions of glycolysis* (p. 208 ff). Alcoholic fermentation and aerobic degradation therefore follow the same reaction sequence up to this point. In the absence of oxygen, pyruvic acid is enzymatically decarboxylated to acetaldehyde. The active enzyme, *pyruvate decarboxylase,* has been found only in plant cells up to now. It is activated by $Mg^{2\oplus}$ and requires thiamine pyrophosphate as coenzyme *(cocarboxylase).* This compound fulfills a similar function in oxidative decarboxylation of pyruvic acid (p. 215 f) as we already know.

Acetaldehyde acts as acceptor compound for the hydrogen which was earlier transferred to $NAD^{\oplus}$ during oxidation of 3-phosphoglyceraldehyde (details of this reaction on p. 210 f). Reduction of

---

* "Ferment" was at that time a term used for the yeast cell or the agent of alcoholic fermentation. The modern term "enzyme" was introduced by Kühne (1878) in order to delimit the soluble components, e.g. from digestive juices, from it. Since the active principle in alcoholic fermentation later also turned out to be soluble and detachable from the cell, the two terms have been used synonymously. Today the term "enzyme" is preferred.

* The term "glycolysis" was originally restricted to anaerobic carbohydrate degradation in the human and animal organism. It was only realized later that most of the reactions involved also take place in aerobic metabolism.

acetaldehyde by reduced NAD is catalyzed by the enzyme *alcohol dehydrogenase:*

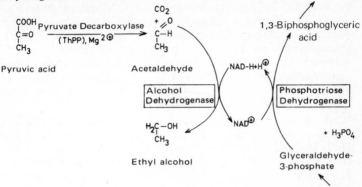

Alcohol dehydrogenase has been isolated and crystallized from yeast. It is a Zn-containing protein with a molecular weight of 151 000. The pure enzyme is used to determine alcohol by means of the "optical test" (cf. p. 15 f).

The validity of the reaction scheme of alcoholic fermentation, originally based on in-vitro findings, was later confirmed in living cells by means of radioactively labeled glucose. Glucose-1-^{14}C was fed to yeast cells and the radiocarbon content of the reaction products formed was determined. 95 % were recovered in the methyl group of ethanol, only traces in the other carbon atoms. About 3 % appeared in CO_2, produced by another system of glucose degradation in yeast cells, the *pentose phosphate cycle* (p. 255 ff).

Through glycolysis we are familiar with the phosphorylation of glucose which also takes place at the beginning of alcoholic fermentation. Since initially acetaldehyde is not available as hydrogen acceptor this function is assumed by a triose phosphate, dihydroxyacetone phosphate. Accordingly, glycerol is formed as reaction product. Once sufficient acetaldehyde has been formed from pyruvic acid, hydrogen from NAD-H + H$^\oplus$ is transferred to this compound. Alcohol formation is then under way. This explains why some glycerol is always found among the end products of alcoholic fermentation (see below).

The earlier statement that one cleavage product of glucose is oxidized by the other in anaerobic glucose dissimilation can be specified for alcoholic fermentation: 3-phosphoglyceraldehyde is oxidized by acetaldehyde. The former is converted to 1,3-biphosphoglyceric acid, the latter to ethyl alcohol.

The yield in chemical energy by this type of anaerobic glucose degradation amounts to 2 moles of ATP per mole of glucose, and is modest in comparison with the 38 moles of ATP produced by aerobic degradation of 1 mole of glucose to CO_2 and H_2O (p. 246). To meet their energy requirements, yeast cells and, in general, all other organisms with a fermentative metabolism, must convert large quantities of substrate. More than 90 % of the theoretically extractable energy remains in the cleavage products and cannot be directly utilized by the organisms. Yeast cells consume up to twenty times more glucose during anaerobiosis than under aerobic conditions. The end product alcohol accordingly accumulates, a process which is highly valued in the production of wine and beer.

When oxygen is supplied to fermenting yeast cells, these facultative anaerobes are now able to degrade glucose aerobically to CO_2 and H_2O. This switch-over in metabolism from fermentation to aerobic dissimilation is termed the "Pasteur effect", after its discoverer. It has been known to exist in animal cells for a long time, and was recently also observed in various tissues of higher plants (roots, bulbs). The Pasteur effect is considered to be a regulatory mechanism whereby glucose degradation and ATP production are adapted to external conditions and internal requirements of the cell (p. 443 f).

After adaptation yeast cells are able to ferment fructose and galactose directly as well. Since the cells contain the enzymes *saccharase (β-fructofuranosidase)* and *maltase (α-glycosidase)*, sucrose and maltose are also utilized after previous enzymatic cleavage to hexoses. On the other hand, direct fermentation of starch is not possible since the *amylases* and *phosphorylases* required for degradation of the macromolecules are lacking. Therefore, germinated barley ("malt") rather than ungerminated barley is preferred in the brewing of beer because the former already contains sufficient fermentable sugars formed during endogenous enzymatic degradation of starch. Even a sufficient oxygen supply does not completely abolish fermentation in yeast cells. Alcohol and CO_2 are produced continuously from hexose via the anaerobic reaction pathway. However, cell proliferation increases rapidly in the presence of oxygen.

The commercial baker's yeast has a relatively low fermentative capacity compared with brewer's yeast since it lacks an active *pyruvate decarboxylase*. This enzyme only becomes active during growth under anaerobic conditions, and then catalyzes the conversion of pyruvic acid to acetaldehyde.

By-products such as glycerol (see above), acetaldehyde, acetic acid, citric acid, succinic acid and fusel oils are also formed in small amounts during alcoholic fermentation. The most important components of fusel oil are propanol, isopropanol, butanol, isobutanol and amyl alcohols. Production of glycerol by means of alcoholic fermentation will be treated later (see below).

Fermentation in Higher Plants. Fermentation processes have also been found in tissues of higher plants; they have been described as "intramolecular respiration" or "anaerobic respiration". Generally, they commence on exclusion of oxygen and are characterized by evolution of CO_2. In some tissues, however, no alcohol is formed under these conditions while in others the amount formed does not quantitatively equal that of hexose consumed. Instead, organic acids appear as end products: lactic acid (see below), oxalic acid, tartaric acid, malic acid and citric acid. Recent findings suggest that *alcohol dehydrogenase* is ubiquitously distributed in higher plants, at least in the seedlings, whereas the occurrence of this enzyme in algae is uncertain.

In most plant tissues anaerobic respiration is exceptional; it starts if supply of atmospheric oxygen is partially or completely blocked. Tissues differ to a large extent in their tolerance to oxygen deficiency: some plants or organs survive a long time, others (e.g. maize seedlings) only 24 hours. Some tissues switch over to anaerobic metabolism even in the presence of oxygen when aerobic respiration is blocked by the addition of specific inhibitors such as cyanide.

Modifications of Alcoholic Fermentation. During fermentation of yeast cells a small amount of glycerol always arises as a by-product besides alcohol. Its quantity may be greatly increased by adding sodium bisulfite to the fermentation mixture:

Acetaldehyde   Sodiumbisulfite   Bisulfite addition compound

This compound reacts specifically with the acetaldehyde formed and thereby abolishes its ability to accept hydrogen from NAD-H + H$\oplus$. The hydrogen is transferred to dihydroxyacetone phosphate instead thus giving rise to formation of glycerophosphate from which glycerol is formed by dephosphorylation:

Dihydroxyacetone   L-α-Glycerophosphate   Glycerol
phosphate

The overall equation is:

$$C_6H_{12}O_6 \longrightarrow C_3H_8O_3 + CH_3CHO + 2 CO_2$$

Acetaldehyde cannot be reduced by NAD-H + H$\oplus$ when fermentation proceeds in an alkaline milieu. Therefore, the end product in this case is also glycerol, formed from dihydroxyacetone phosphate. Acetaldehyde immediately gives rise to formation of acetic acid and ethanol by dis-mutation.

This reaction is probably responsible for the formation of acetic acid and ethanol as described above, observed in fermentation of glucose by some algal species. It is apparently catalyzed by the enzyme *aldehyde mutase* which has been demonstrated in yeast cells, higher plants and liver cells.

The two ways in which glycerol is formed in the course of alcoholic fermentation were discovered by Neuberg and referred to as "second" and "third form of fermentation".

The formation of acetic acid described above must not be mixed up with another mechanism termed *acetic acid fermentation* although this process consumes oxygen:

$$CH_3-CH_2OH \ + \ O_2 \ \longrightarrow \ CH_3COOH + H_2O$$

Ethyl alcohol   Acetic acid

It is not a fermentative process in terms of Pasteur's classical definition since oxygen takes part. According to the modern conception, however, fermentation is a process that delivers a product of relatively high energy content and usually proceeds under anaerobic conditions. Hence, it is reasonable to use the term acetic acid "fermentation". It is typical for several species of bacteria, particularly of the genus *Acetobacter*, which extract energy by oxidative degradation of the energy-rich end product of alcoholic fermentation. This process involves two dehydrogenation steps. First ethanol is converted to acetaldehyde in a reversion of the original formation. The hydrate of acetaldehyde serves as the substrate of a second dehydrogenation which gives rise to formation of acetic acid:

Oxidation of alcohol to acetic acid

The reduced coenzymes transfer the hydrogen or electrons to the respiratory chain where they react with oxygen. The drop in free energy is used to form ATP via the reactions of oxidative phosphorylation. The energy gain from this process ($\Delta G^\circ = -180\ 000$ cal/mole) is accordingly much higher than that in anaerobic degradation processes.

The ability of *Acetobacter* is utilized in industrial production of acetic acid ("wine-vinegar"). Undesirable consequences result when acetic acid fermentation interferes with alcoholic fermentation in wine production and storage, i.e. after infection with these bacterial species.

Lactic Acid Fermentation

The reactions leading to formation of lactic acid as end product of anaerobic glucose degradation, typical for numerous bacterial species as well as for algae, proceed parallel to those of glycolysis up to the formation of pyruvic acid (see reaction scheme). In this case, however, pyruvic acid is not decarboxylated but serves as end acceptor for the hydrogen donated to NAD⊕ in the oxidation of 3-phosphoglyceraldehyde. The enzyme *lactate dehydrogenase* catalyzes this reduction of pyruvic acid to lactic acid.

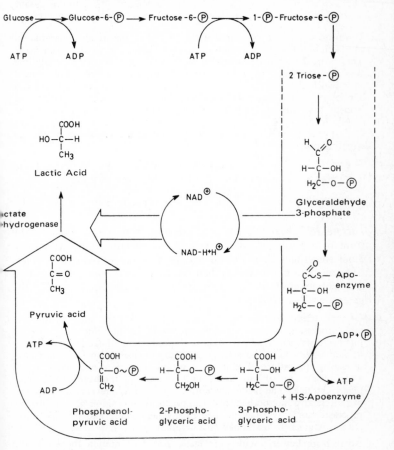

Formally, one C_3 fragment of glucose is oxidized by another one which is thereby converted to lactic acid:

Glyceraldehyde-3-phosphate	Pyruvic acid	1,3-Biphospho-glyceric acid	L-Lactic acid
		(= "oxidized" product)	(= "reduced" product)

The process may be compared with the assembling of an apparatus on an assembly line; for technical reasons one constructive element is removed in between and installed again at the end of the assembly line into the completed apparatus. For this purpose a special tool is used with which the constructive element is not only removed and replaced but to which it is also attached during the entire process. In lactic acid fermentation, the "constructive element" corresponds to hydrogen, the "tool" to the NAD system. Pyruvic acid represents the completely assembled apparatus, glucose is the raw material at the beginning of the assembling process.

The yield of chemical energy is exactly the same as in alcoholic fermentation, namely two moles of ATP per mole of glucose:

$$C_6H_{12}O_6 + 2\ ADP + 2\ H_3PO_4 \longrightarrow 2\ C_3H_6O_3 + 2\ ATP + 2\ H_2O$$

Lactate dehydrogenase has been prepared from muscle tissue in the form of a pure, crystallized protein. Its substrate is L-lactic acid. Since sometimes also D-lactic acid appears among the end products of fermentation in bacteria, their *lactate dehydrogenase* must have a different stereospecificity. – The enzyme has also been found in the green alga *Hydrodictyon*.

Of the bacterial species which form lactic acid, a few are important in milk processing (*Streptococcus thermophilus, Lactobacillus bulgaricus* = joghurt organisms!) and in industrial production of lactic acid (*Lactobacillus delbrückii*). The activity of these organisms also forms the basis of ensilage procedures used in conservation of gherkins and beans as well as in production of sauerkraut and in silage of foodstuffs from green plant materials. The lactic acid formed (0.2–2.5%) prevents development of putrefactive microorganisms. The practical application of fermentation processes can only be mentioned briefly here; more and detailed information is given in textbooks on microbiology.

Apart from bacteria with pure lactic acid fermentation there are also some species which form lactic acid as well as other end products in anaerobic degradation of glucose: ethanol, acetic acid and CO_2. Similar findings have been made in green algae: a few strains of *Chlorella, Ankistrodesmus braunii* and *Scenedesmus* D_3 produce mainly lactic acid as end product of anaerobic dissimilation; in other *Chlorella* strains lactic acid production cannot be demonstrated at all; a third group (of which *Chlorella vulgaris* is the main representative) forms lactic acid along with other products such as acetic acid, ethanol and CO_2. Uncertainty still prevails as to the precise mechanism of fermentation in green algae. The experimental results indicate that the reactions of glycolysis are involved.

Butyric Acid Fermentation

In some species of *Clostridium* the fermentation process is a much more complex one. The starting substance is acetyl-CoA. Butyric acid is formed as an end product or as an intermediate. In the latter case, it is further reduced to butanol. Acetone and isopropanol may be formed as by-products. In a few species of *Clostridium* formic acid appears as an intermediate which is normally decomposed to CO_2 and molecular hydrogen.

Direct Oxidation of Glucose: the Pentose Phosphate Cycle

In plant cells, the major part of glucose (and thus of the carbohydrates) is subjected to aerobic degradation in the consecutive processes of glycolysis, citric acid cycle and respiratory chain. In the course of these processes ATP is synthesized by substrate chain phosphorylation and oxidative phosphorylation. The end products are water and carbon dioxide.

However, glucose is also metabolized, though to a smaller extent, by another dissimilatory mechanism which degrades its carbon chain directly with simultaneous oxidation. By this mechanism hydrogen is removed and donated to $NADP^{\oplus}$ which acts as specific transport metabolite here. Oxygen is not involved in these conversions which may thus also take place in anaerobic organisms. This agrees very well with the hypothesis of substrate dehydrogenation in biological oxidation formulated by Wieland. End products of the direct degradation of glucose are CO_2 and NADP-H + $H^{\oplus}$.

Close analysis of this process has shown that the useful biochemical mechanism of a cyclic reaction sequence is employed here too. Since pentose phosphates play an important role, it is referred to as the "pentose phosphate cycle". The term "Warburg-Dickens-Horecker pathway" (after the biochemists who contributed significantly to its elucidation) is also used.

The reactions of this complex process are in parts identical to those of the photosynthetic CO_2 reduction cycle (p. 145 ff); they proceed, however, in the opposite direction. The intermediate compounds are accordingly sugar phosphate esters. The function they fulfill as the metabolic form of sugars has been discussed earlier (p. 153 f). However, the pentose phosphate cycle is not connected to photosynthesis except that in autotrophic cells the latter process supplies – directly or indirectly – glucose as the fuel. The cycle is active in cells of lower and higher plants as well as in bacteria, yeast, and liver cells. Its intermediates may also enter glycolysis or the pathways of important processes of biosynthesis (p. 260 f).

The direct degradation of glucose is performed in two phases: the first one encompasses the oxidative conversion of glucose-6-phosphate to pentose phosphate, the second one the regeneration of glucose-6-phosphate from pentose phosphate; glucose-6-phosphate enters the cycle again.

Formation of Pentose Phosphate

The starting substance is a phosphoric acid ester of glucose, glucose-6-phosphate. Free glucose or other hexose phosphates must first be converted to this compound (p. 204 f). Oxidation commences at carbon atom 1; by the enzyme *glucose-6-phosphate dehydrogenase* hydrogen is removed and transferred to $NADP^{\oplus}$. 6-phosphogluconolactone formed is spontaneously or enzymatically (by *lactonase*) converted to 6-phosphogluconic acid; $Mg^{2\oplus}$ is required as cofactor:

Glucose-6-phosphate   6-Phosphogluconolactone   6-Phosphogluconic acid

The elucidation of this direct oxidation step by Warburg in 1931 was a milestone in biochemistry. He recognized this step as an enzymatic reaction in which the participation of a coenzyme could be established for the first time. The enzyme involved was named "zwischenferment"; this term is no longer used. Purified *glucose-6-phosphate dehydrogenase* from yeast has recently become an important tool in enzymatic determination of ATP: by means of the "optical test" (p. 15 f) the amount of glucose-6-phosphate formed from ATP and glucose in the presence of *hexokinase* is determined; the quantity measured is NADP-H + H$^{\oplus}$ formed stoichiometrically.

6-Phosphogluconic acid is the substrate for a specific dehydrogenase which removes hydrogen again, this time at carbon atom 3, and reduces NADP$^{\oplus}$. A β-ketonic acid, 3-ketogluconic acid-6-phosphate, is probably formed first. This unstable compound is readily converted to ribulose-5-phosphate by decarboxylation at carbon atom 1. This ketopentose phosphate is enzymatically equilibrated with its aldose form, ribose-5-phosphate, by *ribose-5-phosphate isomerase:*

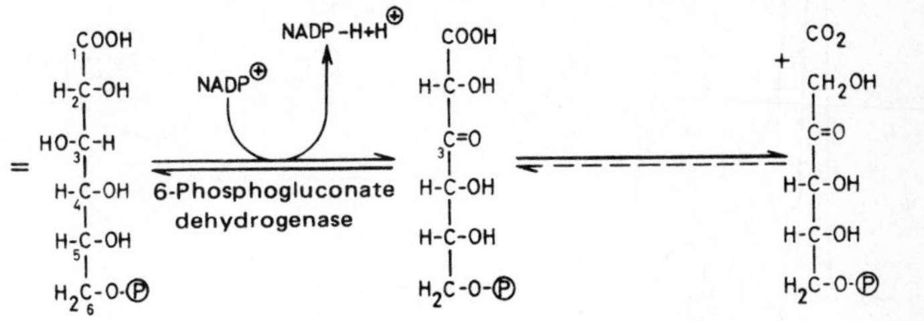

6-Phosphogluconic acid   3-Ketogluconic acid-   Ribulose-5-phosphate
                                6-phosphate

This terminates the first phase, i.e. the oxidation process. Two reduction equivalents are formed, and carbon atom 1 is eliminated as CO_2.

Apart from the oxidation at C-1 of the 6-phosphorylated glucose molecule described above, an oxidation at C-6 of the free sugar also occurs biologically giving rise to formation of glucuronic acid. This reaction does not contribute significantly to hexose degradation except in a few microorganisms. It is, however, biologically important insofar as glucuronic acid formation precedes the synthesis of ascorbic acid in plants and in a number of animal species. Bound to nucleoside diphosphate ("UDP glucuronic acid"), this sugar acid is a key substance in biosynthesis of the polyuronides, the most important compounds of which are the pectic substances (p. 187 ff).

Regeneration of Hexose Phosphate

The pentose phosphates formed are used mainly to regenerate glucose-6-phosphate via a complex process (cf. reaction scheme); its immediate precursor is fructose-6-phosphate. Another pentose phosphate, D-xylulose-5-phosphate, is required besides ribose-5-phosphate. It is formed from ribulose-5-phosphate by an enzyme-catalyzed epimerization at carbon atom 3 (enzyme: *epimerase*).

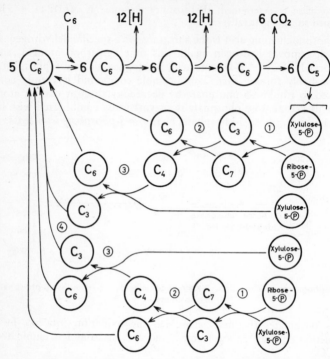

Reaction Sequences in Pentose Phosphate Cycle

From xylulose-5-phosphate and ribose-5-phosphate the compounds sedoheptulose-7-phosphate and 3-phosphoglyceraldehyde are synthesized by *transketolase:* a C_2 fragment (= "active glycolaldehyde") is cleaved from xylulose-5-phosphate and transferred to ribose-5-phosphate by the coenzyme thiamine pyrophosphate (balance: $C_5 + C_5 = C_7 + C_3$). We are familiar with this mode of action of the *transketolase* from the CO_2 reduction cycle in photosynthesis (p. 151 f). In photosynthesis, however, the equilibrium of the catalyzed reaction favors the formation of the pentose phosphates. A C_3 fragment, the dihydroxyacetone moiety, is transferred from sedoheptulose-7-phosphate to an aldose acceptor, 3-phosphoglyceraldehyde, catalyzed by the enzyme *transaldolase.* Erythrose-4-phosphate is left behind, fructose-6-phosphate is formed and converted by *phosphohexoisomerase* to glucose-6-phosphate (p. 165). This molecule may again be subjected to direct degradation (balance: $C_7 + C_3 = C_6 + C_4$):

In contrast to *transketolase, transaldolase* lacks a definite coenzyme. Its action is restricted to cleaving phosphorylated glucose and sedoheptulose by reversal of the aldol condensation, yielding a C_3 fragment, the dihydroxyacetone residue. This remains attached to the enzyme until a suitable acceptor, i.e. an aldose, is available: 3-phosphoglyceraldehyde, erythrose-4-phosphate and possibly also ribose-5-phosphate. Together with *transketolase, transaldolase* brings about the formation of new sugars by displacing or exchanging fragments of their carbon skeleton.

Fructose-6-phosphate is produced from erythrose-4-phosphate by attachment of a C_2 fragment which is donated from xylulose-5-phosphate with the mediation of *transketolase*. From the latter compound 3-phosphoglyceraldehyde originates (balance: $C_4 + C_5 = C_6 + C_3$). This is once again the reversal of a reaction involved in the photosynthetic CO_2 reduction cycle (p. 151 f).

The molecule of 3-phosphoglyceraldehyde left over may either enter glycolysis or be used to regenerate a molecule of fructose-6-phosphate or glucose-6-phosphate, respectively. In the latter case, it must combine with another molecule of triose phosphate via fructose biphosphate as intermediate; *triose phosphate isomerase, aldolase* and *phosphatase* are involved. Three molecules of pentose phosphate must be additionally provided to form this second molecule of triose phosphate via the pentose phosphate cycle. Apart from the C_3 compound, two molecules of fructose-6-phosphate are formed as end products. A total of five molecules of hexose phosphate are synthesized from six molecules of pentose phosphate. In other words, five of the six molecules of glucose-6-phosphate which enter the cycle are recovered, while one is completely degraded to CO_2 and NADP-bound hydrogen:

$$6 \text{ Glucose-6-}(P) + 12 \text{ NADP}^\oplus \longrightarrow 6 \text{ CO}_2 + 12 \text{ NADP-H+H}^\oplus + 6 \text{ Ribulose-5-}(P)$$

$$6 \text{ Ribulose-5-}(P) \longrightarrow 5 \text{ Glucose-6-}(P)
\begin{cases}
4 \text{ Ribulose-5-}(P) \longrightarrow 4 \text{ Xylulose-5-}(P) \\
2 \text{ Ribulose-5-}(P) \longrightarrow 2 \text{ Ribose-5-}(P) \\
4 \text{ Xylulose-5-}(P) + 2 \text{ Ribose-5-}(P) \\
\longrightarrow 5 \text{ Glucose-6-}(P)
\end{cases}$$

$$1 \text{ Glucose-6-}(P) + 12 \text{ NADP}^\oplus \longrightarrow 6 \text{ CO}_2 + 12 \text{ NADP-H+H}^\oplus$$

Six turns of the cycle are accordingly required to decompose a molecule of glucose to CO_2 by stepwise degradation of its carbon chain. The decrease in free energy resulting from the oxidation is not used to conserve chemical energy. The remaining reactions subsequently proceed at about the same energy level.

A conservation of energy during direct degradation of glucose via the pentose phosphate cycle is only possible if the organism or tissue concerned transfers the cleaved and coenzyme-bound hydrogen from NADP-H + H$^\oplus$ to NAD$^\oplus$. Only NAD-H + H$^\oplus$ is able to provide the respiratory chain with hydrogen and to bring about the formation of ATP by oxidative phosphorylation. Since such an enzymatic "transhydrogenation" has only been demonstrated in a few cases so far, energy conservation in the pentose phosphate cycle does not seem to be of much significance.

However, if NADP-H + H$^\oplus$ cannot be oxidized by the respiratory chain or with the help of NAD$^\oplus$, it is easy to see that the level or reduced coenzyme determines the activity of the whole cycle. The catalytic amounts of NADP$^\oplus$ are rather rapidly loaded with hydrogen so that the cycle only continues to turn when the reduced coenzyme is instantaneously re-oxidized and thus activated for uptake of further hydrogen. This is achieved first of all by synthesis reactions which accept hydrogen specifically from NADP-H + H$^\oplus$ (see below). With declining demand, i.e. low synthesis activity, the pentose phosphate cycle gradually comes to a halt (for regulation, see p. 438 f).

An additional oxidation of NADP-H + H$^\oplus$ may also result from the action of a specific *NADP-H-cytochrome c reductase* linked to cytochrome c.

Role of the Pentose Phosphate Cycle in Metabolism. What is the function of the pentose phosphate cycle in metabolism? The findings so far support the concept that it mainly provides reactive hydrogen in the form of NADP-H + H$^\oplus$ for various processes in the cell. A series of syntheses only occur in the presence of these "reduction equivalents", e.g. formation of fatty acids (p. 405 ff) and of mevalonic acid (p. 430 f), or conversion of pyruvic acid to oxaloacetic acid by *malic enzyme*. Moreover, the cycle provides important precursors for the synthesis of nucleotides and nucleic acids (p. 315 ff). When required, also intermediates such as 3-phosphoglyceraldehyde and fructose-6-phosphate enter the process of glycolysis and are thereby aerobically degraded. The cycle thus constitutes an important reservoir for building elements of metabolism which are utilized by quite different pathways of biosynthesis.

The existence and functional efficiency of the pentose phosphate cycle have been demonstrated in cells of both lower and higher plants. Certain problems arise in photosynthetically active cells due to the fact that *transketolase, phosphohexoisomerase* and *phosphopentose epimerase* are also active in the photosynthetic CO_2 reduction cycle. In young yeast cells too, glucose is partially degraded via the pentose phosphate cycle. In most fungi studied so far, its share is about 25%. In many species of bacteria it is the main pathway of glucose dissimilation.

The enzymes of the pentose phosphate cycle are located in the cytoplasm; it is unknown whether they occupy a compartment separated from that of the enzymes of glycolysis.

References

Beevers, H.: Respiratory metabolism in plants. Harper & Row, New York 1961

Caughey, W. S.: Porphyrin proteins and enzymes. Ann. Rev. Biochem. 36: 611, 1967

Chance, B., W. D. Bonner jr., B. T. Storey: Electron transport in respiration Ann. Rev. Plant Physiol. 19: 295, 1968

Green, D. E., I. Silman: Structure of the mitochondrial electron transfer chain. Ann. Rev. Plant Physiol. 18: 147, 1967

Kornberg, H. L.; Anaplerotische Sequenzen im mikrobiologischen Stoffwechsel. Angew. Chemie 77: 601, 1965

Kowallik, W.: Der Einfluß von Licht auf die Atmung von Chlorella bei gehemmter Photosynthese. Planta 86: 50, 1969

Lardy, H. A., S. M. Ferguson: Oxidative phosphorylation in mitochondria. Ann. Rev. Biochem. 38: 991, 1969

Lehninger, A. L.: The mitochondrion. Benjamin, New York 1964

Pullman, M. E., G. Schatz: Mitochondrial oxidations and energy coupling. Ann. Rev. Biochem. 36: 539, 1967

Racker, E.: The membrane of the mitochondrion. Sci. Amer. 218: 32, 1968

Schatz, G.: Wie entstehen Mitochondrien? Umschau 69: 11, 1969

Tolbert, N. E.: Microbodies – Peroxisomes and Glyoxisomes. Ann. Rev. Plant Physiol. 22: 45, 1971

Water and Ions in Metabolism

Water

We have already discussed water as a starting substance in photosynthesis and as an important reactant in metabolism. However, these functions in themselves do not explain why life processes only take place in the presence of water. Practically all metabolic reactions proceed in the aqueous phase. Water serves as indispensable solvent for most inorganic and organic compounds. It is also vital to the function of many proteins, particularly of enzymes. Organisms are accordingly characterized by a high water content; in higher plants it may amount to 90 % and more of the living matter.

In higher plants water also assumes the function of a transport medium outside the cells; it transports nutrient salts from the roots to the aerial organs (p. 276). It is hence understandable that an insufficient water supply impairs or blocks not only photosynthesis but also other elementary life processes. A well-adjusted water balance is therefore required in the individual cell as well as in the whole organism.

Because of the close functional relationship between water economy and ion economy in plant organisms, it seems reasonable to discuss these two factors together.

Properties of Water

The dependence of life processes on water is based on its unique properties. Water is an exceptional compound in physicochemical as well as in biological respects.

Like any other chemical compound it has a melting point and a freezing point. These two values serve as points of reference for our temperature scale. Water is liquid at the temperatures prevailing in large areas of the earth's surface. This liquid state meets the requirements of life processes to a far greater extent than the gaseous or solid states. The relatively high value of the "specific thermal capacity" of water (= the number of calories necessary to raise the temperature of 1 g of a substance by 1 °C) combined with the high water content of organisms (60–90 %) prevents noticeable changes in the temperature of the latter despite the production of heat by their metabolism. Organisms living in water are accordingly protected from extreme fluctuations in the temperature of their biotope. On the other hand, water is also a suitable agent for eliminating surplus heat: evaporation of 1 g of water disposes of more than 500 calories; most other fluids are only half as effective in this respect. Evaporation of water thus constitutes the most important mechanism for eliminating heat. The molar heat of evaporation of water is connected with the "surface tension": the molecules at the surface of a fluid are bound by intermolecular forces to neighboring molecules only on one side, in the interior, however, on all sides (see below). Displacement of molecules at the surface hence requires energy, since some intermolecular bonds must be broken up. The high surface tension of water is of vital significance in maintaining boundaries where aqueous phases border on lipids and proteins in cellular membranes. It also prevents water from being completely drained off from the spaces between soil particles; instead it forms a filmlike coating and can thus be taken up by the plant root (cf. p. 272). Finally, this "cohesion" of water is the basic prerequisite for its movement through the plant cormus, especially in the conducting elements of the xylem which are rather small in diameter (see p. 278 f).

Its high "permeativity" makes water a very suitable solvent for ionized substances. It is an ideal reaction and transport medium for many inorganic and organic compounds. On the other hand, the viscosity of water is relatively low, enabling it to pass rapidly through the conducting vessels. The increase in volume of water when it freezes (= decrease in density!) is also unusual. The maximal density is reached at 4 °C; below this temperature it decreases, i.e. the volume increases. Ice is only 9/10 as dense as water, a circumstance to which organisms living in water owe their survival in a hard winter: lakes freeze over from the surface!

Structure of the Water Molecule

Many of the properties which distinguish water from the analogous dihydrides derive from its molecular structure and intermolecular configuration. The two covalent O-H bonds form an angle of 104 ° in the molecule; each is up to 40 % ionic in character giving the H atoms a weak positive charge, the O atom a weak negative charge (Fig. 82 A). The water molecule is hence a "dipole". Hydrogen bonding is possible because of the two free electron pairs on oxygen, i.e. hydrogen atoms from neighboring water molecules can be attached via weak bonds (Fig. 82 B). This leads to linkage of numerous molecules into a three-dimensional network (= "cluster"). Its size depends on the temperature: about 91 water molecules participate at 0 °, only 8 at 40 °. This accumulation of hydrogen bonds causes the relatively high values of melting point and heat of evaporation as compared with other dihydrides (see above). It also brings about the surface tension described above. It is easy to understand that additional energy is required to release a water molecule bound in this way.

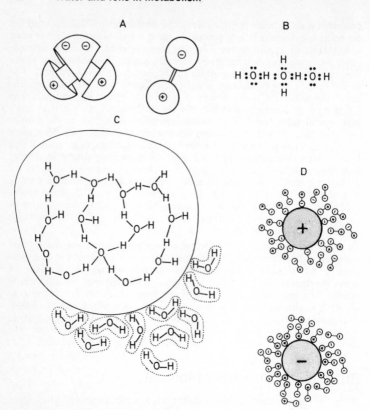

Fig. 82 Structure and properties of water. A, the water molecule as a dipole; B, formation of hydrogen bonds; C, three-dimensional structure ("cluster") formed by water molecules, surrounded by free molecules; D, hydratation envelope surrounding a cation or an anion.

Contrary to expectations, the viscosity is not unusually high despite the accumulation of hydrogen bonds. Two reasons may account for this: 1) the intermolecular hydrogen bonds break off and are formed again at a high rate, and 2) there is a sufficient number of free water molecules available to separate the complex structures from each other (Fig. 82 C) so that no coherent structure can be formed.

The dipole nature of water molecules also gives rise to their characteristic behavior towards ions in that they are attracted and attached to the latter. The charge of the ion determines whether the positive or the negative pole of the water dipole is turned towards the charged particle (Fig. 82 D). As a result of this "hydratation" the ions are surrounded by an envelope of water molecules. Its extent depends not only on the particle size, but also on the number of available charges at the surface. For example, the size of the hydratation envelope of alkali ions decreases markedly with increasing diameter of the ion, although the charge remains the same; this is because the charge density at the surface is decreased. Due to unequal charge distribution electrically neutral particles may exhibit dipole character and also acquire a hydratation envelope, e.g. proteins (p. 378) or membranes.

Dissociation of Water. The hydrogen covalently bound to an oxygen atom occasionally "jumps" to the neighboring oxygen atom as a hydrogen nucleus (= proton). During this "dissociation of water", $H_3O^\oplus$ ("hydroxonium ion") and $OH^\ominus$ ("hydroxyl ion") are formed. The reaction is reversible since the proton combines again with a hydroxyl ion to form water. The concentrations of $H_3O^\oplus$ and $OH^\ominus$ in water are both 10^{-7} molar at 25 °C. At these concentrations water exhibits a "neutral" reaction. It is generally expressed logarithmically; the neutral point has a pH value* of 7.0.

Protons are rapidly transported in water or aqueous solutions with the help of the hydroxonium ion, without actual movement of individual hydroxonium ions or water molecules. A hydroxonium ion transfers a proton onto another water molecule. If this process is repeated several times where water molecules are arranged in a chain and are linked by hydrogen bonds, a proton may travel very rapidly over a certain distance. This mechanism is no doubt very useful in the cell.

Whether a substance reacts as a base or as an acid also depends on the dissociation of water. Both change the concentrations of $H_3O^\oplus$ and $OH^\ominus$ by release of protons (= acid: increased formation of $H_3O^\oplus$) or by attachment of protons (= base: decrease in the amount of $H_3O^\oplus$). Acids are hence "proton donors" and bases "proton acceptors". Both of these processes are reversible; their equilibrium is reached from both sides (cf. p. 6 f).

Water in the Cell

Apart from its function as a solvent and its participation as a reactant in chemical reactions, water assumes additional functions in various parts of the cell. We have already discussed its ability of forming hydratation envelopes around ions (p. 264 f). In much the same way it occupies the surface of proteins and also of polysaccharides via their polar groups (-OH, -COOH, $-NH_2$). Layers of structural water are often formed around these molecules. Although only 5–10 % of the total amount of water in the cell are involved in hydratation, this is of vital importance for the function and

---

* "pH" from "pondus hydrogenii".

maintenance of cytoplasmic structures. Hydratation contributes significantly to "swelling" or "hydration", especially of the cytoplasm. This is a reversible uptake of solvent during which its molecules are deposited into an adequate substance causing an increase in volume. The cell wall is also involved since not only hydratation of polar groups on the polysaccharide molecules takes place but also uptake of water into the intermicellar and interfibrillar spaces. Fluctuations in the pH value and involvement of ions alter the charge conditions, and influence the formation of hydratation envelopes (see p. 378 f); therefore, the extent of hydration is directly dependent on these two factors. The temperature also determines the rate with which water molecules enter the "hydration body" by "diffusion". It is well known that in plant cells particularly $K^{\oplus}$ and $Ca^{2\oplus}$ play a role in this connection. Because of its higher charge $Ca^{2\oplus}$ is superior to $K^{\oplus}$ in discharging proteins and thus in suppressing hydration. Apparently the cytoplasm is also affected by the processes of hydration and dehydration, as indicated by the change of its state from fluid to more solid and vice versa (= change between gel and sol!). The dynamic forces which become apparent here indicate clearly that the activity of the protoplast is shaped not so much by the amount of water available but rather by its thermodynamic state. For this the term *hydratur* has been chosen (Walter). It describes the state of water in bodies, particularly in those of special significance for water economy in the cell: the cell sap as a solution, the cell wall and the cytoplasm as bodies fit for hydration, and the cell and the soil as heterogeneous systems. The *hydratur* is defined as the relative vapor pressure (in bar* or % relative humidity). It is highest when water is available in pure form; the more osmotically active substances are present in a solution, the lower is the *hydratur*. The *hydratur* of air is at its highest when the air is saturated with water vapor.

The introduction of the *water potential* has proved very valuable in the characterization of the state of water in cells or tissues on a thermodynamic basis. Taking a value of zero for pure unbound water under atmospheric pressure and at standard temperature, there is a lower or more precisely, a negative potential for water in solutions or bound to particles or structures; free energy is decreased. This is because the water is now less "available"; the resulting "osmotic potential" or "matrix potential" is generally lower – more negative – than the reference potential of pure, free water. Therefore, the state of the latter can only be reached by supplying energy (measurement is accordingly made in energy units per volume: $erg \times g^{-1}$ or $erg \times cm^{-3}$; often expressed in units of pressure: $10^6 erg \times cm^{-3} = 1$ bar; see also p. 270 f).

---

* 1 bar = 10^6 dyne $\times cm^{-2}$ = 0.987 atmosphere.

This "depression of water potential", ΔW or Ψ, is also convenient for describing the thermodynamic state in the cell provided that it is in equilibrium, and the water potential is the same throughout the cell: in the cytoplasm, in the vacuole and in the cell wall. In the vacuolized plant cell the *hydratur* of the cytoplasm is decisively influenced by the *hydratur* of the cell sap, i.e. by the osmotic value or "osmotic potential" (cf. p. 270) of the contents of the vacuole. From the latter the depression of water potential for the cytoplasm can be measured, at least approximately. This system of cytoplasm-vacuole (= "plasmic *hydratur*") must be distin-guished from that of the cell wall or cell surface (= "external *hydratur*"). Since the latter is practically free from dissolved substances the state of water is largely determined by the fine structure (see p. 266). A "matrix potential" brought about by hydrostatic forces in the capillary spaces is almost exclusively responsible for the depression of water potential in this region of the cell. However, there is also a matrix potential in the cytoplasm to be considered which results mainly from binding of water to colloids; its value is rather low. It is absent altogether in the vacuole.

State of Water in the Plant Cell

Assuming that life first developed in the sea, the "invention" of a membrane by which precursors of cells or organisms were separated from the external environment with its relatively high salt con-centration (~ 0.5 M) was certainly an important prerequisite. The cells or organisms were thereby enabled to create their own "inner environment" and an intricate chemical system of cellular metabo-lism. A barrier was erected against the influx of external sub-stances, which at the same time maintained a controlled and selec-tive uptake of vital compounds as well as permitting excretion of waste products. In a broad sense, this principle also applies to all the cellular organelles enclosed by membranes in more highly de-veloped cells. The difference in chemical composition between ex-ternal medium and internal environment of the living system, which is characteristic for all organisms, is ultimately based on the presence of a cytoplasmic membrane. Despite the fact that it only measures about 100 Å, it obviously meets the diverse demands (see above) in an optimal way. These qualifications of the membrane must be closely connected with the molecular structure.

In connection with the structure of **biological membranes,** the classical model of a "unit membrane" which does not provide for the active trans-port function has recently been replaced by models stressing the special function of a membrane. A good example is the thylakoid membrane which we have already discussed (p. 74 ff). Instead of regular layers of protein and lipid molecules ("protein-lipid-protein membrane") – a concept based on the largely uniform appearance of many biological membranes in electron micrographs – a more differentiated, sometimes even variable structure is visualized in such "functional membranes", with specific proteins and different molecules supplementing the typical components. Various models have accordingly been developed which are based in part on the classical unit membrane and in part on a more complex ultra-structure with dynamic properties, i.e. with a certain mobility of its components. Fig. 83 represents two current models.

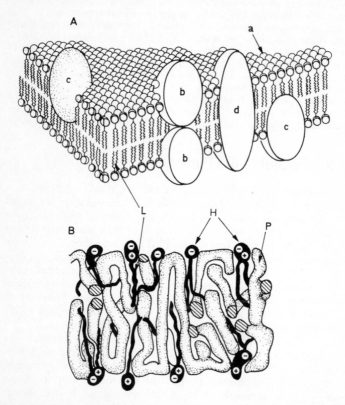

Fig. 83 Hypothetical models of the structure of a biomembrane. A: "Liquid mosaic" (after Capaldi); smaller proteins (a) are embedded in a bilayer lipid phase and cover the surface in a coherent layer; large proteins penetrate into the membrane either reaching the central lipid phase (b, c) or extending from one surface to the other (d); they may be involved in transport across the membrane ("tunnel proteins"). Both protein units are thought to be mobile within the lipid phase. B: Model after Benson; the proteins (P) extend through the membrane; the lipid molecules (L) embedded among the protein mole-cules reach the membrane surface with their polar "heads" (H). The lipid molecules are arranged in a bilayer with their hydrophobic "tails" meeting in the center in model A whereas their "heads" form the membrane surface, together with the small protein units (a).

Biomembranes appear where "compartments" (p. 5) must be separated from one another, be it the cytoplasm of the cell from its external environment, or a cellular organelle from the cytoplasm. Biomembranes accordingly assume multiple functions. As a physiological barrier they do not only prevent swamping by all sorts of substances, but also selectively enable certain substances to pass either by a "passive" or by an "active" mechanism, i.e. against a concentration gradient (cf. p. 271). The membranes of thylakoids and mitochondria belong to the type of biomembranes that carry enzymes and redox catalysts the arrangement of which enables a reaction sequence to proceed directionally across a membrane (p. 128 ff).

The separation barrier represented by a membrane creates special conditions for the passage of substances in aqueous solution. These conditions depend on the permeability of the membrane as well as on the *hydratur* or water potential of the cell (p. 266 f). We shall briefly consider the various possibilities.

Diffusion occurs if a permeable membrane separates two solutions of different concentrations: the membrane offers little resistance to the solvent and the solutes when they pass, equilibrating the concentrations. Diffusion is "restricted" if the passage of solutes and of solvent is limited by the size of the pores in the membrane and by the membrane thickness.

"Facilitated diffusion" occurs when individual substances are transported through a membrane selectively and at a relatively high rate, e.g. ions, while most of the other solutes diffuse very slowly. A specific *carrier* probably binds the molecule or ion, forming a carrier-substance or carrier-ion complex at the external surface of the membrane. The complex traverses the membrane thus bringing the molecule or ion to the inner side of the membrane. Here the complex breaks down, releasing the carried molrcule or ion into the inner compartment (details p. 271). This transport mechanism is also termed *catalytic permeation*.

Osmosis. A different process occurs when the membrane is semipermeable. In this case, practically only the solvent (as a rule water) and hardly any solutes (ions or molecules) penetrate through the membrane to the solution of higher concentration (= *osmosis*). The latter, since it exerts a "pull" on the water molecules on the other side of the membrane, is diluted and increases in volume; a high hydrostatic pressure develops within, the *osmotic pressure*. It can be determined with the help of an osmometer and depends mainly on the concentration of the dissolved particles. If the solutes are electrolytes, the number of ions formed in aqueous solution and the activity coefficients at moderate or weak dilutions must be taken into account in calculating the *osmolarity*, i.e. the osmotically active concentration.

Within certain limits the plant parenchyma cell constitutes an osmotic system since its cytoplasm is enclosed by semipermeable membranes and surrounds a large central vacuole (the elastic cell wall permits free entrance of water and ions!).The cell sap, containing a relatively high concentration of dissolved substances, exhibits a high osmotic value; the cytoplasm is altogether semipermeable. Therefore, water molecules tend to pass from the dilute solution in the cell wall into the vacuole, following the "diffusion pressure". The vacuole will take up water molecules until the elastic cell wall limits its expansion. The "pull" or "absorption force" (negative pressure) of the vacuole approaches zero, the resistance of the wall or *turgor pressure* corresponds to the osmotic pressure (or "osmotic value"). It is easily derived from the equation of the osmotic state $S = \pi^* - P$ (S = absorption force, π^* = osmotic value or "potential osmotic pressure", P = cell wall pressure). Conversely, if there is no cell pressure, i.e. $P = 0$, it follows that the driving potential or "pull" of the cell has the same value as the potential osmotic pressure: $S = \pi^*$.

Where the single cell is part of a tissue complex, a "tissue tension" comes into effect, i.e. the pressure of the cell wall is shaped by an "external pressure" ($= A$) of the neighboring cells which is either positive or negative: $S = \pi^* - (P \pm A)$.

The water potential or, more precisely, the depression of water potential (ΔW, see p. 266) may also be used to describe the osmotic state. The equation for the water potential should be formulated as follows: the potential of the water in the cell (Ψ_c) equals the osmotic potential (Ψ_π, negative value!) plus the pressure or turgor potential (Ψ_p, positive value!) extended by a matrix potential (Ψ_τ, negative value, see p. 266):

$$\Psi_c = \Psi_\pi + \Psi_p + \Psi_\tau$$

The more depressed the water potential of the cell, the higher is its absorption force or "pull". The corresponding diagram (Fig. 84) indicates that in a cell exerting full turgor pressure, i.e. in the state of water saturation, the water potential is zero, and that the absolute value

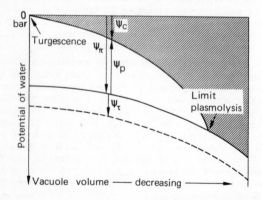

Fig. 84 Changes in the water potential of a vacuolized plant cell brought about by various conditions (see text). Ψ_c, potential of water in the cell; Ψ_p, turgor potential; Ψ_π, osmotic potential; Ψ_τ, matrix potential (after Larcher).

of the turgor potential is equal to the sum of the osmotic potential plus the matrix potential:

$$\Psi_\pi + \Psi_\tau = \Psi_p; \; (\Psi_c = 0)$$

Turgor potentials are normally in the range of 5–10 atmospheres, but there are vast differences among the various species; values of up to 100 atmospheres may occur.

If a single turgid cell is placed in a hypertonic external medium its water potential is subject to a typical change: the vacuole will become smaller and the pressure on the cell wall will accordingly decrease until $\Psi_p = 0$; now the protoplast shrinks away from the cell wall. In this state of "limit plasmolysis" the water potential of the cell is exclusively determined by the osmotic potential and the matrix potential:

$$\Psi_c = \Psi_\pi + \Psi_\tau.$$

Active Transport always occurs when uncharged molecules or ions are conveyed through a membrane not by diffusion or osmosis, but directionally ("vectorially") against a concentration gradient, with consumption of energy. This concept includes various kinds of uptake mechanisms for anumber of ions (see p. 283 f) and organic compounds (sugars, organic acids, amino acids). A common feature is the dependence on chemical energy, usually in the form of ATP. The involvement of $Na^\oplus$- and $K^\oplus$-dependent *ATPases* in the transport system is therefore also under discussion. This would involve *carriers* (see above), the chemical nature of which (protein or lipid?) is still a matter of much controversy. It is also uncertain whether carriers actually traverse the membrane or whether they are proteins embedded more or less firmly in the membrane structure. An argument against the occurrence of mobile carriers is the disproportion between the speed of transport and the relatively slow movement of macromolecules in a membrane. Therefore, more recent views stress the possibility of changes in the conformation of proteins which penetrate right through the membrane in the form of "tunnel" or "bridge" proteins and take over the presumed function as carriers of solutes from one side of the membrane to the other. The details of this process are still obscure.

Since carriers behave in many ways like proteins or enzymes, they are often – usually inappropriately – termed *permeases* or *translocases*. Valinomycin, a cyclic peptide (p. 376) serves as a convenient model substance transporting $K^\oplus$ ions across the inner mitochondrial membrane.

Absorption of Water

Generally, plant organisms use their whole surface for the absorption or uptake of water. In land plants, however, only the roots have constant access to the water available in the soil, representing a highly specialized and efficient organ of uptake.

Poikilohydric plants (bacteria, blue-green algae, lower green algae, fungi and lichens) are highly dependent on their environment for water uptake, i.e. on the water content of their substratum or of the air. Absorption takes place by hydration (p. 265 f), accompanied by release of heat. The amount of water gained by means of this mechanism is rather large and may considerably exceed the amount of dry matter. Insufficient water supply affects these organisms seriously since their cells lack a central vacuole which could be used for water storage. Therefore, various metabolic activities are stopped when the water supply is insufficient. However, irreversible interruptions of important functions do not result, since metabolism and growth proceed normally again once absorption of water has been resumed.

Homoiohydric plants are less dependent on external conditions in respect of their water supply, since their cells generally have a central vacuole. Its water reserve has a stabilizing effect on the *hydratur* of the cell. However, most homoiohydric plants do not survive desiccation. Consequently, at least all the highly developed species are equipped with devices at the surface (e.g. cuticle!) which minimize or actively regulate the inevitable transpiration (stomata!). Since at the same time a close functional connection exists with the roots, major changes in water supply can be endured without impairment of vital functions. On the other hand, the formation of a protective barrier against transpiration vastly reduces the absorption of water via the surface of the plant. This is now restricted to areas without cuticle (hydathodes, p. 280; sites of hair attachment) or to special differentiations (water-absorbing hairs in epiphytic *Bromeliads*). Thus, the main entry site for uptake of water into the plant is the root.

Water in the Soil. According to the nature of the soil the absorbed water will be distributed as "adhesive water" or as "sinking water". The latter normally reaches the ground water very rapidly; because of the relatively great depth only a few plants equipped with specialized root systems reach down that far. The adhesive water, on the other hand, constitutes the actual source of water for most plants. As indicated by the name it is held in capillaries between the soil particles ("capillary water"). The "water capacity" of a specific type of soil is determined by its composition and the size of pores present. These exert capillary forces up to a diameter of $60\ \mu$. A high content of organic substances and colloids has an enhancing effect. Apart from capillary binding, water is also attached to soil colloids by hydratation (p. 264 f) and, to a small extent, to ions. The latter indicate that salts are dissolved in the soil water (see p. 282 ff for absorption of these by plants).

Again the water potential (p. 266) serves as an appropriate measure for describing the state of water in the soil. Since it is partially bound by various forces and structures and since it contains ions, adhesive water is neither fully free nor pure in the soil. By analogy with the state of water in a plant cell (cf. p. 269 f) this leads to a more or less marked depression of water potential. Since the osmotic potential and also the hydrostatic pressure potential are in general relatively low and therefore neglectable, the "capillary potential" or matrix potential (p. 266) contributes decisively to binding of adhesive water, and thereby to the apparent absorption force or negative pressure of soil (negative values for ΔW!).

Absorption of Water by the Roots

As indicated above adhesive water is only absorbed by the root when its absorption force (p. 269 f) is greater than that of the soil. In fact, the water potential of the cells involved in absorption (see below) does have the required low (= negative!) values. They are in the range of a few atmospheres and only slightly higher than the driving potential of the soil but in most cases this suffices for an adequate absorption of water. Since the size of the absorbing surface is a decisive factor, the root system generally shows many subdivisions and branches. It often expands in the search for water; in zones where water is abundant parts often atrophy or die off. Roots may also become large and fleshy, and may contain a large quantity of stored water as in xerophytes. In hydrophytes roots atrophy or disappear as a result of adaptation to the biotope.

The composition of the soil determines via its content of adhesive water which plants will thrive or what adaptations will be necessary for survival. Soil with fine pores requires a significantly higher absorption force of the absorbing cells than sandy soil since the water is more firmly held by the former. Its potential often increases markedly when a relatively small amount of water is removed: this is replaced rather slowly due to limited flow through the narrow capillaries. The water columns may even break. The state of "permanent wilting" has proved to be a useful criterion. It is brought about in hydrophytes by a driving potential of the soil of 5–8 bar (p. 266), in agricultural crop plants between 10 and 20 bar, in plants adapted to dryness and in woody plants between 10 and 30 bar.

The actual sites of absorption are the root hairs, lateral protuberances of the root epidermis (= "rhizodermis") the walls of which lack thickening and covering with cutin. The root hairs appear only in a definite zone of the surface, a few cm behind the root tip which coincides with the metabolically most active part of the root tissue. Root hairs are short-lived, functioning only for a few days or weeks. Their structure enables them to penetrate into the spaces between the soil particles and to tap the water in the capillaries. The surface of other parts of the root is often covered with a layer of cork cells and becomes progressively impermeable for water and mineral salts. This impermeability is, however, by no means absolute.

Initial entry of water (and of ions dissolved in it, cf. p. 282) into the root may occur via two routes of radial transport across the cortex towards the stele: 1. Through the cell walls by diffusing freely into the intermicellar and interfibrillar spaces (cf. p. 266), using this coherent system to spread relatively fast to the endodermis. Here this *apoplastic transport* through the *outer free space* (see p. 282 f) or "free diffusion space" comes to an end: the flow is impeded by the Casparian strip, an impregnation of the radial and transverse walls with hydrophobic substances which makes them impermeable and prevents spreading of the water through the cell walls. It has to enter the cytoplasm and the vacuole in order to reach the stele, preferentially using the "transfer cells" whose water potential becomes the crucial factor. This kind of movement is typical for the second way of entry as well. 2. Absorption and conduction of water by osmotic movement across the plasmalemma and into the protoplasts and vacuoles of the cortex cells. This is referred to as *symplastic transport;* the "symplasma" or "symplast" (after Münch) is the functional continuum of the protoplasts of a tissue all interconnected by the plasmodesmata (Fig. 85). Water and solutes may move along this route to the stele provided an adequate gradient exists in the radial direction. In actual fact, the absorption force is highest in the innermost parts because the vessels in the vascular tissue remove relatively large amounts of water from these cells; their water potential is accordingly relatively low, i.e. highly negative. The absorption force gradually decreases outwards, and the values of the water potential become less negative. Accordingly, the water follows directionally along this gradient.

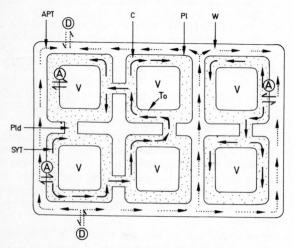

Fig. 85 Symplastic transport (SYT →) and apoplastic transport (APT – – →) in a plant tissue (schematic). A : Active processes involved in membrane transport of water or substances between the "outer free space" and the "symplast"; D : Uptake and release of substances from the "outer free space" by diffusion. C = cytoplasm; Pl = plasmalemma; Pld = plasmodesmata; To = tonoplast; V = vacuole; W = cell wall (after Lüttge).

The protoplasts may of course compensate for a water deficit from the surrounding water-filled cell walls, too, i.e. from the "external free space". This will depend mainly on their matrix potential (p. 266). The interconnection of symplastic and apoplastic water transport proves that it is not the single cell but the whole tissue complex – in this case the root cortex – that acts as a functional unit in water economy. The root cortex, in its turn, is subject to superior control by the whole plant. The cortex often assumes the function of a collecting system or reservoir which makes the plant more independent of sudden and short-term disturbances in the water flow. Symplastic transport is controlled by the metabolism of the cells; moreover, in many cases it proceeds at a speed higher than that of apoplastic transport. However, the significance of the latter should not be underestimated; this is evident in the transport of ions which will be discussed later (p. 282 f).

The movement of water from the endodermis, especially from the transfer cells, into the stele and thus into the long-distance system leading upwards has not yet been definitively elucidated. The rapid removal of water from the transfer cells cause a strong absorption force, i.e. a more negative water potential than that of the cortex cells; the resulting potential difference may well play an important role in the flow of water into the vascular tissue and the vessels. On the other hand, an "active", i.e. an energy-dependent component, cannot be excluded. It is demonstrated in the form of *root pressure,* causing exudation of sap from the cut surface in decapitated plants (main stem cut off above the ground). The root pressure which amounts to up to 6 atmospheres and can easily be demonstrated by fitting a manometer in the place of the main stem, supports the "pull" or absorption force (see below) active in water transport and brought about by transpiration. In fact, it replaces this potential when transpiration is low (during high atmospheric humidity!) or zero (leafless trees in spring!). Since the vessel sap exuded contains a relatively large number of ions, these could be the actual cause of the increased osmotic potential and the hydrostatic pressure resulting from it. It is still disputed whether these ions are secreted "actively", i.e. with consumption of energy, or "passively" from the root tissues into the vessels. The dependence of the root pressure on energy supply is emphasized by the finding that it can be abolished by specific inhibitors of respiration (p. 201) or by blocking the oxygen supply.

Limited movement of water, however, may occur through the phloem and even through cells outside the vascular tissue. There is also evidence that a rapid lateral (= radial) movement of water takes place via the vascular rays, i.e. towards the periphery of the stem.

Conduction of Water in Land Plants

The flow of water in the xylem of the vascular system which con-
ducts water absorbed by the roots to the leaves is distinguished as
"conduction of water" or "long-distance transport of water" from
the processes in the root cortex (= "displacement of water" or
"middle-distance transport" after Lüttge). The construction of the
conducting system permits a rapid and proportionate water supply
to all parts of the cormus. The speed of water conduction is generally between one and several meters per hour, in exceptional
cases more than a hundred meters per hour (lianas!). The total
amount of water which a plant absorbs and transports during its
life is very large; however, only a small fraction of it remains in the
cells or is actually utilized in metabolism. Most of it leaves the plant
as water vapor by transpiration (see below). This is because the
cormus of land plants with its intermediate water potential (cf. p.
278 f) extends with some of its parts into regions of varying *hydratur:* with the root into the water-soaked soil (= lower water potential!), with the shoot system into the air, with a generally low
water content (= high water potential!). The water balance is accordingly characterized by a continuous dissipation of water into
the surrounding atmosphere via the shoot system which must be
compensated for by absorption from the soil and upward transport against gravity. Apart from small diurnal fluctuations a dynamic equilibrium is usually established. With these considerations
we have already outlined the basic processes as well as the problems
underlying water economy in the plant.

Transpiration. In enlarging upon the above considerations we shall
confine ourselves to a brief review of "transpiration", the most
important mechanism of water release in higher plants (cf. also
p. 278 f). It consists of liberation of water vapor by the aerial parts
of the shoot, mainly through the stomata of the leaves; this may
account for up to 90 % of transpiration (= "stomatal transpiration"). The rest is liberated by the surface of the leaves and the
shoot. In the epidermal cells the rate of water loss is drastically cut
down due to the layer of cutin coating on their external walls;
however, diffusion cannot be completely prevented (= "cuticular
transpiration").

Transpiration in small plants and isolated parts of a shoot can be calculated from the loss of weight resulting from release of water vapor.
The amount of water simultaneously absorbed can be determined with
the potometer provided that it compensates exactly for the loss by
transpiration.

The intensity of transpiration depends on various factors such as light intensity, temperature, movement of air, humidity, but also on the structure and functional efficiency of the stomata. The highest values were found in plants living in water or on very wet soil. Among the land plants the herbaceous plants in sunny locations are the most active ones. In trees and shrubs the rate of transpiration is comparatively low; the rates for poplar and birch exceed those for oak and beech under similar conditions.

Only about 1–2 % of the leaf surface is exposed by normal opening of the stomata which occur mainly in the lower epidermis and are often confined to this surface (100–1000/mm²). Nevertheless, transpiration amounts to 50–70 % compared with evaporation from a water surface of the same size. This astonishingly high rate is explained by the "boundary effect": the diffusion field formed over each stomate is larger than over an equal area of an open water surface; consequently more water molecules pass per unit of time. It is known that stomatal transpiration is regulated by the guard cells, a process in which not only the water content of the air plays a role but also such factors as illumination, temperature, partial pressure of CO_2 and the *hydratur* of the plant. The stomata accordingly represent complex systems of regulation controlling several different processes. Of these, transpiration is governed above all by physical laws. It is closely correlated with photosynthesis since the traffic of CO_2 and O_2 proceeds mainly via the stomata and may temporarily become a limiting factor: closing of the stomata at noon in order to reduce transpiration necessarily blocks the uptake of CO_2; hence photosynthesis is reduced despite maximal illumination.

Mechanism of Water Transport

In the leaf a "middle-distance transport of water" (cf. p. 276) can be discerned similar to that in the root cortex: the mesophyll cells continuously lose water by transpiration to the intercellular spaces which form the means of communication with the atmosphere when the stomata are open. The factor responsible for this is the difference in water potential or the *hydratur* gradient between air (at 50 % relative humidity about 1000 atmospheres!) and the cells (10–100 atmospheres, cf. p. 276). The walls of the cells involved are affected first: their water potential will be markedly lowered leading to a change in the equilibrium of the cell potential (cf. p. 270 f); the latter tends to equilibrate again by mobilizing water from the cytoplasm and the vacuole. This affects mainly the osmotic potential with the result that the cell eventually has a strongly negative water potential. Now, because of its absorption force (negative pressure), it removes water from the adjacent cell in which the same sequence of events occurs. This cell consequently "pulls" water from the next one, and so on, until the connection with an element of the conducting system in the leaf has been formed. The same processes underlie cuticular transpiration: the external walls of the epidermal cells continuously lose water to the surrounding air despite the presence of a cuticle. This deficit is compensated for by "pulling" water from the cytoplasm and from the vacuole of the epidermal cell via the other cell walls. The water potential of the epidermal cell is accordingly depressed markedly; the resulting difference in potential leads to removal of water from the mesophyll cell bordering on the epidermal cell. The same gradient of water potential is formed as described above.

In this way water is continuously lost to the atmosphere, especially via the cell walls (= "apoplastically", p. 273 f). How much symplastic transport contributes to transpiration is as yet unknown; still, a "short-distance transport of water" by osmotic movement does take place. The overall "pull" or water-lifting power of transpiration exerts a suction force on the sap in the vessels which is drawn upwards. Here is the actual origin of the ascending movement of water in the vessels from the endodermis to the endings in the leaf.

This mechanism has an important consequence: in its long distance transport of water and ions the higher land plant largely makes use of the particular physical conditions of its environment, namely the *hydratur* gradient between soil and atmosphere to which it has successfully adapted. Its own energy budget is practically not drawn on by water transport except for the small amount of energy used to produce the root pressure (p. 275).

The velocity of the "transpiration flow" constituting the "long-distance transport of water" in the plant is influenced mainly by the *hydratur* gradient or the gradient of water potential from the roots to the leaves (Fig. 86). It also depends on the number and the diameter of the vessels or tracheids present, on the "conducting area", and on various forces offering resistance to conduction: gravity, filtration resistance, frictional resistance, etc. The velocity is determined by the intensity of transpiration provided water absorption through the roots is unimpeded. Fluctuations in the rate of transpiration affect the transpiration flow rapidly and vigorously. The flow may proceed at different speeds in various regions of the cormus according to the anatomical features.

Despite high filtration resistance in the narrow vessels the sap columns from the leaves to the endodermis do not break – even at maximal tension or "pull" by transpiration (about 40 atmospheres) and at extreme length (more than 100 meters in trees). This tensile strength is caused by the cohesion of water molecules which is typical for gas-free aqueous solutions, and by adhesion of the water column to the walls of the ves-

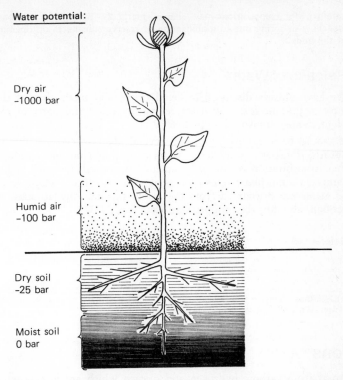

Fig. 86   Gradient of water potential (in bar) from the roots to the leaves (schematic; after Kausch).

sels. Due to their thickened walls, the latter are only slightly compressed even at extreme tension or "pull". Nevertheless, this slight reduction in diameter suffices to reduce the diameter of a tree trunk measurably in periods of intense transpiration. On the other hand, adhesion considerably increases the frictional resistance to the movement of the vessel sap.

The magnitude of the tensile strength or pressure of the sap can be experimentally determined in a pressure bomb using cut twigs of higher plants which are put under gas pressure. The water column in the vessels, having moved inwards after the cutting, is forced to its original position with increasing external pressure. The pressure which has to be applied in the bomb corresponds to the tensile strength of the sap. The values measured amount to about 30–50 atmospheres. This is sufficient to supply water to trees up to a height of about 140 meters; in fact, taller trees have not been found.

The sap usually contains small amounts (0.1–0.4 %) of inorganic and organic compounds, among them substances formed by the roots (alkaloids). Its composition may change markedly during the seasons, especially in spring due to mobilization of reserve substances on opening of the buds.

Release of Water

We have already discussed in detail origin and mechanism of the most important form of water release by the aerial parts of the plant, namely transpiration (p. 276 f).

It should be mentioned, too, that several plant species are able to absorb, transport, even excrete water as well as vital ions although their transpiration is low or completely depressed due to a highly saturated atmosphere. This process of *guttation* during which drops of water are discharged from the leaves through "hydathodes" is brought about by the root pressure (details p. 275).

References

Kuiper, P. J. C.: Water transport across membranes. Ann. Rev. Plant Physiol. 23: 157, 1972

Larcher, W.: Ökologie der Pflanzen. UTB Eugen Ulmer, Stuttgart 1973

Lüttge, U.: Stofftransport der Pflanzen. Springer, Berlin–Heidelberg–New York 1973

Philip, J. R.: Plant water relations: some .physical aspects. Ann. Rev. Plant Physiol. 17: 245, 1966

Ions

Apart from carbon, hydrogen and oxygen which are incorporated into organic compounds during photosynthesis the plant organism also requires other elements for its metabolism, particularly for the biosynthesis of own compounds: the non-metals nitrogen, sulfur and phosphorus, and the metals potassium, calcium, magnesium and iron. In addition, most plants need "trace elements" for normal development: manganese, zinc, copper, boron, molybdenum and cobalt. As indicated by their collective name, substantially smaller quantities of these elements are required compared with those of the former group. All these elements are regarded as essential since each is indispensable for normal plant development. In case of absence or insufficient supply, typical deficiency symptoms will arise (p. 289). These requirements of the plant are usually met by certain mineral salts ("nutrient salts") which are absorbed from solutions in the form of the cations $K^{\oplus}$, $Ca^{2\oplus}$, $Mg^{2\oplus}$ and $Fe^{2\oplus}$ (or $Fe^{3\oplus}$) and of the anions $NO_3^{\ominus}$, $SO_4^{2\ominus}$ and $PO_4^{3\ominus}$ (or $H_2PO_4^{\ominus}$). In algae and plants living in water the whole surface is involved in absorption of these ions, in land plants (cormophytes) only the root. Cormophytes may, however, absorb ions also through their leaves; in agriculture and horticulture "leaf fertilization" is sometimes practised.

For many plants the presence of all these essential elements in an appropriate ionic form in the soil is not sufficient. Their concentrations must be in the appropriate balance: in respect of the osmotic potential the total ionic strength must not exceed a certain value. Often the pH value significantly affects the absorption of ions (see below).

In order to ascertain the absolute requirement for certain elements in a plant, and the optimal quantities of the various types of ions, the roots of the plant are immersed in a nutrient solution containing several inorganic salts dissolved in water. This experimental device, developed by the great plant physiologist Julius Sachs, showed that a plant requires larger quantities of certain elements in an appropriate ionic form. These "macronutrients" or "macroelements" comprise N, S, P, K, Ca, Mg besides C, H and O; Fe is sometimes added to this group. When these were supplied to plants in sufficient quantities as highly purified salts dissolved in distilled water, deficiency symptoms (e.g. bleaching of chlorophyll, brown discoloration of the leaves) occurred nevertheless. These symptoms indicate that the plant needs still additional elements, however, in such small quantities that the inevitable impurities of the less pure salts originally used had been sufficient to meet the requirement for these "micronutrients" or "microelements": Mn, Zn, Cu, Mo, B, Cl and, in exceptional cases, Co. Accordingly, the first nutrient solution for higher plants developed by Knoop about 1880 (components: $Ca[NO_3]_2$, $MgSO_4$, KH_2PO_4, KNO_3, $FeSO_4$) had to be supplemented by the salts of the micronutrients ($MnCl_2$, $ZnSO_4$, $CuSO_4$, H_3BO_3, Na_2MoO_4, $CoCl_2$) because of the high purity of the chemicals today. Since this solution contains sufficient and well-balanced amounts of all the essential macro- and micronutrients, most plant species will grow as well as in soil when other conditions such as illumination and CO_2 supply are favorable. In such "solution culture" or "hydroculture" the nutrient solution in which the roots are immersed is usually aerated by bubbling air through it since the absorption of ions depends on the oxygen supply (see below).

Under natural conditions a plant must absorb the ions from the soil through the root. The composition of the soil determines the extent of plant growth. Its liquid phase, the soil solution, contains free ions in a form suitable for absorption. The solid phase with its inorganic and organic constituents serves mainly as a reservoir of nutrients which releases the ions by weathering or decomposition. As mentioned before, a good oxygen supply increases the absorption of ions by the root. This is ensured where the soil constitutes a crumbly substrate readily permeable to air due to the presence of abundant organic substances and numerous microorganisms. The leaching of certain ion species in humid climates caused by a deficiency in colloids is of great importance for the content of nutrient salts in the soil. The ions most affected are $Ca^{2\oplus}$, $Mg^{2\oplus}$ and $NO_3^{\ominus}$; $PO_4^{3\ominus}$ is the least affected. Both this phenomenon and the constant depletion resulting from plant growth cause certain fluctuations in the ion concentration of the soil solution, but the buffer capacity of the soil generally prevents one-sided accumulation of a single ion species which would have a toxic effect on the plants. Thus, hydrogen ions are exchanged with other cations adsorbed on the surface of soil colloids, preventing an excessive acidification of the soil which would impair the microflora, decrease the stability of the soil colloids and limit plant growth. Some plant species, however, tolerate acidic soil; lupines, with an optimal growth at a pH value of 5.0, belong to this group of "acidophilic plants". In contrast, the "acidophobic plants" like lucerne and barley prefer alkaline soils (pH values 6.5–8.0). Another group has an intermediate position, or is insensitive to changes in the pH value of the soil. It should be noted that in solution cultures a pH value between 4.5 and 6.0 is optimal for most plant species.

After repeated crops, soils suffer particularly from depletion of nutrient salts. This deficiency must be compensated for by means of fertilizers in order to prevent limitation of plant growth. "Mineral" fertilization directly replaces the nutrient salts removed from the soil by the crop. This practice was introduced by Justus von Liebig (1803–1873), the founder of modern agricultural chemistry. "Organic" fertilization with manure, compost etc. primarily increases the organic components of the soil thus supporting the growth of microorganisms; later on nutrient salts are released by decay (see above).

Absorption of Ions

When discussing absorption of water by the root (p. 273 ff) it was noted that also ions permeate the "outer free space" by free diffusion. Its volume amounts to about 0.10–0.15 ml per gram fresh weight; the plasmalemma represents the inner boundary of the "outer free space" since the latter is increased by plasmolysis of the cells involved. The plasmalemma as the limiting biomembrane (p. 267 f) is usually highly impermeable to ions and organic molecules.

Processes in the "outer free space" are additionally affected by anions which occupy the surface of the cell wall and the plasmalemma after ionization (= release of protons!) of the carboxyl groups of proteins and pectins and of the phosphate groups of phospholipids. These charges cause a preferential, but reversible adsorption of divalent cations which may, however, be exchanged in part with univalent cations.

This can be observed clearly when isolated tissue or an isolated root is immersed for several minutes in a solution of a salt, e.g. $CaCl_2$, then taken out and intensively rinsed with water: contrary to expectations, a certain amount of the salt is retained by the tissue and is only released upon subsequent immersion of the tissue in a solution of another salt, e.g. $MgSO_4$. The fraction retained inspite of rinsing consists of $Ca^{2\oplus}$, i.e. of cations, which were held by the negative sites of an exchange material in the tissue until displaced by cations of a different kind, in this case by $Mg^{2\oplus}$.

The space in which cations are held by negative sites or negative charges and are hence confined has been termed the "Donnan free space". If no distinction is made between freely diffusing ions in the "outer" space and those bound to negatively charged sites of the cell wall, the term "apparent free space" is used, in which all free and bound ions are contained. – This kind of "exchange absorption" is not directly related to the actual ion absorption into the cytoplasm via the plasmalemma.

Further movement of ions occurs within the *symplast* (p. 274); after initial uptake in the cells concerned the ions cross the plasmalemma into the cytoplasm. Once there, they move within the cytoplasm through plasmodesmata to the parenchyma cells of the stele. This "symplastic movement" leads directly to the conducting elements of the xylem, i.e. the ions leave the symplast by traversing the plasmalemma and penetrate into the "outer free space" of the cell walls before entering the vessels.

This "middle-distance transport" (p. 276) of ions occurring in the root cortex which precedes the "long-distance transport" is a rather controversial problem which is far from clear. The fact that certain cations and anions within the vacuoles of many plant cells reach concentrations far higher than their concentrations outside (particularly striking in aquatic plants!) indicates that selection, absorption and accumulation of ions is an "active" process in the sense that it does not occur spontaneously but against the existing diffusion gradient, with concomitant expenditure of energy. The terms "active absorption" and "active transport" are used to refer to this kind of movement dependent on metabolic activity and energy production of the cells (cf. p. 271). If the process of ion absorption by a tissue or an isolated root is followed as a function of time (for methods, see below) two phases are easily discernible: a short-term phase with a high absorption rate followed by a long-term phase with a lower but constant rate of ion absorption. During the first phase ions apparently penetrate into the "outer free space" by diffusion; during the second phase, however, "active", i.e. metabolically controlled absorption of ions into the protoplasts takes place; this is indicated by the finding that the process is dependent on temperature and supply of oxygen, and is sensitive to specific inhibitors, e.g. cyanide, 2,4-dinitrophenol (p. 243).

The selectivity of this process is more difficult to explain. The "carrier" hypothesis (p. 271) has been useful in accounting for this and other features of ion transport. According to this concept the ions move across the membrane bound to mobile organic molecules (= "carriers"). The carrier specifically binds an ion at the external surface of the membrane; the complex traverses the membrane and the ion is brought to the inner surface where it is released into the "inner space", i.e. the cytoplasm or a cell compartment. The carrier is then regenerated, probably by a change in molecular configuration which restores its capacity for ion transport. The whole process shows characteristics of a biochemical reaction.

Chemically very similar ions such as $K^\oplus$ and $Rb^\oplus$ or $Ca^{2\oplus}$ and $Sr^{2\oplus}$ (but not $K^\oplus$ and $Na^\oplus$ or $Ca^{2\oplus}$ and $Mg^{2\oplus}$) may compete for the "active site" or "binding site" of the carrier. Members of such ion pairs compete with each other in the process of absorption; both the ions obviously fit into a common binding site of the carrier. This "ion competition" is an important argument in favor of the carrier hypothesis. However, so far a carrier has not been directly demonstrated nor has its structure been elucidated (see also p. 271).

"Active" absorption of cations through biomembranes may cause anions to follow in order to equilibrate the electrical potential difference which has been created. This tendency of ions to restore electrical neutrality also applies when anions have been "actively" absorbed in excess. This shows that obviously only one type of ion has to be "actively" transported while the other one with opposite electrical charge will "passively" move through the membrane pores into the cytoplasm, following the electrical potential gradient. In fact, certain observations do indicate the existence of such a "passive" transport of cations; thus intensive absorption of $NO_3^\ominus$ is often accompanied by an increased uptake of cations. This correlation, however, does not exist universally.

Another hypothesis of ion absorption is based on the concept that very small hydratated ions may traverse the plasmalemma or other biomembranes without specific binding to a carrier. They use the water-filled pores in the membrane the diameter of which is about 10 Å. This ion movement is not brought about by energy-providing metabolic processes, but probably by chemical and electrical potential gradients. The negative charge of the cytoplasm and the vacuole would cause the cations to migrate into the interior of the cell. The resulting "passive" accumulation of ions ends when equilibrium is reached, which is predetermined by the potential involved. Additional absorption requires an "active" mechanism i.e. by means of carriers, with consumption of energy. This hypothesis cannot satisfactorily explain the selectivity of ion absorption; nevertheless, the concept of "passive" absorption ("influx") or release ("efflux") of ions in plant cells is no doubt justified.

Finally, we must bear in mind that biomembranes like the plasmalemma are highly but not completely impermeable to ions. This may be observed clearly when the external concentration is high compared to the internal, i.e. intracellular, relatively low concentration. A certain portion will inevitably penetrate into the cytoplasm – even if the ions concerned are useless ($Na^\oplus$, $Cl^\ominus$) or poisonous (heavy metal ions). The cell thus generally fails to exclude unwanted ions quantitatively.

So far we have not mentioned the ability of root cortex cells to accumulate ions in their vacuole which were "actively" absorbed previously. However, only a small portion of these is involved since the capacity of the vacuole is limited. Therefore, most of the ions "actively" absorbed will enter into the "symplastic transport" and eventually reach the stele.

Methods. For analyzing the movements of ions in cells and tissues of plants the employment of unstable isotopes like ^{35}S, ^{32}P, ^{36}Cl, ^{45}Ca, ^{86}Rb (cf. p. 142 f) has proved to be extremely useful. From appropriate compounds containing one of these isotopes solutions are prepared in which roots – isolated or as part of an intact plant – as well as tissues from various plant organs are immersed. By means of autoradiography (p. 144) absorption, movement or accumulation of various ion species can be studied in tissue slices subsequently prepared from the plant material.

Long-Distance Transport of Ions

The mechanism of transfer of ions into the conducting elements of the xylem, i.e. their delivery into the long-distance transport of root and stem, is still largely obscure. Since xylem exudates usually show a higher ion concentration than the milieu outside the root, an "active" transport seems to operate between the two.

As far as the location of this transport is concerned, several hypotheses exist. According to an early concept (after Crafts and Broyer) the active transport step is represented by the absorption of ions into the cytoplasm of the cortical cells which are assumed to have a higher salt concentration than the solution in the xylem vessels. Thus the movement of ions in the symplast would be passive, following a diffusion gradient towards the stele. Here ions leak out of the cytoplasm by traversing the plasmalemma of the stelar parenchyma cells and reach the lumen of the vessels via the "outer free space" of the cell walls and the vessels. The premise of this hypothesis that the plasmalemma of the stelar cells is "leaky" thus permitting ions to leave the cytoplasm and reach the vessels has recently been doubted. According to another concept the parenchyma cells of the stele are the terminals of the symplastic pathway which "actively" transfer ions into the extracellular space of the stelar cell walls and the xylem vessels. Having reached the latter, the ions are carried upwards by the mechanisms already described in the long-distance transport of water (p. 278): the "root pressure flow" and the "transpiration flow". The solvent as well as the ions and organic compounds ascend (= "solution flow" or "mass flow"; cf. p. 191 f).

By means of this long-distance transport the ions reach all regions of the aerial plant organs as well as regions in older parts of the root. On the other hand, cells close to xylem elements in stems, branches etc. may selectively withdraw ions or organic molecules from the vessel sap by an "active" or "carrier" mechanism. Thus they decrease the vessel sap and often profoundly change its ionic composition. The same effect is brought about by withdrawal or secretion of water, processes that also affect the flow in the vessels, independent of ion transport. As mentioned previously, the ions selectively withdrawn from the vessels by associated parenchyma cells may travel through the cytoplasm of the tissue cells, i.e. in the "symplast", reaching the receiving sites farthest away, the mesophyll cells of the leaves. Diffusion via the "outer free space" is also possible; in this case the protoplasts of the tissue cells are "bathed" by the ion solution from which essential ions can be absorbed by an "active" process.

A special feature of the vessels and tracheids consists in the fixed negative charges they bear along the inner surface of their cell walls so that a reversible exchange mainly of divalent cations from the sap takes place. Since these are alternately bound and displaced according to their concentration and affinity, an upward migration of cations along the surface occurs which may be compared to the events taking place during cation exchange chromatography (p. 380 f). Monovalent cations obviously exhibit much less affinity for the exchange sites. The composition and concentration of the ascending vessel sap, however, are not significantly changed by these ion exchange reactions because the ions absorbed displace others which are released in equivalent amounts into the solution.

It should be mentioned here that the xylem sap, apart from the ions $K^\oplus$, $Ca^{2\oplus}$, $Mg^{2\oplus}$, $PO_4^{3\ominus}$, $NO_3^\ominus$, and $SO_4^{2\ominus}$ also contains small amounts of amino acids and other nitrogen-containing compounds as well as organic acids and even sucrose; these compounds are formed in the root. Particularly in trees nitrogen is often transported in the form of amides and amino acids, to a lesser extent as $NO_3^\ominus$ or $NH_4^\oplus$. The amount of dissolved substances varies, depending on the physiological state of the plant. The pH value of the sap is generally in the range of 5.0–6.0.

Excretion of Salt. Plants in locations with a relatively high concentration of salt (coastline, saline desert; = "*halophytes*") have mechanisms at their disposal to excrete excessive or unwanted salts (see p. 284). These "salt glands" which are common in species of the families *Plumbaginaceae* and *Frankeniaceae*, but occur also in those of other genera (*Avicennia, Aegialitis, Spartina, Tamarix, Limonium*) often cover the surface of the leaves in vast numbers; accordingly a layer of salt crystals is observed. This excretion of ions is also an "active" and selective process which depends on the metabolic activity and energy production of the gland cells.

Another kind of ion release by the leaf surface is called "leaching"; this is important in the salt balance of plants exposed to continuous wetting by water, e.g. in the tropics or during prolonged spraying or sprinkler irrigation. In this case, the water is believed to be in communication with the solution of ions in the "outer free space" in which the protoplasts of the mesophyll cells of the leaf are usually bathed. Since the water is constantly removed and renewed the ions permeate the outer cell walls and are washed off from the leaf surface – in spite of the cuticle and waxy excretions of the leaf which are not entirely impermeable. The process of leaching is apparently a "passive" one brought about by diffusion.

Metabolism of the Anions

Of the anions which reach the cytoplasm, only phosphate is directly utilized in metabolism; nitrate and sulfate must first be reduced before the elements nitrogen and sulfur are available for biosynthesis of numerous cellular components such as proteins and nucleic acids. Although chemically quite different, nitrogen and sulfur have rather similar properties which are biologically important. The metabolic processes of these two elements are therefore discussed together.

Autotrophic organisms rely exclusively on the oxidized compounds of nitrate and sulfate to meet their requirements for nitrogen and sulfur. Autotrophy for nitrogen and sulfur is hence associated with carbon autotrophy. Apart from higher plants, ferns, mosses and algae, a number of bacteria and fungi also utilize nitrate (in some cases also ammonia or molecular nitrogen) and sulfate as sole source for their nitrogen and sulfur supply. Some species of bacteria, protozoa, mammals and also man utilize the two elements exclusively in the reduced form, i.e. as organic compounds. However, all stages of transition between strict autotrophy and strict heterotrophy for nitrogen and sulfur exist in nature. There is, moreover, a pronounced metabolic plasticity in certain types of cells or organisms insofar as they adapt their form of nitrogen and sulfur absorption to the conditions prevailing in the biotope.

A common feature of the two elements is that each participates in redox reactions resulting in a transfer of 8 electrons: the nitrogen in $NO_3^\ominus$ is thereby converted from the pentavalent positive form to the trivalent negative form of NH_3 ($N^{5\oplus} \rightarrow N^{3\ominus}$), the hexavalent positive sulfur in $SO_4^{2\ominus}$ to the divalent negative state ($S^{6\oplus} \rightarrow S^{2\ominus}$). With this reduction, i.e. the uptake of electrons provided by appropriate metabolic processes, the assimilation of the two elements commences. Formally, the two anions act as electron acceptors which are only incorporated into cellular compounds after reaching their lowest state of oxidation. We shall later go into the details of this process.

"Dissimilatory reduction" ("nitrate" and "sulfate dissimilation") is distinguished from "assimilatory reduction" in respect of the introduction of nitrate and sulfate into metabolism. In the former, the two salts merely serve as electron acceptors; the compounds formed, namely sulfide, NH_3, N_2 or N_2O, are released into the medium. "Assimilatory" reduction comprises reactions by which nitrogen and sulfur are incorporated into cellular organic compounds. In spite of apparent differences in the reaction mechanisms of dissimilatory and assimilatory reduction, the products of the former may also serve as precursors in biosyntheses while assimilatory reduction may also remove excess electrons if necessary.

Both nitrogen and sulfur are subject to a characteristic "biogeochemical" cycling, i.e. between living and non-living matter. The schemes below show details. Originating from salts both elements are incorporated into organic material; during its decomposition they are released in the form of free gases or bound in inorganic salts. In both forms they are again at the disposal of microorganisms and plants.

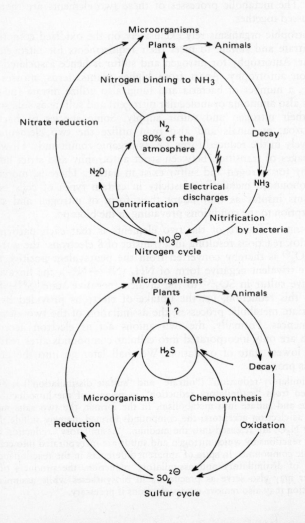

NO$_3^{\ominus}$

Nitrate is the most important source of nitrogen for autotrophic organisms; it is, however, present in relatively small amounts in the soil and in rivers, lakes etc. When the supply is inadequate (see below), lower and higher plants may change over to the nitrogen of ammonium (NH$_4^{\oplus}$) though absorption of this compound has the great disadvantage that the soil is thereby acidified: NH$_4^{\oplus}$, OH$^{\ominus}$ and SO$_4^{2\ominus}$ ions are absorbed from the mixture of ions resulting from NH$_4$ salts (NH$_4^{\oplus}$, Cl$^{\ominus}$, SO$_4^{2\ominus}$) and from water (H$^{\oplus}$, OH$^{\ominus}$), while Cl$^{\ominus}$ and, above all, H$^{\oplus}$ ions remain in the soil solution thus changing its pH value.

Several algal species may in exceptional cases utilize organic nitrogen compounds such as urea, amino acids or amides to meet their nitrogen requirements.

Depletion of nitrate or ammonium salts in the medium gives rise to deficiency symptoms in the plant organism. These affect various metabolic activities more or less seriously. Since the element nitrogen is involved in numerous reactions of the cellular metabolism, impairment or blocking of its supply affects primarily the enzymes, the formation or renewal of which consumes relatively large quantities of nitrogen. It is still an open question, however, whether all the physiological and morphological changes resulting from nitrogen deficiency can be explained in this way.

In most plants advanced nitrogen deficiency results in a degradation of chlorophyll (= "chlorosis"), of proteins and of certain components of RNA. On the other hand, the lipid content of the cells increases significantly, particularly in unicellular green algae (*Ankistrodesmus, Chlorella, Chlamydomonas*). This is often accompanied by formation of secondary carotinoids (see also p. 54); the cells are then orange-red in colour. The products of photosynthesis which is still active in the beginning, probably flow into carbohydrate metabolism at first before they are utilized exclusively in lipid synthesis. With nitrogen not available, amino acids and nucleotides cannot be formed as precursors of proteins and nucleic acids; thus the autotrophic cell regulates its substance production in this way.

Experimentally induced nitrogen deficiency has turned out to be a successful method for studying nitrogen absorption and utilization in green algae; when such deficient cells are transferred into normal nitrogen-containing culture medium and simultaneously illuminated, the deficiency symptoms rather rapidly disappear. During this phase photosynthesis and respiration (among other processes) change in a characteristic way due to increased absorption of nitrate, nitrite or NH$_4$ salts.

Nitrate Reduction. As mentioned previously nitrate must be reduced to NH$_3$ before its nitrogen can enter cellular metabolism. The overall process of this nitrate reduction is given by the following equation:

$$HNO_3 + 8\,[H] \longrightarrow NH_3 + 3\,H_2O$$

In the cell, however, the conversion of nitrate to ammonia proceeds by a sequence of several reactions each transferring two electrons ("2-electron transport"; cf. p. 22 f). A total of eight electrons are required:

$$\overset{(+5)}{NO_3^{\ominus}} \xrightarrow{+\,2e} \overset{(+3)}{NO_2^{\ominus}} \xrightarrow{+\,2e} \overset{(+1)}{X_1} \xrightarrow{+\,2e} \overset{(-1)}{X_2} \xrightarrow{+\,2e} \overset{(-3)}{NH_3}$$

The main donor for the electrons or the hydrogen required is either NADP-H $+$ H$^{\oplus}$ which is probably provided by photosynthesis in photoautotrophic organisms, in others by dissimilatory processes (pentose phosphate cycle, degradation of fatty acids), or NAD-H $+$ H$^{\oplus}$.

The conversion of nitrate to nitrite has been studied in detail; the identity of the end product with NH_3 has been established by several lines of evidence. On the other hand, the chemical nature of "X_1" ($=$ intermediate with valency $+ 1$) and "X_2" ($=$ intermediate with valency $- 1$) is unknown. The possibility of the conversion of nitrite to NH_3 proceeding without free intermediates is still under discussion (p. 291).

The demonstration of intermediates of nitrate reduction in green cells of higher plants and algae is hampered by the fact that they never accumulate in noticeable amounts. By an artifice, however, the involvement of nitrite as an intermediate can be demonstrated in cultures of the green alga *Ankistrodesmus:* on raising the H$^{\oplus}$ ion concentration of the culture medium (pH 3.0–3.5) in darkness the cells excrete relatively large amounts of nitrite into the medium. The nitrite is further metabolized by changing the pH value of the medium to 7.0 (Kessler). In *Scenedesmus* excretion of nitrite already starts below a pH value of 7.0. In several strains of *Chlorella* nitrite is mainly accumulated in the medium under anaerobic conditions, in darkness as well as in light. 2,4-Dinitrophenol, an uncoupling compound ("DNP"; p. 243), completely inhibits nitrate reduction at the stage of nitrite in *Ankistrodesmus* and in the blue-green alga *Anabaena cylindrica,* during illumination; the formation of nitrite from nitrate is not affected. Obviously, only the reduction of nitrite is closely associated with the photosynthetic electron transport and the ATP formation coupled to it.

It is interesting to note that nitrite accumulated by cells of the green alga *Ankistrodesmus* (see above) is very rapidly absorbed by these upon illumination. This "photochemical nitrite reduction" also takes place in an atmosphere of pure N$_2$; 1.5 moles of oxygen are formed per mole of nitrite reduced.

There is a corresponding reaction in *Anabaena cyclindrica,* spinach chloroplasts and wheat leaves. In all cases known, however, the nitrite-reducing system is saturated by relatively low light intensities.

The Biochemical Mechanism of Nitrate Reduction. The main component in the reaction system of nitrate reduction is *nitrate reductase,* an enzyme rather widely distributed. It was discovered as early as 1904 in potatoes.

The purified enzyme from higher plants, algae, fungi and bacteria contains molybdenum as metal component and probably also iron since in several preparations of the enzyme *(Chlorella, Neurospora)* a cytochrome was found. *Nitrate reductase* of *Chlorella* (molecular weight about 500 000) consists of a flavoprotein component *(NAD-H diaphorase)* and the true *nitrate reductase* moiety with active SH groups (= *molybdoflavoprotein).* The specific hydrogen donor is $NAD-H + H^{\oplus}$. In contrast, *nitrate reductase* of *Neurospora* has a molecular weight of about 228 000; it also contains flavin, molybdenum, cytochrome and SH groups. Hydrogen or electrons are supplied by $NADP-H + H^{\oplus}$, and flow via FAD, cytochrome and Mo to nitrate (p. 293).

The activity of *nitrate reductase* in vitro depends on its redox state; oxidation (by ferricyanide) usually favors activation, reduction (by $NAD-H + H^{\oplus}$; NH_3 plus illumination), however, inactivation.

Though many preparations of *nitrate reductase* from green plant organisms have turned out to be specific for $NAD-H + H^{\oplus}$, some have also been found which are active with $NADP-H + H^{\oplus}$. Similar preparations from fungi *(Aspergillus, Torulopsis),* however, were always NADP-H-dependent.

Another important component of the *nitrate reductase* reaction system is *nitrite reductase* which catalyzes the conversion of nitrite formed during the first reaction step. This enzyme which has been detected in numerous autotrophic and heterotrophic organisms, has a molecular weight of about 63 000 in the case of *Chlorella* and contains iron. In the presence of reduced ferredoxin it converts nitrate to NH_3, consuming a total of 6 electrons; free intermediates do not appear. In vitro ferredoxin-dependent *nitrite reductase* apparently functions without addition of ATP or another energy-rich compound; in barley leaves and in other in-vivo systems, however, compounds uncoupling phosphorylation inhibit the reduction of nitrite. In heterotrophic cells without ferredoxin *nitrite reductase* is believed to be present as a FAD-dependent metalloproteid (p. 382) depending on $NADP-H + H^{\oplus}$ or $NAD-H + H^{\oplus}$ as primary hydrogen donor.

As to the location of *nitrate reductase* and *nitrite reductase* in the cell, uncertainty prevails. In leaf cells *nitrate reductase* is assumed to be present only in the cytoplasm, the *nitrite reductase,* however, in the chloroplasts. Other findings indicate an attachment of both enzymes to the "microbodies" (peroxisomes; p. 227) of leaf cells. In roots *nitrite reductase* has been found bound to particles (proplastids, etioplasts?), whereas the *nitrate reductase* has been found as soluble enzyme in the cytoplasm.

Corresponding enzymes active in "dissimilatory nitrate reduction" ("nitrate dissimilation"; p. 287) have been isolated from various bacterial species. The "dissimilatory *nitrate reductase*" of *Micrococcus* is an iron-sulfur protein (cf. p. 237) with labile sulfide and molybdenum (molecular weight about 160 000). An enzyme of the "dissimilatory *nitrite reductases*" has been isolated from *Achromobacterium.* It catalyzes the reduction of nitrite to NO; its molecular weight was estimated to be about 69 000; the enzyme probably contains copper.

Nitrate reduction in *Neurospora crassa* has been studied by means of biochemical mutants. Several of these also excrete nitrite into the culture medium besides hydroxylamine. Other mutants completely lack the ability to reduce nitrate; they depend on nitrite as source of nitrogen. Similar observations have been made with mutants of *Aspergillus* and *Chlorella.*

Cell-free extracts of one mutant exhibited active NADP-H-dependent *nitrate reductase* activity when a molybdenum-containing moiety of another enzyme, e.g. *nitrogenase* (p. 296), had been added. These enzymes obviously contain a common, exchangable component independent of their origin; it is probably identical with a protein of low molecular weight containing molybdenum.

Even the formation of NH_3 can be demonstrated under modified culture conditions, e.g. in illuminated cells of *Chlorella* inadequately supplied with CO_2. They are obviously unable to incorporate the reduced nitrogen into organic compounds because the appropriate precursors are lacking. In leaves NH_3 has been demonstrated as end product of nitrate reduction by means of the nitrogen isotope ^{15}N (introduced as $^{15}NO_3^{\ominus}$).

In darkness nitrate reduction obviously proceeds with concomitant consumption of carbohydrates; this additional respiration can be detected experimentally due to the "extra CO_2". Two moles of CO_2 are formed per mole of NH_3 formed. This finding is based on studies with blue-green algae and unicellular green algae. Under anaerobic conditions (N_2 atmosphere) the process is usually inhibited.

Numerous bacterial species, as well as several green algae *(Ankistrodesmus, Scenedesmus,* some strains of *Chlorella)* are equipped with *hydrogenase* (p. 139) and are thus able to use molecular hydrogen for reducing nitrate, nitrite or hydroxylamine to NH_3 in darkness. The algal species mentioned preferentially convert nitrite and hydroxylamine in this way:

$$HNO_2 + 3 H_2 \longrightarrow NH_3 + 2 H_2O$$

The electron transport involved in nitrate reduction probably proceeds via the following steps:

Probably the valency of molybdenum changes reversibly between $5^{\oplus}$ and $6^{\oplus}$.

Apart from this electron transport a second one has been described which is perhaps characteristic for "dissimilatory nitrate reduction". Its key enzyme either contains cytochrome or is closely coupled to it ("cytochrome-coupled nitrate reduction"). Thus, the "dissimilatory *nitrate reductase*" (p. 292) of *Micrococcus* formed under anaerobic conditions is bound to particles which contain cytochrome. The latter is thought to play an important role in providing electrons.

In **induction** of the enzyme system active in nitrate reduction, NH_3 apparently plays a more important role than nitrate or nitrite. NH_3 is believed to act as *repressor* (p. 446) in the absence of which enzyme synthesis is initiated (= "de-repression" p. 447). The promoting action of light in this process as observed in several green plant organisms (*Chlorella, Chlamydomonas, Phaseolus*) has not yet been elucidated.

Nitrate is only one of the forms occurring in the cyclic conversion process of nitrogen in nature. Another form is molecular nitrogen which is released from inorganic or organic binding by chemical or biological processes. In this form it cannot be directly utilized by most organisms. The biologically necessary balance between free and fixed nitrogen is only maintained through the activities of a number of microorganisms which absorb N_2 from the atmosphere and convert it first of all to an inorganic compound. The nitrogen thus bound is now at the disposal of these organisms for the synthesis of cellular substances. The profound biological importance of this process of nitrogen fixation is emphasized by two figures: for the symbiotic form (*Leguminosae*) a yield of 100–200 kg N per hectare and year has been calculated; for N_2-binding organisms living free this value is believed to be about 5 kg N/hectare and year.

Binding of Free Nitrogen. Organisms which make use of free atmospheric nitrogen have so far only been found among prokaryotes. Here, free-living organisms are to be distinguished from symbionts. The first group comprises aerobic and anaerobic bacteria including phototrophic bacteria (*Rhodospirillum, Chromatium, Chlorobium* and *Rhodomicrobium*) and blue-green algae (*Nostoc, Tolypothrix,*

Anabaena). To the second group belong some species of blue-green algae as well as the "root nodule bacteria" *(Rhizobium)* and the *Actinomycetales* which live as symbionts in root cells of higher plants, particularly in those of the legumes, and give rise to formation of the typical root nodules.

Plants thus infected benefit from this process in that they utilize the products of bacterial nitrogen fixation. Until recently it was common belief that rhizobia are unable to fix nitrogen outside the host cells, but lately several investigators demonstrated *nitrogenase* activity in species of rhizobia grown in pure culture. The activity was particularly high in slow-growing "cowpea" strains (see also p. 300). Root nodule bacteria retain the capacity of nitrogen-fixation for a limited period in isolated root nodules as well.

Root Nodules. More refined experimentation has established that N_2-binding in root nodules is exclusively brought about by symbiotic bacteria present in their cells. Cell-free extracts prepared from these *bacteroids* (in contrast to free-living forms the cells of symbionts are enlarged, roundish and even branched) exhibit N_2-fixation and contain the key enzyme of this process, *nitrogenase* (see below). In vivo, the organic N-compounds formed by the bacteroids are delivered to the host cells, penetrate into the conducting system (see Fig. 87, b) and are thus at the plant's disposal. Obviously, the bacteroids do not have to be lyzed before setting free these compounds. In the long-distance transport of the latter (mainly amino acids and amides, e.g. asparagine, glutamine; p. 358) the xylem is probably involved, the elements of which are usually present in the nodules together with those of the phloem (Fig. 87, b).

An important prerequisite for nodule formation and establishment of a true symbiosis is the compatibility of the particular plant species and the bacterial strain; thus, several species of *Rhizobium* exist according to their host specifity. The process of infection is preceded by a congregation of bacteria around the root which probably results from excretion of certain organic compounds by the root. They cause the root hairs to curl in a characteristic fashion. The bacteria now invade preferentially young root hairs by means of enzymes degrading the cell walls; after strong proliferation they form threadlike aggregates surrounded by a membranous structure containing pectins and cellulose: the "infection thread". It is continuously enlarged and lengthened, making its way from the site of infection through the root hair into the cortical cells (Fig. 87, e): the cell walls are probably passed via the pits. When the infection thread reaches a tetraploid cell its wall is dissolved and the bacteria are released into this cortical cell. It begins to divide, with the adjacent diploid cells following suit. Branches of the infection thread meet other tetraploid cells in the neighborhood, inducing further divisions. With the mediation of phytohormones a local new meristem is thus established which gives rise to formation of a root nodule. During this phase the characteristic changes in the shape of the bacterial cells take place leading to "bacteroids" (cf. above; Fig. 87, d). They are obviously not in direct contact

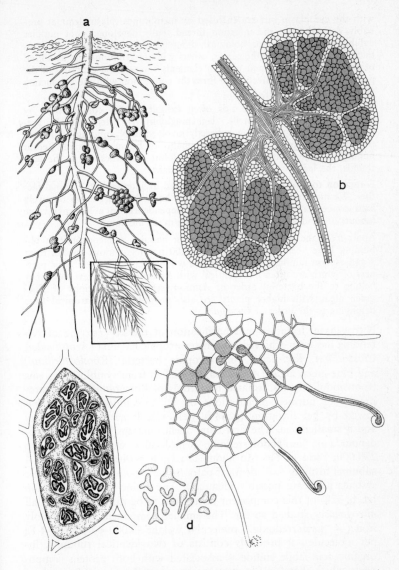

Fig. 87 Symbiotic N₂-binding in root nodules of the legumes. a: pea root with nodules; b: section through a fully developed nodule; c: section through a cell invaded by *Rhizobium*; d: morphology of the "bacteroids" in a nodule cell; e: penetration of bacteria by means of an "infection thread" which traverses the root cortex. (Schematic; by courtesy of H. G. Schlegel).

with the cytoplasm but are enclosed by membranes (plasmalemma) probably derived from the infection thread. Fully developed root nodules exhibit a surprising morphological differentiation: while the cells containing bacteroids form the center, others without symbionts act as a kind of endodermis; xylem and phloem elements traverse the central and the peripheral part, respectively, and form the connection with the conducting system of the root (Fig. 87, b).

The cells containing bacteroids often exhibit a pink color caused by enrichment of a hemoproteid, "leghemoglobin" (= "legoglobin"). Presumably, this compound is not directly involved in N_2-binding; its role in the binding or transfer of oxygen remains to be substantiated. Recent findings indicate that although leghemoglobin is synthesized in the bacteroids, the legume is the genetic determinant.

Formation root of nodules is also influenced by internal and external factors. We may mention some of the latter here: dependence on relatively high concentrations of $Ca^{2\oplus}$ and Co, dependence on a low pH value (about 5.0), and inhibitory effect of nitrate in the medium.

Apart from the findings in legumes, root nodules and concomitant N_2-binding have been observed in about 100 non-leguminous species from 7 angiosperm families. All of these are woody plants, the symbiotic partners of which are often not identical with *Rhizobium;* in a few cases they belong to the bacterial order of *Actinomycetales*. Associations of blue-green algae with higher plants are also possible, e.g. in tropical trees belonging to the order of *Cycadales*.

Nitrogenase. Isolation and purification of the key enzyme of N_2-binding has been achieved from heterotrophic bacteria *(Azotobacter, Clostridium, Klebsiella),* phototrophic bacteria *(Rhodospirillum)* and blue-green algae *(Anabaena)* as well as from symbiotic systems (root nodule tissues, isolated bacteroids of *Rhizobium).* Chromatography of purified preparations on DEAE-cellulose (p. 381) followed by gel filtration (p. 48 f) yielded a larger protein moiety and a smaller one. The first component contains iron and molybdenum (= "molybdoferredoxin"; molecular weight 200 000 to 270 000; ratio of Fe: Mo about 17 : 1); it probably consists of 4 subunits forming two identical pairs (in *nitrogenase* from symbiotic systems all 4 are possibly identical). The amino acid composition (cf. p. 378) of this component turned out to be rather similar in all *nitrogenases* studied so far. The smaller component contains only iron (= "azoferredoxin"; molecular weight 40 000–67 000; 2–4 Fe per molecule); it probably consists of two identical subunits. Inorganic acid-labile sulfide is associated with both protein components which therefore do not correspond in their structure to bacterial ferredoxins but rather resemble the flavodoxins, i.e. "iron-sulfur proteins" (= "nonheme-iron proteins", p. 237) occurring in bacteria and algae. The ratio between the large component and the smaller one required for optimal activity of isolated *nitrogenase* from *Clostridium pasteurianum* has been estimated to be 1 : 2.

All preparations of *nitrogenase* studied so far require an electron donor, ATP and $Mg^{2\oplus}$ for the reduction of N_2 or another appropriate substrate like acetylene (C_2H_2) under anaerobic conditions. The latter compound has turned out to be a very suitable substrate for measuring *nitrogenase* activity in vitro and in vivo. In the absence of reducible substrate $H^{\oplus}$ ions may function as electron acceptors: molecular hydrogen is released. In contrast to ATP which cannot be replaced by any other nucleoside triphosphate, $Mn^{2\oplus}$, $Co^{2\oplus}$, $Fe^{2\oplus}$ or $Ni^{2\oplus}$ may well compensate for $Mg^{2\oplus}$ though resulting in a decrease in activity of the *nitrogenase* system.

Nitrogenase preparations are more or less inactivated in the presence of oxygen even if they originate from aerobic organisms, e.g. *Azotobacter vinelandii*. The smaller enzyme component appears to be particularly sensitive in this respect. Therefore, activity measurements must be carried out under anaerobic conditions (see above).

According to a recent model the smaller enzyme component ("azoferredoxin") is reduced first and forms a complex with ATP and $Mg^{2\oplus}$ from which electrons then flow via the larger enzyme component to N_2. For a better understanding of the reaction catalyzed by *nitrogenase* chemical models have been developed which reduce N_2 and other appropriate compounds (C_2H_2, cyanide, N_2O, nitriles isonitriles) with dithionite or borhydride as hydrogen donor. Systems with Mo bound to cysteine, glutathione (p. 375), thioglycerol or aminoethanthiol were rather efficient in reducing N_2 or any of the substrates mentioned above when supplied with ATP, $Fe^{2\oplus}$ and $Mg^{2\oplus}$.

N_2-Binding in Vivo. Little is known about the origin of the hydrogen or electrons utilized in N_2-binding in vivo. In aerobic N_2-binding organisms and in bacteroids NADP-H + $H^{\oplus}$ probably serves as the primary hydrogen donor while ferredoxins or similar compounds like flavodoxins (p. 305; "azotoflavin" in *Azotobacter*) hold a key position in the electron flow to *nitrogenase*. Oxidative phosphorylation or substrate chain phosphorylation (p. 211), as well as photophosphorylation in autotrophic species, would provide ATP. For blue-green algae a direct photoreduction of ferredoxin by *photosystem I* (p. 106 ff) with a subsequent electron transfer to *nitrogenase* has been proposed (see below). Anaerobic organisms like *Clostridium pasteurianum* obviously rely on the metabolism of pyruvic acid which supplies ATP as well as hydrogen or electrons (p. 298). Acetyl phosphate provides the energy which is used to phosphorylate ADP to ATP.

The function of *hydrogenase* (p. 136 f) often present in free-living and symbiotic N_2-binding organisms *(Rhizobium)* is evidently restricted to the removal of H_2 formed by *nitrogenase* in the absence of N_2 or other reducible substrates. A reaction of the released H_2 with oxygen giving rise to the formation of ATP might well occur.

The first stable product of N_2-binding and hence key intermediate in the biosynthesis of cellular nitrogen compounds is NH_3. Evidence for this conclusion has come mainly from experiments where the incorporation of ^{15}N into intact cells, root nodules and cell-free *nitrogenase* systems was observed: ^{15}N appeared after short-term incubation mainly in accumulated ammonia – provided the formation of α-ketoacids, the specific NH_3 acceptors (= conversion to amino acids and amides! cf. p. 362 f), had been experimentally blocked.

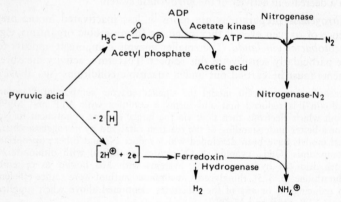

Little is known in detail about the intermediates and reactions involved in conversion of N_2 to NH_3. Since six electrons are required for the conversion of one molecule of nitrogen ($N_2 \rightarrow 2\ NH_3$) the reaction must proceed via several steps. Hydroxylamine and hydrazine, formerly regarded as probable intermediates, certainly have no physiological significance in nitrogen fixation.

An acceptable working hypothesis is based on the concept that N_2 initially forms a complex with the large component ("molybdoferredoxin"; p. 296) of *nitrogenase;* a stepwise reduction without cleavage of the bonds between the two nitrogen atoms is thus possible. Finally, enzyme and NH_3 are released.

The inhibitory effect of oxygen on N_2-binding even in aerobic organisms has long been known. Accordingly, the organisms concerned must possess natural protecting mechanisms for their *nitrogenase* system; this holds particularly for blue-green algae because of their photosynthetic O_2 production. One possible mechanism would be an increase in respiration, consuming surplus O_2 before it reaches the *nitrogenase (Azotobacter!);* N_2-binding is indeed often accompanied by an increase in O_2 uptake.

Another type of protective mechanism appears to operate in those species of blue-green algae which form *heterocysts:* these are supposed to lack *photosystem II* (p. 114 ff) and hence O_2 evolution in light; as a consequence access of oxygen to the N_2-binding system is prevented. This does not exclude the possibility that the undifferentiated cells in heterocystous species also contain *nitrogenase;* its activity, however, would be inhibited in light due to O_2 evolution of *photosystem II* and would only be restored in darkness at low pO_2 or under anaerobic conditions.

NH_3 apparently plays an important role in regulating N_2-binding since its presence prevents the synthesis of *nitrogenase* ("repression"; p. 446); on the other hand, synthesis of *nitrogenase* is initiated when the supply of NH_3 is exhausted ("de-repression"; p. 447). It should be noted that the two protein components of *nitrogenase* (p. 296) react in conformity in degradation and synthesis of the enzyme. In *Clostridium* carbamylphosphate probably exerts a regulating influence on *nitrogenase* activity and hence on N_2-binding.

The Agricultural Significance of Nitrogen Binding.

According to recent estimates biological nitrogen binding amounts to about two-thirds of all fixed nitrogen, yielding 175×10^6 metric tons per annum. Since the amount of fixed nitrogen in nature is limited a serious restriction is imposed on the capacity of world agriculture. Therefore, increasing amounts of supplementary nitrogen have to be provided in the form of chemical fertilizers containing nitrogen fixed by energy-consuming industrial methods, e.g. by the Haber process. Since 1950 a tenfold increase in the use of these fertilizers has been registered. About 44×10^6 metric tons per annum of fixed nitrogen now come from this source. There are, however, considerable differences in the relative contributions of biologically and chemically fixed nitrogen to agriculture in the various parts of the world: Australia is leading with 99 % biologically fixed nitrogen for plant production; in the United States 68 %, in India 57 % and in Britain 40 % of nitrogen are fixed biologically. With world population increasing and agricultural techniques improving, especially in the underdeveloped regions of the world, the demand for chemically fixed nitrogen will continue to increase. On the other hand, for chemical fixation of nitrogen an irreducible quantity of energy must be supplied to make the nearly inert diatomic gas (N_2) combine with other compounds. Since the "energy crisis", costs have escalated thus affecting the production and distribution of nitrogenous fertilizer and leading to a crisis in world agriculture. Therefore, there is an obvious need to find new ways of supplementing or improving present methods of nitrogen fixation. As far as this process as a chemical reaction is concerned a substantial reduction in cost is likely to result from advances in the techniques by which the raw materials and energy are supplied. Recent research in these processes seems to open up new possibilities which we shall discuss briefly.

Among the prokaryotic organisms which carry out biological nitrogen binding the blue-green algae (p. 293 f) have turned out to be agriculturally important too. They live on the surface of the soil and tolerate wide ranges of temperature and oxygen supply. Recent studies have revealed that they are very useful in rice paddies since they can fix up to 20 kg nitrogen per hectare and year. Moreover, blue-green algae in combination with various higher organisms form very efficient nitrogen-fixing systems (cf. p. 294), e.g. *Anabaena* with the floating fern *Azolla,* or *Nostoc* with liverworts or the angiosperm *Gunnera*. In subarctic regions symbiotic associations between blue-green algae and fungi give rise to a significantly large group of nitrogen-binding lichens. The selection of free-living, nitrogen-binding species of blue-green algae and breeding of other successful symbiotic associations between nitrogen-binding blue-green algae and agriculturally important crops may help to increase plant production at low costs.

As discussed previously root nodule symbiosis is obviously restricted to a limited number of plant families generally considered as primitive (p. 296); the reasons for this symbiosis are far from clear. The question has been raised whether these families developed some special qualities, or whether members of advanced families lost some attributes essential for the establishment of symbiosis. Nevertheless, artificially induced associations of *Rhizobium* ("cowpea strains") with non-leguminous plant cell cultures *(Bromus, Brassica, Pisum, Nicotiana, Vigna)* leading to nitrogen fixation and *nitrogenase* activity in the non-host plant cells have been successfully performed. These results together with other evidence (see below) may indicate that the genetic information for *nitrogenase* is encoded in *Rhizobium* and that its expression apparently depends on a diffusible factor excreted by the plant cells. These studies might be an important step towards the establishment of nitrogen-binding systems consisting of rhizobia and agriculturally important crops and forage plants. On the other hand, these attempts to develop new nodular symbioses made up of nitrogen-binding bacteria and plants originally lacking them, raise the significant question as to the effects these artificial developments may have on the terrestrial nitrogen cycle (p. 288). One has to consider the restriction of root nodule symbiosis to a few plant families during evolution and the reasons for it. Bearing in mind possible disastrous consequences it seems more realistic to make best possible use of the natural nitrogen-fixing systems already existing, namely nodulated legumes and rhizosphere associations, by improving the techniques of inoculation and by breeding of desirable genetical characteristics in both legume hosts and rhizobial strains for maximal fixation of nitrogen.

Valuable knowledge has been obtained recently by studies of the genetics of *nitrogenase* synthesis in free-living bacteria (p. 296). *Nitrogenase* genes (collectively referred to as "nif") were successfully transferred from *Klebsiella pneumoniae* into a mutant strain of this organism lacking the capacity for nitrogen binding, as well as into non-nitrogen-binding *E. coli,* thus establishing the capacity of nitrogen binding in these cells. This manipulation of the *"nif"* genes undoubtedly shows up the interesting possibility of transmission of these genes into eukaryotic plant cells as well. However, apart from the problems arising regarding incorporation and protection of the genes, the main obstacle in such a transfer is the establishment of an adequate physiological and genetic environment within these cells: protection of the *nitrogenase* from cellular oxygen (cf. p. 298 f), supply with ATP, establishment of electron carriers, organization of a regulatory system. Successful genetic transfer of *nitrogenase* to higher plants remains unrealistic as long as we are unable to explain why eukaryotes have resisted the acquisition of *nitrogenase* despite permanent contact of their genoms with those of rhizobia, blue-green algae and nitrogen-fixing bacteria during millions of years.

$SO_4^{2\ominus}$

Sulfur is absorbed and transported in the form of sulfate in most plant organisms. Under conditions of deficiency, however, several organic sulfur compounds such as cysteine, cystine or glutathione (p. 375) may assume this function.

Two pathways are open to sulfate absorbed by the cells: 1) direct incorporation into cellular compounds after an initial "activation", and 2) "assimilatory reduction" or "dissimilatory reduction" (= "sulfate dissimilation"; p. 287). The products of the first reaction sequence are sulfuric acid esters of phenols, steroids, polysaccharides (agar-agar, carragon in algae!), choline and others. Probably the sulfonic acids (direct bondage of S to C!) which are constituents of sulfolipids (in thylakoid membranes) and of taurines (in red algae) are also formed in this way. The enzyme system involved, *sulfotransferase,* has been detected in animals as well as in several plant organisms (mangrove, red algae, fungi); 3-phosphoryl-5'-adenosinephosphorylsulfate ("PAPS" or "active sulfate") serves as substrate.

With consumption of ATP and mediation of a *sulfurylase (ATP-sulfate adenyl transferase)* the anhydride compound from adenylic acid and sulfuric acid, adenosine-5'-phosphoryl sulfate ("APS"; structure p. 195) is initially formed in an endergonic reaction.

The reaction mechanism has been studied in bacteria, yeast, higher plants and animal organisms. Active *sulfurylase* was found in extracts of bacteria, algae, chloroplasts and animal tissues.

"Active Sulfate"

Since the equilibrium of the strongly endergonic reaction ($\Delta G^\circ = +11\,000$ cal/mole) favors ATP and sulfate, it is coupled to two other reactions in order to increase the yield of APS. In the first one, the pyrophosphate formed is cleaved by a *pyrophosphatase* thus removing a reaction end product from the reaction mixture and shifting the equilibrium to the right, i.e. in favor of the formation of APS. The reaction also partially compensates for the negative energy balance ($\Delta G^\circ = -6000$ cal/mole). Full adjustment, however, is only attained by the second coupled reaction in which an *APS kinase* (see below) transfers a phosphoric acid residue from ATP to position 3' of the ribose in APS in an exergonic reaction ($\Delta G^\circ = -5000$ cal/mole). The sum of all changes in free energy in the reactions involved is hence zero or slightly negative. The end product PAPS (see above) or "active sulfate" is formed with a good yield. The enzyme involved is termed *adenosine phosphorylsulfate kinase (ATP: adenylsulfate-3'-phosphotransferase)* "Active sulfate" has been demonstrated in algae and isolated chloroplasts, *APS kinase* in chloroplast preparations and several bacterial species.

The second mechanism supplying sulfate to the cellular metabolism (cf. p. 301) is far from clear. **"Assimilatory sulfate reduction"** in *Chlorella* appears to rely exclusively on APS as substrate which is probably formed from PAPS by a specific dephosphorylation step catalyzed by *3'-nucleotidase* (see reaction scheme below). Since the formation of PAPS is favored, particularly in the presence of pyrophosphate and ATP (p. 302) this compound might serve as a reservoir of sulfate in the cell, providing APS via the *nucleotidase* reaction whenever it is required. This concept is substantiated by the fact that PAPS is also involved in the transfer of sulfate into

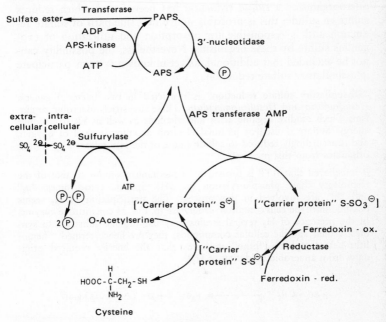

Probable reaction sequence during assimilatory sulfate reduction (after SCHIFF and HODSON)

an ester bond; an efficient regulatory control of both sulfate-consuming processes could be exerted via the two "pools" of APS and PAPS.

A second enzyme system *(APS reductase complex)* catalyzes the subsequent conversion of sulfate to sulfite, thiosulfate and sulfide, respectively. It has been detected in numerous microorganisms and algae *(Chlorella, Chondrus)* as well as in spinach chloroplasts. In *Chlorella* the system consists of a *transferase,* a "carrier" protein of low molecular weight with SH groups, and of a *reductase* (see reaction scheme on p. 303). The *transferase* brings about the transfer of $SO_4^{2\ominus}$ from ASP to the carrier protein; the complex formed is subject to the action of the *reductase* which, in the presence of electron donors (reduced ferredoxin), converts acetylserine to cysteine.

As in the case of nitrate reduction, the system of sulfate reduction seems to proceed without release of intermediary products (sulfite, thiosulfate, sulfide). Findings to the contrary effect are presumably unspecific and result from nonphysiological conditions. In some microorganisms a *sulfite reductase* has been found which reduces sulfite to sulfide; this is probably active in a side-path of sulfate reduction and is responsible for absorption and utilization of exogenous sulfite by these organisms. Nevertheless, the possibility cannot be excluded that additional, as yet unknown systems participate in assimilatory sulfate reduction.

"Dissimilatory sulfate reduction" is restricted to two bacterial genera: *Desulfovibrio* and *Desulfotomaculum.* These are strict anaerobic organisms which oxidize organic acids and alcohols as well as H_2 to produce energy. Sulfate is used as terminal electron acceptor instead of oxygen, and is accordingly reduced to sulfide (most of the H_2S released by sewage originates from this process!).

It is believed that ATP is formed by a mechanism similar to that of the respiratory chain phosphorylation (p. 241). Specific cytochromes, "c_3" and "b", are involved in this process. The participation of H_2 seems possible since the cells contain *hydrogenase* (p. 136) as constitutive enzyme. In the presence of H_2 several strains may use organic substrate to synthesize their specific cellular compounds; they are hence termed "chemo-litho-heterotrophic" ("litho" indicates that the energy required originates from anaerobic oxidation of H_2).

$$H_2SO_4 + 4 H_2 \longrightarrow H_2S + 4 H_2O \quad (\triangle G^\circ = -45{,}6 \text{ kcal})$$

The probable reaction sequence starting with APS is shown in the following scheme:

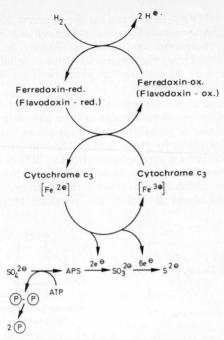

Probable conversions during dissimilatory
sulfate reduction (after ROY and TRUDINGER)

The amino acids cysteine, homocysteine and methionine represent important sulfur compounds in metabolism. To some extent they exist free, for the most part, however, they are bound in proteins. Thus sulfur is also an essential component of all enzymes. In the form of free SH groups it activates the "thiol enzymes". Moreover, the cell contains a number of sulfur compounds which, acting as coenzymes, assume a key position in metabolism. In some of these sulfur is firmly bound to the molecular structure (thiamine pyrophosphate, biotin); in others it forms a free SH group, representing the active site of the molecule (coenzyme A, lipoamide).

Impaired or blocked absorption of sulfur has serious and multiform effects on metabolism. The deficiency symptoms which appear greatly resemble those of nitrogen deficiency; this is not surprising since both elements are components of proteins and enzymes.

$PO_4^{3\ominus}$

The element phosphorus is absorbed by plants in the oxidized form of orthophosphate (under acid conditions as $H_2PO_4^{\ominus}$, under neutral conditions as $HPO^{2\ominus}$). In contrast to the elements nitrogen and sulfur which enter the metabolism of the plant cell only after having been reduced, phosphorus can participate directly as phosphoric acid or phosphate. Hence the compound active in cellular metabolism is the same as the one used in absorption and transport of this element by plant organisms.

During periods of an abundant supply or increased uptake of phosphate, many plant organisms are able to store phosphates. In lower plants this function is probably assumed mainly by the polyphosphates, in higher plants by phytic acid (= "phytin", see structure). In the polyphosphates a varying number (up to 10^6) of phosphoric acid residues are bound as anhydrides. The structural formula shows that the molecules consist of a chain-like anion in which two phosphorus atoms at a time are connected by one oxygen atom:

Polyphosphate

Inositol hexaphosphate
= phytic acid

The members with low molecular weight of this class of compounds are termed mono-, di-, tri- and tetraphosphates. The monomeric and the dimeric compound are conventionally called orthophosphate and pyrophosphate, respectively.

Recent studies indicate that cyclic condensed inorganic phosphates (= "metaphosphates") containing 2 to 7 phosphoric acid residues occur in cells of the blue-green alga *Anacystis nidulans*.

Since formation of polyphosphates is an endergonic process, it is coupled to other energy-supplying processes of cell metabolism. The reaction proceeds with an astonishing speed in yeast and in a number of bacterial species. Little is known in detail of this process as yet. Apart from the function as phosphate reservoir, a function as energy reservoir has been ascribed to the polyphosphates as well: it is believed that a surplus of ATP in the cell is used to establish energy-rich bonds in polyphosphate. This concept is supported by the observation that an enhanced polyphosphate formation takes place in various algae in light when the CO_2 supply is inadequate or lacking. Indeed, the free energy of hydrolysis of the bond $-P-O-P-$ amounts to $7000 - 10\,000$ cal/mole, thus being in the same order of magnitude as the energy-rich phosphate bonds in ATP.

Another possible function of polyphosphates under discussion is the formation of complexes with cations such as $Ca^{2\oplus}$, $Mg^{2\oplus}$ and $Na^{\oplus}$ which may be released from the complex when required in metabolism.

The role of phytic acid (hexaphosphoinositol; see structure above) as a phosphate reservoir is indicated by its accumulation in seeds where phosphorus practically occurs solely as the Ca-Mg salt of this compound. During germination a steep rise in the activity of phosphatases which cleave phytic acid *(phytases)* is observed together with a release of phosphoric acid.

The element phosphorus contributes significantly to the construction of important cellular compounds. Phosphoric acid esters play a prominent role in many metabolic processes: photosynthesis, glycolysis, the pentose phosphate cycle, nucleic acid synthesis. But these phosphoric acid esters can only be formed when functioning of the ADP/ATP system is maintained by an adequate supply of phosphoric acid or inorganic phosphate. As a constituent of phosphatides and phospholipids the element participates substantially in the structure of cellular membranes.

As in the case of the elements nitrogen and sulfur, phosphorus deficiency has extensive consequences and results in a variety of symptoms. A clear decision whether these represent direct or indirect effects, is usually not possible.

The Role of Cations in Metabolism

$Mg^{2\oplus}$: Magnesium is not only a constituent of the chlorophyll molecule (p. 39 f) but also an integral component of the ribosomes (p. 383). Moreover, this ion activates a number of enzymes; among these are *RNA polymerase* (p. 343 f) and *polynucleotide phosphorylase* (p. 343). In this function magnesium may be replaced to some extent by $Mn^{2\oplus}$.

$Fe^{2\oplus}$: Iron is required in the synthesis of all those enzymes which contain heme or hemin as prosthetic group: cytochromes (p. 237 ff), *peroxidases* and *catalases* (p. 418). It is also a component of the "iron-sulfur proteins" (ferredoxins, p. 112; flavodoxins, p. 237).

$Ca^{2\oplus}$: As a structural constituent calcium links the macromolecular protopectins in the middle lamella (p. 186). It is also found in an insoluble form in calcium phosphate, calcium oxalate and calcium carbonate. The ability to dehydrate cytoplasmic proteins seems to be of greater significance; here potassium ions act as antagonists (cf. p. 266).

$K^{\oplus}$: Apart from its effect on cytoplasmic hydration mentioned above, the precise function of potassium is still unknown although it is absorbed by higher plants in far larger quantities than the other cations. Chemical bonding with organic molecules has not been demonstrated so far.

To give examples of the mode of action of **microelements** we shall mention a few of these which enter metabolism partly as anions and partly as cations. *Nitrate reductase* (p. 291) and *nitrogenase* (p. 296) are activated by *Mo. Cu* is bound is plastocyanin (p. 122) and in the phenol oxidases *(phenolase, tyrosinase;* they cause the brown coloring when plant tissues are cut: potatoes, fungi and fruit). *Mn* is part of the O_2-evolving system of photosynthesis (p. 116). *Co* is an essential factor in N_2-binding of root nodules (p. 296).

References

Beevers, L., R. H. Hageman: Nitrate reduction in higher plants. Ann. Rev. Plant Physiol. 20: 495, 1969

Bergerson, F. J.: Biochemistry of symbiotic nitrogen fixation in legumes. Ann. Rev. Plant Physiol. 22: 121, 1971

Bollard, E. G., G. W. Butler: Mineral nutrition of plants. Ann. Rev. Plant Physiol. 17: 77, 1966

Broda, E.: Aktiver Transport durch biogene Membranen. Naturwiss. Rundschau 21: 483, 1968

Dixon, R. O. D.: Rhizobia. Ann. Rev. Microbiol. 23: 137, 1969

Hardy, R. W. F., R. C. Burns: Biological nitrogen fixation. Ann. Rev. Biochem. 37: 331, 1968

Kessler, E.: Untersuchungen zum Problem der photochemischen Nitratreduktion in Grünalgen. Planta 49: 505, 1957

Kessler, E.: Anorganischer N-Stoffwechsel. Fortschr. Botan. 29: 119, 1967

Laties, G. G.: Dual mechanisms of salt uptake in relation to compartimentation and long-distance transport. Ann. Rev. Plant Physiol. 20: 89, 1969

Leggett, J. E.: Salt absorption by plants. Ann. Rev. Plant Physiol. 19: 333, 1968

Mengel, K.: Ernährung und Stoffwechsel der Pflanze, 4th Ed. Fischer, Stuttgart 1972

Salzer, E. H.: Methoden der Hydrokultur. Kosmos, Stuttgart 1965

Scheffer, F., P. Schachtschabel: Lehrbuch der Bodenkunde, 7th Ed. Enke, Stuttgart 1970

Schiff, J. A., R. C. Hodson: The metabolism of sulfate. Ann. Rev. Plant Physiol. 24: 381, 1973

Sitte, P.: Biomembranen: Struktur und Funktion. Ber. dtsch. botan. Ges. 82: 329, 1969

Stewart, W. D. P.: Nitrogen fixation in photosynthetic microorganisms. Ann. Rev. Plant Physiol. 24: 289, 1973

Thompson, J. T.: Sulfur metabolism in plants. Ann. Rev. Plant Physiol. 18: 59, 1967

Weigl, J.: Ionentransport und Membranstruktur in Pflanzen. Ber. dtsch. botan. Ges. 82: 445, 1969

Metabolism of the Cellular Components

The substrate which the autotrophic cell utilizes to construct its own chemical components originates from carbohydrate at the hexose level as it is produced by photosynthesis or chemosynthesis by reduction of CO_2. From this simple "precursor" together with the elements nitrogen, sulfur and phosphorus the "building block" molecules are synthesized, which are then assembled to form the characteristic molecular components as well as the subcellular structures, i.e. the organelles. The cell has to carry out "chemical work" (p. 3) in order to achieve these biosynthetic reactions comprising the enzyme-catalyzed formation of precursors and building blocks with relatively low molecular weight as well as of complex molecules. The energy required comes from the end product of autotrophic CO_2 reduction, too; it is degraded by dissimilation yielding "energy equivalents" in the form of ATP. The principle underlying its participation in biosynthetic reactions is the "substrate activation" by phosphorylation (p. 19 ff), an effective chemical device to ensure that these reactions proceed to completion.

In order to maintain their "species specificity" a number of cellular constituents must be identically reproduced. The construction plans are encoded in the genetic material of the cell and transmitted to the sites of synthesis by mobile informational molecules. The biosynthetic activity of the cell is thus determined by the transmission and preservation of genetic information. This finally manifests itself in the specific activities and functional structures of a differentiated cell. Molecular structure and biochemical function are based on the same fundamental biological principle which further emphasizes their close relation.

The biosynthesis of cellular compounds depends primarily on the growth rate; it is hence particularly intensive in rapidly growing unicellular organisms or in dividing tissue cells of higher organisms. Biosyntheses also take place in cells which neither divide nor increase in total mass. The reason for this is that all cellular components are subject to a continuous process of degradation and building-up, resulting in a "dynamic steady state": the rate of synthesis equals the rate of breakdown.

Among the chemical constituents of the cell the rate of this metabolic turnover varies considerably. Marked differences may also exist in this respect among the different cell types of an organism or an organ. Even the apparently stable components of cell organelles (mitochondria, nucleus, chloroplasts) with their numerous enzymes undergo a metabolic turnover.

The Macromolecules

Numerous substances, especially molecules of low molecular weight, are widely distributed constituents of plant cells at whatever level of differentiation. They accomplish important functions in primary and secondary metabolism. This category comprises all those compounds which are intermediates of photosynthesis, dissimilation and the pentose phosphate cycle. Polysaccharides serving as reserve products like starch or glycogen also belong to this group. Apart from these compounds the cell contains others whose structure is specific for an individual cell or organism only and is an expression of *species specificity*. The specific features of an organism thus originate from a unique molecular structure of certain classes of compounds which possess a positive quality, i.e. *information*. For realization of this quality the construction principle of the *macromolecule* is at the disposal of the cell: from a limited number of precursor substances of low molecular weight chainlike structures are formed by polymerization or polycondensation. Differences in the specific order (= *sequence)* and in the number of the components result in an enormous number of variations which, considering the vast number of organisms, are necessary to preserve the species specificity. A great advantage of this construction principle is that the apparatus of synthesis can meet demands for production of information-containing molecules at a minimal expense of substrate and energy. That they are synthesized with painstaking accuracy is shown by the fact that the biosynthesis of nucleic acids and proteins, i.e. typical macromolecules containing information, is a very complex process – not because the establishment of the chemical bond linking the building blocks is particularly complicated but because it is of enormous importance that they are inserted into the chainlike molecules in the specific order or *sequence* demanded by the genetically determined "blueprint" of the organism concerned. Even the smallest deviation from this pattern may have fatal consequences for the vitality of the cell or organism because it would impair or destroy the *biological activity* of the macromolecule.

The macromolecular construction principle is also applied in compounds representing the typical reserve products or fibrous substances of the cell. In these the aspect of species specificity naturally is of minor importance (p. 174).

Before discussing the biosynthesis of macromolecules we shall consider the formation of building blocks from the elements carbon, hydrogen, oxygen, nitrogen and sulfur.

Nucleic Acids

The macromolecules of nucleic acids differ from those of the poly-saccharides discussed previously (p. 174 ff) in three fundamental aspects. The first one relates to their structure: the molecular building blocks, the "nucleotides", do not originate from a single class of compounds, but comprise components of different chemical nature. The second difference concerns their function: nucleic acids are exclusively designed to store or to transfer biological information that is ultimately expressed in protein macromolecules. A division of labor exists among the nucleic acids insofar as a specific type assumes the function of an informational carrier or memory: *deoxyribonucleic acid* ("DNA"; contains D-deoxyribose!)*. These molecules are able to accomplish this task because of certain structural premises. The third deviation consists in the participation of the elements nitrogen and phosphorus in the construction of nucleic acids.

The transfer or *transcription* of information, however, devolves upon nucleic acids belonging to the class of *ribonucleic acids* ("RNA"; contains D-ribose!). They comprise the nucleic acids which participate in the complex process of protein biosynthesis and which are constituents of certain cell structures.

Although only four different building blocks in the form of nucleotides are available for the construction of any nucleic acid macromolecule, there is an enormous variability with respect to molecular structure. This variability is essential for the maintenance of specificity which is expressed particularly in the order or *sequence* of the individual nucleotide building block, and also to some extent in the chain length of the macromolecules containing between about 80 and more than 100 000 nucleotides. The biosynthesis of nucleic acids is thus rather more complex than that of polysaccharides. Adherence to the specific sequence is of vital significance when the mononucleotide units are arranged in the chain-like macromolecules of nucleic acids: any deviation impairs the biological function and

---

* In some viruses, ribonucleic acid (RNA) constitutes the genetic material: tobacco mosaic virus and a few bacteriophages.

may give rise to transmission of an incorrect information with serious consequences on the protein being synthesized. Artificial induction of such defects in the cell has become an important method for analyzing the structure and function of nucleic acids.

Historical Aspects. Nucleic acids were discovered by Friedrich Miescher, a co-worker of Hoppe-Seyler, in Tübingen (1869) when he isolated "nuclein" from suppurated dressings. This substance, rich in phosphoric acid, exhibited an acidic reaction and was soluble in alkali. Later on, he obtained the same material from fish sperm together with a basic protein which he named "protamine". The term "nucleic acids" dates from R. Altmann (1889). In the years 1879–1900 Kossel and his colleagues succeeded in demonstrating organic bases in nucleic acids. Levene and Jacobs (1909) identified ribose as the characteristic sugar of "yeast nucleic acid", and in 1930 deoxyribose as the sugar moiety of "thymus nucleic acid".

This is the origin of the old terms "yeast nucleic acid" for RNA and "thymonucleic acid" for DNA. The demonstration of nucleic acids within the cell and the establishment of its location became possible after development of the following methods: the DNA-specific "Feulgen reaction" developed by Feulgen and Rossenbeck in 1924, the "ribonuclease test" which Brachet used to identify a portion of the basophilic material of the cell as RNA, and Caspersson's microspectrophotometric technique which allows a quantitative determination of nucleic acids in histological sections by measuring their specific UV absorption. Avery and his co-workers (1944) for the first time convincingly proved the biological function of DNA as genetic material in their "transformation" experiment. Their findings were subsequently confirmed by numerous experiments, particularly with viruses and bacteriophages. The properties and functions of the other nucleic acid species have only been recognized in the last ten years; they have been experimentally confirmed at least for microorganisms.

Let us now consider the chemical constitution of nucleic acids beginning with the monomeric unit of their macromolecules, the nucleotide.

Nucleotides as Building Blocks of Nucleic Acids

From our discussion of the coenzymes we are already familiar with the principle underlying the construction of nucleotides: an organic nitrogen-containing base is linked with a pentose to which a phosphoric acid residue is attached. With the base component nitrogen as a new element enters the macromolecule (its origin in plant cells has already been treated; p. 289 ff). The bases present in nucleotides are derived from pyrimidine and purine:

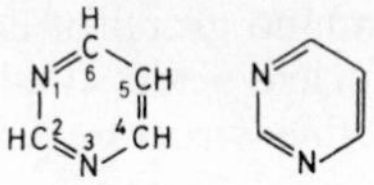

Pyrimidine

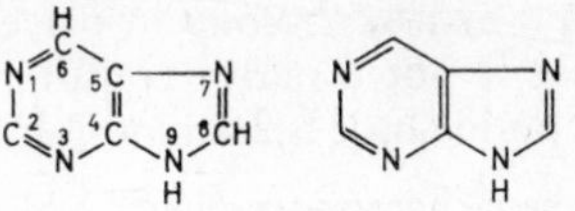

Purine

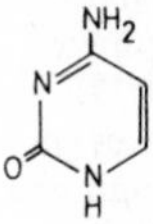

Cytosine

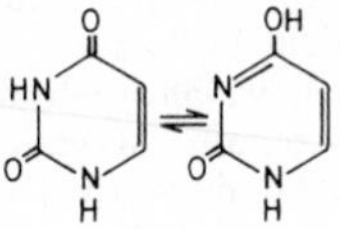

Uracil

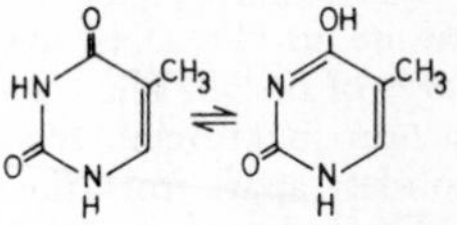

Thymine

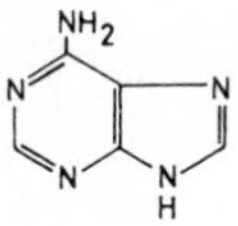

Adenine

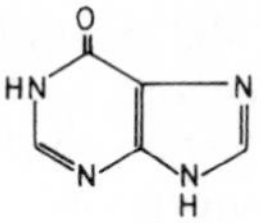

Hypoxanthine

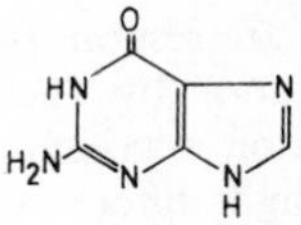

Guanine

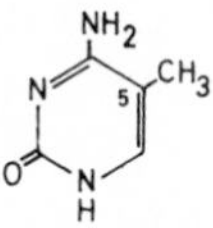

5-Methyl cytosine

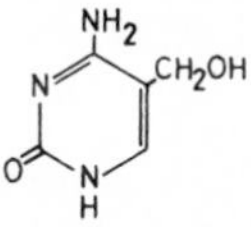

5-Hydroxymethyl cytosine

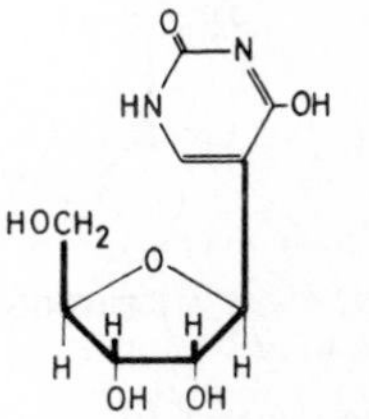

Pseudo uridine

Pyrimidine has an aromatic ring structure consisting of four carbon and two nitrogen atoms. Substitution in position 2 (oxygen), position 6 (OH or NH_2 group) and position 5 (CH_3-) leads to the pyrimidine bases: cytosine, uracil and thymine. While uracil occurs practically only in RNA, thymine is a typical component of DNA.

In the double ring system of the purines* we recognize the pyrimidine structure attached to another nitrogen-containing ring structure. We have already encountered adenine (6-aminopurine) as a constituent of ATP and other nucleotide coenzymes. Hypoxanthine

---

* The name originates from the term "purum uricum".

contains a hydroxyl group in place of the amino group of adenine. This base is not a native constituent of nucleic acids. Guanine as another purine base is 2-amino-6-hydroxypurine (structure p. 313).

Besides these characteristic bases additional "rare" or unusual bases have been discovered in several nucleic acid components in recent years. The DNA of higher plants generally contains a small amount (3–6 %) of 5-methylcytosine. Cytosine is completely replaced by 5-hydroxymethyl cytosine in the DNA of a particular class of bacteriophages: T-even phages of *E. coli*. The *transfer RNA* (p. 338 f) which assumes an important function in protein biosynthesis contains several rare bases in the nucleotides: apart from the methylated derivatives of the main bases and 4,5-dihydrouracil the unusual nucleosides "pseudouridine" (for structure see p. 313), isopentenyladenosine, 4-thiouridine, inosine and ribothymidine (p. 338).

The bases are responsible for the characteristic absorption of ultraviolet radiation by nucleotides and nucleic acids. The latter have an absorption maximum at 260 nm which is a mean value resulting from the slight differences in absorption of the individual bases.

The pentose sugar in nucleotides is either β-D-ribose or β-D-2-deoxyribose:

β-D-Ribose   β-D-Deoxyribose

This is the reason for the chemical and functional differences between ribonucleic acids (RNA) and deoxyribonucleic acids (DNA; containing deoxyribose). In DNA thymine replaces uracil; thymine is absent in RNA.

The compound formed from a base and a sugar is termed a "nucleoside". A link is created between carbon atom 1 of the pentose and the nitrogen in position 3 of the pyrimidine base or in position 9 of the purine base, respectively, one molecule of water being released. Strictly speaking, this is a glycosidic bond (p. 167 f), i.e. a "N-glycosidic bond". The nucleosides have trivial names which are derived from the bases; pyrimidine derivatives have the ending "-idine", those of purine the ending "-osine". With the attachment of a phosphoric acid residue a nucleoside is converted into the corresponding nucleoside monophosphate or "mononucleotide". This occurs via an ester bond involving carbon atom 5' of the pentose (its carbon atoms are designated 1'–5'):

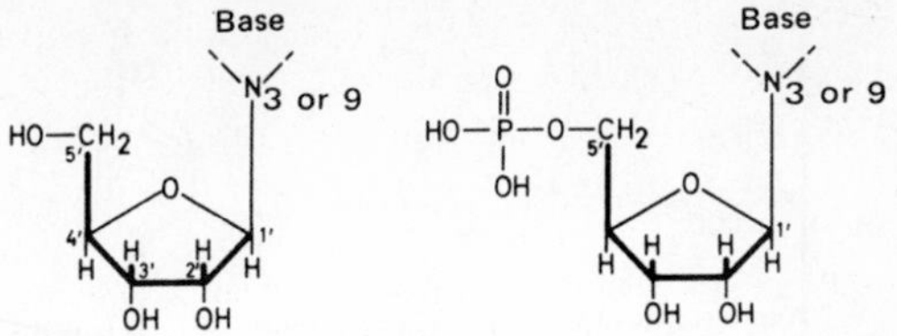

Ribonucleoside (section)   5'-Nucleoside monophosphate

We already know from the adenylic acid system as well as from nucleotide coenzymes that the 5'-phosphoric acid ester can be extended by an additional phosphoric acid or pyrophosphoric acid residue. These "nucleoside diphosphates" and "nucleoside triphosphates" are the reactants of nucleic acid biosynthesis; they represent an energy-rich, "activated" substrate (cf. p. 19 f). Let us now outline their formation in the cell (for isolation of nucleotides, see p. 337 f).

Biosynthesis of Nucleotides. The mechanism of nucleotide biosynthesis has been studied mainly in bacteria and animal cells. Only few results have been obtained with plant cells so far; they seem to show a pathway identical to that found in the former organisms.

The key precursor of pyrimidine nucleotides is orotic acid which must be synthesized first (scheme p. 316). From aspartic acid, an amino acid with two carboxyl groups (p. 358), and carbamyl phosphate the intermediate carbamyl aspartic acid (also termed "ureidosuccinic acid") is formed by mediation of *L-aspartate carbamyltransferase*.

Carbamyl phosphate, a mixed acid anhydride, acts as the "activated", i.e. energy-rich component of this reaction. In bacteria it is formed by the following reaction catalyzed by a *carbamyl phosphate kinase*:

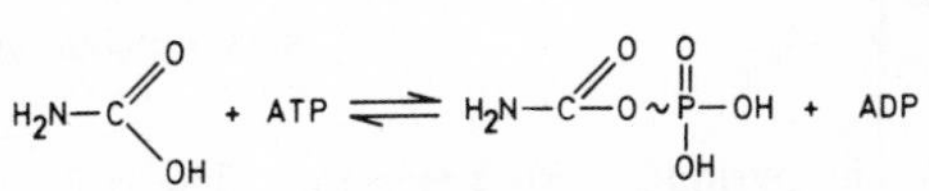

Carbamyl phosphate

The enzyme has been found in mitochondria from *Phaseolus aureus* seedlings, too.

Carbamyl aspartic acid is converted to dihydroorotic acid by a cyclo-dehydration: the carboxyl group reacts with the amino group

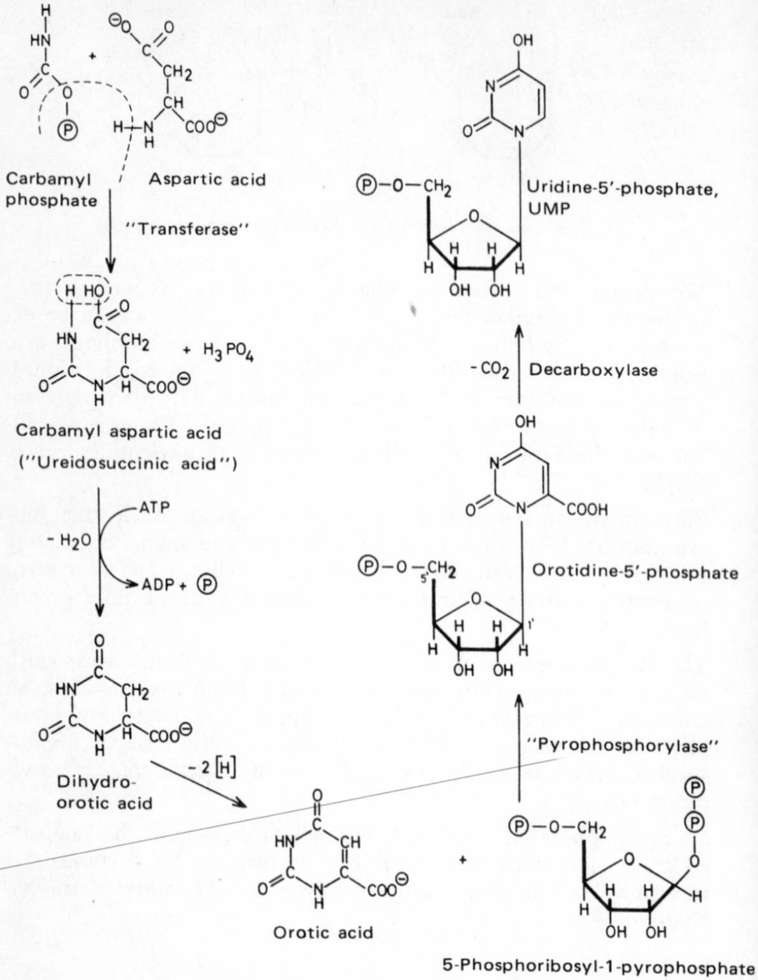

Carbamyl phosphate   Aspartic acid

"Transferase"

Carbamyl aspartic acid ("Ureidosuccinic acid")

+ H₃PO₄

- H₂O

ATP

ADP + (P)

Dihydro-orotic acid

- 2 [H]

Orotic acid

5-Phosphoribosyl-1-pyrophosphate

"Pyrophosphorylase"

Orotidine-5'-phosphate

- CO₂   Decarboxylase

Uridine-5'-phosphate, UMP

giving rise to the pyrimidine ring structure. The reaction is endergonic and consumes ATP; it is catalyzed by *dihydro-orotase*.

The dihydroorotic acid formed is subsequently oxidized to orotic acid by *dihydroorotic acid dehydrogenase ("L-4,5-dihydroorotate : NAD oxidoreductase")*; this enzyme has been isolated and purified from bacteria and demonstrated also in extracts of wheat embryos. It is probably a flavoprotein which transfers the hydrogen to $NAD^{\oplus}$.

The conversion to the nucleotide, orotidine-5'-phosphate (= "orotidic acid"), is accomplished by N-glycosidation; 5-phosphoribosyl-1-pyrophosphate is the activated substrate of this condensation reaction and *orotidine-5'-phosphate pyrophosphorylase* (= *"orotidine-5'-phosphate: pyrophosphate phosphoribosyl transferase"*) the active enzyme.

The ribose for the construction of nucleosides and nucleotides is generally supplied from the "pool" of the pentose phosphate cycle; as explained previously, this serves as a reservoir of precursors for the anabolic (= "building") metabolism. In photosynthetically active cells, the CO_2 reduction cycle may represent a second "pool" for ribose.

A specific decarboxylase which has been isolated and partly purified from etiolated leaves of *Phaseolus vulgaris* converts orotidine-5'-phosphate to uridine-5'-monophosphate (UMP), setting free CO_2. This compound and other uridine nucleotides are intermediates in the synthesis of cytidine nucleotides; in uridine triphosphate (UTP) the hydroxyl group at carbon atom 6 is enzymatically replaced by an amino group, leading to formation of cytidine triphosphate (CTP).

In vivo an equilibrium is probably immediately established enzymatically between the UMP synthesized and uridine or uracil. The compound required for the formation of pyrimidine nucleotide is removed from this equilibrium mixture. Besides de-novo synthesis of the pyrimidine ring system with orotic acid as intermediate many cells are able to convert free bases, e.g. uracil, into a nucleoside by linking with ribose-1-phosphate; the nucleoside may subsequently be phosphorylated by ATP to a nucleotide. To some extent nucleotides are used as substrate, too. This explains why leaf tissue as well as unicellular algae *(Chlorella, Anacystis)* incorporate added uracil, uridine, cytidine, thymine and thymidine rather rapidly into their nucleic acids, although these compounds are not intermediates in the biosynthetic pathway of uridine-5'-phosphate or cytidine-5'-phosphate (CMP). Results of tracer experiments performed with leaves and seedlings using ^{14}C-orotic acid or its labeled precursors indicate that orotic acid has a corresponding key position in pyrimidine nucleotide synthesis in plants.

Some details in the biosynthesis of the two deoxyribose derivatives of pyrimidine, thymidine-5'-monophosphate and d-cytidine-5'-monophosphate, are still unknown. Specific reductases probably carry out the reduction of the pentose at the level of uridine diphosphate (UDP) and cytidine diphosphate (CDP), respectively. Thymidine-5'-phosphate is formed from d-UMP by the introduction of a methyl group at position 5. The methyl donor is "active formaldehyde", a derivative of tetrahydrofolic acid.

Tetrahydrofolic acid is another one of those coenzymes which are concerned with activation and transfer of C_1 fragments ("C_1 metabolism"), i.e. the hydroxymethyl group (= "activated formaldehyde") and the formyl group (= "activated formic acid") in the cell. Folic acid belongs to a group of chemically closely related compounds; it was first discovered as a vitamin in yeast, liver cells and leaves (hence the name!) and as a nutritional factor for microorganisms (lactic acid bacteria). The molecule of folic acid is composed of a pterine ring, p-aminobenzoic acid and glutamic acid:

Pteridine   p-aminobenzoic acid   Glutamic acid

Folic acid

Pterines are naturally occurring pigments, e.g. in blue-green algae and in insects (in butterfly wings, hence the name!). The second component, p-aminobenzoic acid, is required as a growth substance by a number of microorganisms. It is well known that this is where the sulfonamides, compounds of a very similar structure, show their inhibitory effect. The various members of the folic acid group differ from one another mainly by the varying number of glutamic acid residues bound in the molecule. While the substance acting as coenzyme contains only one, the "folic acid conjugates" may contain up to seven. As a donor for "activated formaldehyde", tetrahydrofolic acid has a hydroxymethyl group attached to N^{10} which probably leads to an additional ring formation involving N^{10} and N^5:

N^5, N^{10}-Methylene-tetrahydrofolic N^{10}-Formyl-tetrahydrofolic acid
acid

("active formaldehyde")   ("active formic acid")

The principal source of the hydroxymethyl group is the amino acid serine, which is subject to a reversible transfer of the β-carbon atom (cf. p. 358). The formyl group (= "activated formic acid") is also attached to

N^{10} of tetrahydrofolic acid (= "coenzyme F"). The existence of an energy-rich anhydrous form, N^5–N^{10}-methenyl-tetrahydrofolic acid, is under discussion. The C_1 fragment is derived from various metabolic reactions and certainly not from free formic acid. It is required for the biosynthesis of purine nucleotides (see below) as well as for other reactions.

By subsequent ATP-consuming phosphorylation thymidine-5'-phosphate and d-cytidine-5'-phosphate are converted to thymidine-5'-triphosphate (TTP) and d-cytidine-5'-triphosphate (dCTP), respectively; these compounds assume the function of precursors in DNA biosynthesis (p. 329 ff).

In contrast to the synthesis of pyrimidine nucleotides, the synthesis of purine nucleotides starts from ribose-5-phosphate: the purine ring system is formed by the addition of various small precursors or molecular groups to the sugar moiety. The amino acid glycine acts as the starter substance; to its C-C-N skeleton other carbon and nitrogen atoms are attached. The following reaction scheme summarizes the main reactions of biosynthesis of purine nucleotides; we shall discuss them briefly.

Initially, ribose-5-phosphate is "activated" by ATP and a specific kinase yielding 5-phosphoribosyl-1-pyrophosphate which we know already from the synthesis of orotidine-5'-phosphate (p. 317). This compound reacts with glutamine to form 5-phosphoribosylamine. It is subsequently conjugated with glycine in an ATP-dependent reaction giving rise to formation of an amide, glycineamide ribonucleotide, which contributes three atoms to the ring formation. An additional carbon atom comes from tetrahydrofolic acid in the form of an activated formyl group (see above). However, before the ring of imidazole is formed, a second amination takes place in which glutamine serves as amine donor. The energy required for the reaction comes from ATP, just as in the subsequent ring-closure which is an ATP-dependent dehydration. The following construction of the second ring system (pyrimidine!) begins with the addition of a carboxyl group originating from bicarbonate to the aminoimidazole ribonucleotide in a reversible reaction. The "active" form of this C_1 fragment is carboxybiotin (the function of biotin, another coenzyme involved in C_1-metabolism, will be discussed in connection with the biosynthesis of fatty acids, p. 405 f). The compound formed (= 5-amino-4-imidazolecarboxylic acid ribonucleotide) reacts with aspartic acid to form an amide, with consumption of ATP. After expulsion of fumaric acid, a further ring atom in the form of an amino group is left over. The introduction of the last carbon atom and final closure of the purine ring is brought about once again by a "transformylation" in which N^{10}-formyl tetrahydrofolic acid (see above) acts as the donor. The ring-

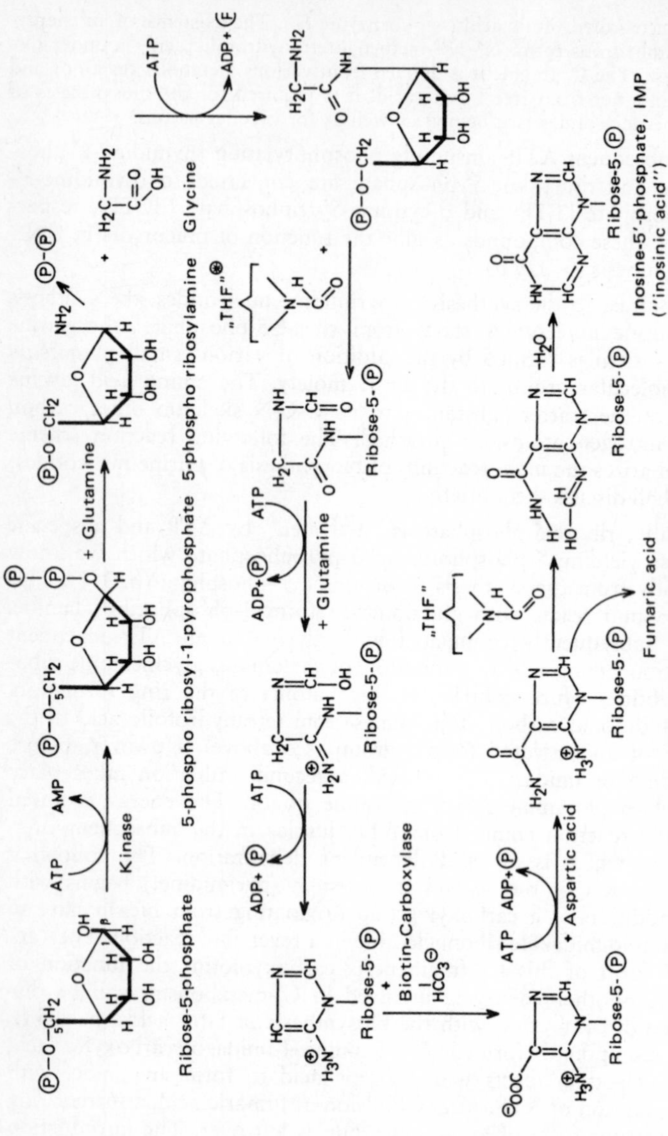
Ribose-5-phosphate
ATP
AMP
Kinase
5-phospho ribosyl-1-pyrophosphate
+ Glutamine
5-phospho ribosylamine
Glycine
ATP
ADP+ Ⓟ
„THF"
Ribose-5-Ⓟ
ATP
ADP+Ⓟ
+ Glutamine
Biotin-Carboxylase
$HCO_3^{\ominus}$
ATP
ADP+Ⓟ
+ Aspartic acid
Fumaric acid
„THF"
$-H_2O$
Inosine-5'-phosphate, IMP
("inosinic acid")

Synthesis of purine nucleotides

(⊛ "THF" = Tetrahydrofolic acid)

closure with dehydration yields a purine derivative, inosine-5'-phosphate (inosinic acid, IMP; with the base hypoxanthine!) which is converted to the typical purine nucleotides of the nucleic acids by additional enzyme-catalyzed reactions.

The complex reaction sequence leading to IMP, and the mode of action of the enzymes involved have been elucidated mainly in bacteria and animal cells. There are only few results from experiments with plant cells.

Guanosine-5'-phosphate (guanylic acid, GMP) is formed from IMP in two reaction steps (see scheme). First, hydrogen is cleaved by a NAD-dependent *inosine-5'-phosphate dehydrogenase;* then the product, xanthosine-5'-phosphate, is aminated in position 2. The NH_2 group originates from the amide group of glutamine or from $NH_4^{\oplus}$. Two reactions are also required to convert IMP into adenosine-5'-phosphate (adenylic acid, AMP). In the first reaction aspartic acid is bound to the purine ring at position 6, requiring GTP as energy equivalent. The second reaction is very similar to that applied in the construction of the purine ring, i.e. the synthesis of 5-amino-4-imidazole carboxamide ribonucleotide (p. 319); in a

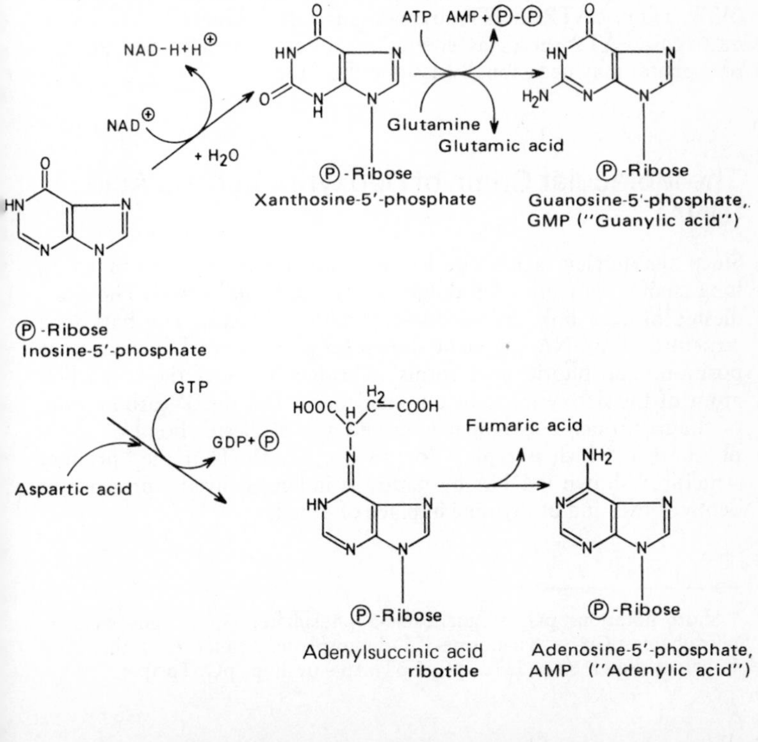

Xanthosine-5'-phosphate

Guanosine-5'-phosphate, GMP ("Guanylic acid")

Inosine-5'-phosphate

Adenylsuccinic acid ribotide

Adenosine-5'-phosphate, AMP ("Adenylic acid")

similar reaction the enzyme *adenylosuccinase* splits off fumaric acid from the adenylsuccinic acid ribo(nucleo)tide thus leaving an NH_2 group in position 6 attached to the ring. The reaction product is AMP. *Inosine-5'-phosphate dehydrogenase* has been demonstrated in pea seedlings, *adenylosuccinase* in yeast cells.

By analogy with the pyrimidine nucleotides, purine nucleotides are also formed from the free bases and ribose-5-phosphate by more or less specific *nucleotide phosphorylases* and by ATP-dependent *kinases*. The *nucleoside pyrophosphorylases* convert the bases directly to nucleotides, using 5-phosphoribosyl-1-phosphate as precursor.

The deoxyribose derivatives of purine are formed by reduction of the corresponding ribonucleoside diphosphates, probably in a reaction similar to the one occurring in the conversion of the corresponding compounds derived from pyrimidine nucleotides. Deoxyribonucleosides are also formed by the action of *nucleoside phosphorylase* in the presence of deoxyribose-1-phosphate, though it is not certain whether this pathway is of significance in the cell.

Before they can act as precursors of RNA and DNA the purine nucleotides must be phosphorylated to the nucleoside triphosphates (ATP, GTP, dATP, dGTP) by two successive kinase reactions. In most cases ATP serves as energy equivalent; other nucleoside triphosphates may occasionally assume this function.

The Molecular Chain of Deoxyribonucleic Acid (DNA)

Since the nucleic acids consist of many nucleotides arranged in long chains, they are accordingly termed *polynucleotides*. Phosphodiester bridges link the successive building blocks. The backbone structure of a DNA fragment shown on p. 323 reveals their precise position. Phosphoric acid forms a bridge between the 5' carbon atom of the deoxyribose of one nucleotide and the 3' carbon atom of the next one. This is hence termed a 3'-5'-diester bond of phosphoric acid which is typical for nucleic acids*. That the "primary structure" shown is DNA by nature is indicated by the presence of deoxyribose, and of thymine in place of uracil.

---

* Short notation: pG = guanosine-5'-phosphate; Gp = guanosine-3'-phosphate; pGp = guanosine-3',5'-diphosphate. Sequence of the DNA fragment shown on p. 323: – pCpGpTpAp – or d(–pCpGpTpAp–).

Deoxyribonucleic acid, DNA
(fragment)

A precise determination of the molecular weight of individual native DNA species is as yet not possible. During isolation from eukaryotic cells in which they are bound to proteins as "nucleoproteids" (p. 382) the long chain-like molecules very easily break into fragments. Nevertheless, it is known that the molecular weight must be very high, since values of 10^9 have been obtained, corresponding to several million nucleotide units. Previously obtained values of 1–10 million are suspect, since most probably molecule fragments were used, which resulted from inadequate isolation methods. DNA macromolecules are thus of an order of magnitude which makes them visible in an electron microscope with high resolving power. The DNA of a bacteriophage, for instance, is visible as helical thread of about 50 μ length.

The high molecular weight further increases the already large scope for variability in the structure of DNA; as mentioned previously, it is caused by changes in the nucleotide sequence. Unfortunately, there are as yet no methods permitting the determination of the nucleotide sequence for DNA. All the genetic information which a

DNA molecule contains is encoded in the specific sequence of the nucleotides over the entire length of the chain. It is obvious that with the employment of four different nucleotides (dAMP, dGMP, dCMP, TMP) an enormous number of combinations and, thus, of different sequences in DNA molecules is potentially possible. We shall later examine in some detail how the information laid down and encoded in the molecular structure of DNA is transmitted and finally expressed in protein synthesis (p. 386 ff).

Secondary Structure of DNA

Quantitative determination of the components of isolated DNA preparations of high molecular weight has always yielded a molar ratio of 1 for the bases adenine and thymine, as well as for the bases cytosine and guanosine. Accordingly, purine and pyrimidine bases are present in equal numbers. While animal DNA is relatively rich in adenine + thymine, guanine + cytosine dominate in the DNA of a number of bacterial species.

The quotient $\frac{A + T}{G + C}$ gives an indication as to the origin of a DNA sample. On the basis of the striking fact that the base components in the nucleotides of all DNA species are present in pairs of equal amounts and on the basis of other experimental data Watson, Crick and Wilkins* proposed a new structural model for DNA. It has been confirmed by experimental data so convincingly that it is generally accepted today. It is based on base pairing between thymine and adenine, and between cytosine and guanine, caused

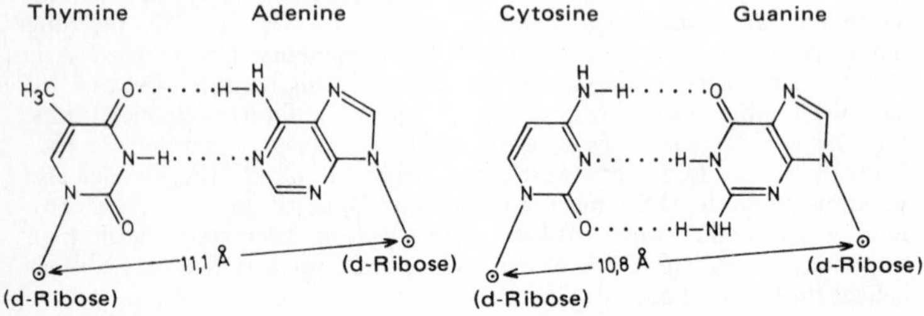

Thymine   Adenine   Cytosine   Guanine

Base pairing

---

* For this contribution they received the Nobel Prize in 1962.

by formation of hydrogen bonds (p. 324). Other pairings of bases are not possible.

The existence of these bonds between the uniquely fitting base components of single nucleotides enables two polynucleotide chains or "strands" to form a common structure if the base sequence of one strand is *complementary* to that of the other, i.e. each nucleotide base reacts with the specific counterpart on the same level in the second strand.

The structural model of DNA meets this requirement: two polynucleotide strands are held together by hydrogen bonds between the fitting pairs of bases over the entire length. In other words, the base sequence of one strand definitively determines that of the other since only thymine and adenine or cytosine and guanine can pair, i.e. form stable hydrogen bonding with each other. This *complementarity* is illustrated by the following diagram:

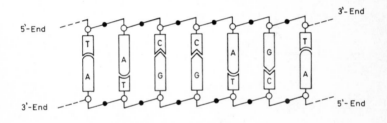

The Two-Stranded Structure of DNA (Fragment)
(○ = deoxyribose; ● phosphoric acid)

The two polynucleotide strands of DNA form a right-hand twisted double "helix" structure: the two strands are intertwined helically around each other (see schematic representation p. 326). While the deoxyribose molecules linked by phosphoric acid form the true backbone, the base pairs are arranged perpendicularly to the axis of the helix. The two base pairs "fit" exactly into the free space between the two chains. Other pairings are not possible because they would not fit into the helix. The two strands of the DNA molecule run in opposite directions, i.e. the 3'-end of one strand is opposite the 5'-end of the other (see scheme above). The one strand thus runs from 5' to 3', the other from 3' to 5'.

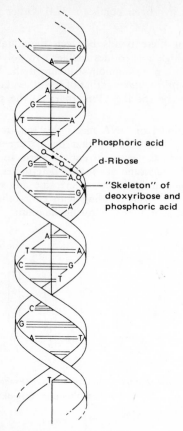

Watson-Crick Model of DNA

The double-stranded nature of a DNA preparation, i.e. its "native" state, can be tested by placing it in a dilute salt solution, gradually raising the temperature and continuous measuring of the optical density at 260 nm. Fig. 88 shows the resulting "melting curve". Decay of the double helix into single strands and fragments takes place between 80° and 90 °C and will be made evident by a striking change in optical density. This "T_m-value" is defined as the temperature at the transition midpoint. It is specific for each DNA species and can be used for characterization. The single strands formed during heating recombine, at least to some extent, to a double-stranded structure on slow cooling ("renaturation").

Native DNA can be demonstrated more precisely by means of "isopycnic" gradient centrifugation, which separates macromolecules exclusively on

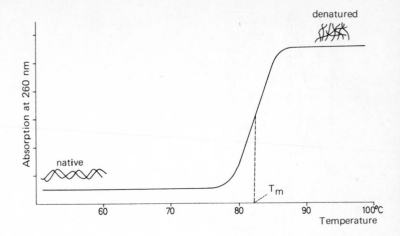

Fig. 88 Change in ultraviolet absorption (260 nm) during heat denaturation of native (double-stranded) DNA. T_m: "melting" temperature (= "transition mid-point").

the basis of their density. It differs therefore from the sucrose gradient centrifugation, a method of "zone centrifugation" discussed previously (p. 95 f). In isopycnic gradient centrifugation the dissolved DNA is layered on top of a solution of cesium chloride of appropriate concentration in a centrifuge tube or mixed with this solution before filling the tube. The tube is subsequently centrifuged at high speed in an analytical ultracentrifuge. An equilibrium is established between sedimentation and diffusion, and a concentration gradient is formed. At the bottom of the tube there is a higher density which gradually decreases towards the surface. The large molecules will come to rest – after sedimentation in the field of gravity or after rising against it – in a band which corresponds to their buoyant density in the solution. One or more bands will have been formed when equilibrium has been reached. By means of a built-in optical system (UV at 260 nm), the position of these bands can be followed during centrifugation by continuous measuring of light absorption. The photometric evaluation of the graph obtained is shown in Fig. 89 for various DNA preparations; a specific density can be assigned to each. Conclusions as to the molecular weight of the macromolecular compounds can be drawn from the position of the individual bands (p. 336).

The selectivity of this "cesium chloride equilibrium centrifugation" (according to Meselson, Stahl and Vinograd) is so high that even two DNA samples of the same kind are easily separated into two distinct bands when one of these contains predominantly the heavy nitrogen isotope ^{15}N and the other one the normal nitrogen ^{14}N (Fig. 90). This combination with isotope techniques has proved particularly useful in studies of the mode of DNA replication (p. 332 f).

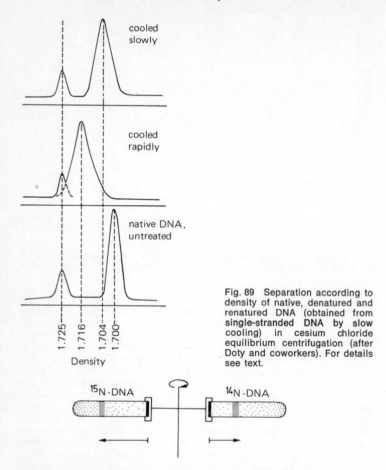

Fig. 89 Separation according to density of native, denatured and renatured DNA (obtained from single-stranded DNA by slow cooling) in cesium chloride equilibrium centrifugation (after Doty and coworkers). For details see text.

Fig. 90 Movement of "heavy" DNA, labeled with the nitrogen isotope ^{15}N, and of the same DNA species with normal ^{14}N during centrifugation in a density gradient. ^{15}N-DNA forms a band in the region of higher density, i.e. it accumulates closer toward the bottom of the tube, compared to ^{14}N-DNA.

The double helix is the native form of the DNA molecule. This also holds true for DNA of plant cells. Among the few exceptions, the most wellknown is the DNA of the bacteriophage Φ X 174; it consists of only one strand which is probably closed to a ring (molecular weight 1.7 million). Circular DNA in which both strands are closed to a ring is present in cells of E. coli (molecular weight about 3×10^9).

As to the amount of DNA per cell the following rule applies: the greater the amount of DNA, the more an organism has advanced in evolution. However, one must not jump to the conclusion that the information content has increased to the same extent. On the contrary, it has been found that the DNA of highly developed eukaryotes contains long segments without structural genes. Renaturation experiments (p. 326) have shown that these consist of sequences of a few hundred up to a thousand of nucleotide pairs which, recurring many times over, make up considerable sections of DNA. It has been estimated that between a thousand and a hundred thousand copies per genome may occur with identical nucleotide sequence or else with a more or less varying sequence. The occurrence of these "repetitive" DNA segments and those practically present in a unique version is characteristic for individual eukaryotic organisms. The "satellite DNA" seems to be particularly rich in repetitive sequences which may amount to 5% (HeLa cells) or even 10% (mouse) of the total DNA in the genome.

It has been experimentally confirmed that chloroplasts and mitochondria contain small quantities of DNA. Compared to nuclear DNA this "extranuclear DNA" has a different base composition and a different density; accordingly they can be separated from each other by density gradient centrifugation (p. 326 f). Just as in the case of bacteria and blue-green algae, the DNA of the cellular organelles is free, i.e. unlike nuclear DNA it is not bound to histones and other proteins. Moreover, its double helix forms an open or highly twisted (= "supercoil") ring. This conformity between prokaryotes on the one hand, and mitochondria and chloroplasts on the other is also evident in the structure of the ribosomes (p. 383 ff), in the protein-synthesizing system (p. 387 f), and in the specific inhibitory effects of antibiotics (p. 395). These findings provide new arguments in favor of an old hypothesis according to which originally free prokaryotes invaded higher cells during evolution and developed into mitochondria and chloroplasts via endosymbiosis.

Mitochondrial DNA, amounting to less than 1% of the total cellular DNA, has a molecular weight of about 10^7. Per mitochondrion 2–6 molecules of DNA may occur. Chloroplasts may contain up to 6% of the DNA in a cell. The molecular weight ($1–2 \times 10^8$) is significantly higher than that of mitochondrial DNA. Per chloroplast 40–80 molecules of DNA have been estimated which, in contrast to nuclear DNA, does not contain 5-methylcytosine (see p. 313). The question whether and to what extent the DNA of mitochondria and chloroplasts stores genetic information for the establishment of the complex structure of these cellular organelles (see p. 155) cannot be answered at present since only a few relevant facts are known so far.

DNA Biosynthesis: Identical Reduplication

Replication of DNA in the cell must aim at copying exactly the nucleotide sequence of its molecules in order to maintain specificity. DNA biosynthesis always takes place when cell division requires duplication of the genetic material. In the end, each of the two daughter cells formed has been provided with the same complete double-stranded DNA as the "mother" cell. The fact that genetic information can be transmitted in this way from generation to generation is due to the unique ability of DNA of *identical reduplication* or *replication*, i.e. net synthesis in accordance with given directions. This emphasizes the fundamental difference to the synthesis of other macromolecules such as starch or cellulose where there is no preservation of species specificity or information.

Apart from structural aspects the enzyme-catalyzed process of identical reduplication of DNA shows clearly why the double-stranded structure is the most convenient for accomplishing the biological function of DNA. The two existing polynucleotide chains act as building plans in the formation of new *complementary* nucleotide sequences: each chain unequivocally determines the sequence of the complementary chain because of base pairing (cf. p. 325).

The *polymerases* which act as specific enzymes of DNA synthesis were isolated for the first time from bacteria*. The four typical

Deoxyribonucleoside triphosphate

---

* Kornberg and Ochoa received the Nobel Prize in 1959 for their discoveries in the field of nucleic acid biosynthesis.

building blocks of deoxynucleotide triphosphates (dATP, dGTP, dCTP and TTP) are required as substrate. The test mixture must also contain a small quantity of native DNA as "primer" (= "template") of the reaction, and Mg$^{2\oplus}$ ions. The polymerase causes the sequential formation of new 3'-5' phosphodiester linkages by splitting off the two terminal phosphoric acid residues of the precursor nucleoside triphosphate as pyrophosphate and adding the monophosphate moiety to the terminal deoxyribonucleoside via its 3'-hydroxyl group (p. 330).

Two energy-rich bonds (2 × 7000 cal/mole) are required for the insertion of each mononucleotide unit. Since the energy of hydrolysis of the phosphodiester bond formed is $\Delta G° = -6500$ cal/mole the process of synthesis is energetically favored. The energy used to activate substrate, i.e. to convert the mononucleotide to the nucleoside triphosphate, is derived from ATP, the universal transport metabolite for cellular energy. The ATP "pool" is fed by the processes of dissimilation. In photoautotrophic cells an additional source of ATP is provided by photophosphorylation.

According to the present state of knowledge, the DNA added as "primer" is responsible for the arrangement, i.e. for the sequence of the nucleotides in the newly synthesized deoxyribopolynucleotide. It is to be noted that this enzyme-catalyzed reduplication process will only proceed if a DNA polynucleotide strand is present to provide the "building plan" for the enzyme. This strand, serving as *template,* is converted into a double helix by the addition of a mononucleotide the base of which fits the corresponding base in the template in each reaction step. Thus the new complementary strand grows continuously along the "template strand". Where the latter has adenine in the nucleotide, only a thymine nucleotide is allowed at the opposite position in the new strand. Thus, guanine fits with cytosine, thymine with adenine and cytosine with guanine.

Such a mechanism implies that the DNA double helix of the Watson-Crick model is initially unwound into two single strands. To each a complementary strand is then synthesized giving rise to the

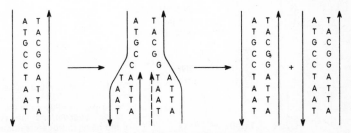

Semiconservative DNA replication

formation of two complete new double helices (see scheme). Each double helix has the same nucleotide sequence and composition in the single strands as the original "template" molecule. The best proof for this identity of "template" and product would be a comparison of the base sequences. Determination of the latter, however, is experimentally too difficult to achieve at the present time. One adequate approach to this problem is the base ratio analysis, i.e. $\frac{A + T}{G + C}$ which in most cases shows the same value for synthesis product as for "template" DNA.

In accordance with the mode of formation each of the newly formed (= "daughter") double strands consists of a parent strand and a newly synthesized polynucleotide strand. This kind of replication is termed "semiconservative"; it seems to be typical for plant cells, too. This is indicated by studies with callus cells of *Nicotiana tabacum* which were carried out according to the classical experiment of Meselson and Stahl with *E. coli*, using the heavy nitrogen isotope ^{15}N (Fig. 91). After growing several generations of cells in the presence of $^{15}NO_3^\ominus$ in a liquid medium, they are transferred to a fresh culture medium now containing normal (= ^{14}N) salts. When the cells have divided once more their DNA is isolated and subjected to CsCl density gradient centrifugation. Its density (1.703) is at the midpoint between that of ^{15}N-DNA (from cells grown in presence of $^{15}NO_3^\ominus$; 1.711) and that of ^{14}N-DNA (from cells grown in normal medium with $^{14}NO_3^\ominus$; 1.696). After the next cell division in

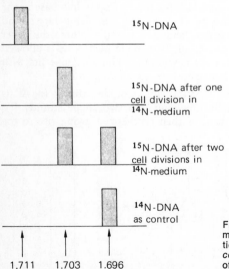

^{15}N-DNA

^{15}N-DNA after one cell division in ^{14}N-medium

^{15}N-DNA after two cell divisions in ^{14}N-medium

^{14}N-DNA as control

1.711   1.703   1.696
Density

Fig. 91   Demonstration of semiconservative DNA replication in plant callus cells *(Nicotiana tabacum)* by means of the heavy nitrogen isotope ^{15}N. For details see text.

normal medium about half of the DNA was "light"; the rest, however, was "half-heavy" (see Fig. 91) as is to be expected from semiconservative replication.

In *HeLa* cells the mechanism of semiconservative DNA replication of eukaryotes has been studied in detail. The cells were labeled with ³H-deoxythymidine; subsequent autoradiography (p. 144) of the isolated DNA showed that synthesis of new DNA (indicated by the silver grains along the DNA thread) certainly does not begin at one end and progress continuously over the entire length, but starts at several points. Therefore, "replicative units" exist, each of which is independently and asynchronously replicated starting from a "growth point". These segments are of various length and may comprise up to a hundred thousand base pairs. The pieces formed as products of this "discontinuous" DNA synthesis must subsequently be linked by the enzyme *ligase*. Hence eukaryotes would be equipped with a mechanism of DNA replication similar to that of bacteria. In fact, Okazaki found small DNA fragments for the first time which could be identical with the early products of DNA replication.

The mechanism of a discontinuous DNA synthesis would also explain how the implicated elongation of the polynucleotide chain from the 5′ end in the Watson-Crick model might occur (scheme p. 331). So far only growth from the 3′ end has been experimentally confirmed. Taking this as basis, the new complementary strand could arise if first of all several small fragments are synthesized contrary to the growth direction of the

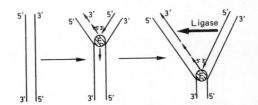

Hypothetical course of the "discontinuous" DNA synthesis

whole strand, i.e. from the growth point (E) in the "wrong" direction, namely from 5′ to 3′. After their completion the fragments would have to be joined by a *ligase*. Perhaps even both the new complementary strands are constructed according to this mechanism.

Cells of *E. coli* obviously dispose of at least three different *DNA polymerases*. In contrast to the original concept the task of the enzyme first isolated by Kornberg *(polymerase I)* is probably restricted to renewal of DNA segments which have lost their functions, e.g. after transcription has occurred (p. 347). The *DNA polymerases II* and *III* seem to be responsible for the actual replication; they are probably only active when bound to the cell membrane.

The initial unwinding of the double helix and the subsequent helix formation in the two daughter molecules are among the unsolved problems of DNA synthesis.

Enzymes of DNA Degradation

In a number of plant species enzymes have been found which cleave DNA molecules in the same way as the *deoxyribonucleases (DNases)* from bacteria, i.e. by hydrolyzing the internucleotide bonds. They belong thus to the *phosphodiesterase* group. The cleavage products, oligonucleotides and large, non-dialyzable chain residues, are difficult to analyze more precisely. Two types of *phosphodiesterases* can be distinguished by the different way in which they cleave the phosphodiester bond. The one type is represented by a pancreatic

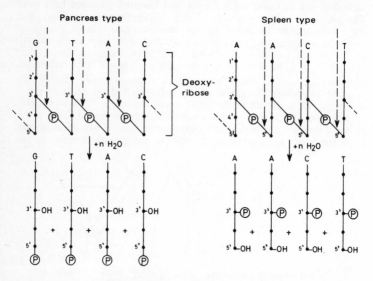

Mode of action of Deoxyribonucleases

enzyme for which the splitting of the 3'-phosphoester bond is characteristic (cf. scheme above); the fragments accordingly have a phosphoric acid residue in the 5' position. The *DNase* from spleen represents the other type; it cleaves the 5'-phosphoester bond and gives rise to formation of fragments phosphorylated at the 3' end.

There are also a number of non-specific *phosphodiesterases* which not only attack the macromolecules of nucleic acids, but also compounds of other classes provided that these possess a phosphoric acid group attached by diester bond. Such enzymes have been found in intestinal mucous membranes and in snake venoms.

Specific *5'-nucleotidases* which occur in muscle and human seminal fluid dephosphorylate the 5'-mononucleotides resulting from complete degradation of DNA. These are *phosphomonoesterases* to which the relatively non-specific *"acid" phosphatases* (pH optimum around 5.0) and the *"alkaline" phosphatases* (pH optimum 7–8) belong as well. This group of enzymes utilizes 3'- and 5'-monophosphates of different compounds as substrate. *Acid phosphatases* have been isolated from plant cells, e.g. from the green algae *Hydrodictyon* and *Acetabularia*. Another enzyme from this group may be active in the photosynthetic CO_2 reduction cycle (p. 150 ff).

Structure and Properties of Ribonucleic Acid (RNA)

The constructional principle underlying DNA is realized in RNA, too: unbranched chainlike macromolecules, the building blocks of which are joined by phosphoric acid via the pentose unit of the nucleotides. In this case, D-ribose molecules are linked by 3'-5' diester bonds. In the structure formula (p. 336) we note the lack of thymine and the appearance of uracil as nucleotide base.

At present, little is known about the secondary structure of RNA. In contrast to DNA, a double helical structure resulting from base pairing seems to play a minor role (p. 339 f). Another significant difference is that RNA occurs in three different forms in the cell. Since the classification according to chemical criteria coincides with the different functions, the RNA classes are termed: *transfer* RNA, *ribosomal* RNA and *messenger* RNA.

Methods. Gentle disruption of the cells must precede the extraction of RNA from algae or plant tissues. The homogenate obtained is mixed with buffer-saturated phenol or a phenol-cresol mixture and shaken to dissociate the nucleic acids from protein (= "nucleoproteids", p. 382). On centrifugation the mixture yields a phenol phase and an upper aqueous phase. Practically all the RNA components and part of the DNA are present in the aqueous phase; this layer is recovered. Alcohol is added and the solution is left standing in the cold, whereupon the nucleic acids precipitate from the aqueous solution. After washing with ether or alcohol the total amount of nucleic acids thus obtained is ready for separation into the various components. The following methods are applied:

1. Sucrose gradient centrifugation, a method of "zone centrifugation" with which we are already familiar (p. 95 f); an example is shown in Fig. 94 (p. 341).

2. Column chromatography on a mixture of methylated serum albumin and kieselgur ("MAK").

3. Sedimentation in the centrifugal field of an analytical ultracentrifuge (up to 300 000 g!). Since they are heavier than their solvent, i.e. water, the RNA macromolecules slowly migrate to the bottom of the centrifuge

Ribonucleic acid,
RNA (section)

tube which rotates perpendicularly to the axis. They move at different speeds depending on their size. From the speed measured a sedimentation coefficient "s" may then be calculated for the various classes of RNA molecules, the migration of the macromolecules being continuously monitored during centrifugation (cf. p. 341). The value of "s" is closely correlated with the molecular weight: the greater the molecular weight, the higher is "s". This procedure is thus also suitable for the determination of molecular weight, and is used particularly by protein chemists (p. 377).

4. Separation on polyacrylamide gel by "disc" (= discontinuous) electrophoresis. The separation of macromolecular polyelectrolytes (nucleic acids, proteins) in an electrical field by this analytical method (for quantities of 1–100 µg) is substantially improved by additional application of the concentration and molecular-sieve effect (p. 48 f). The concentration effect is brought about by establishing zones of different pH values or/and different acrylamide concentrations within the gel column to be used for separation. Before being applied onto the separation gel with small pores, the sample is usually trapped in gel with relatively large pores ("concentrating" or "collecting" gel). In this way the substances to be separated are collected in a narrow zone before entering the separation gel

where electrical charge and molecular size cause the individual components to move at different speeds and to accumulate in distinct bands. Their position is ascertained by staining or by direct UV photometry scanning. For detection of labeled substances (^{3}H, ^{14}C) the gel column is frozen and then sliced into discs 0.6–1.0 mm thick; these discs are then treated with an appropriate solvent and the radioactivity released is measured. The original pattern of the labeled components in the gel can be reconstructed from the data obtained (see Fig. 95, p. 350).

The base composition of an RNA species is often used for characterization. For its determination the macromolecules must first be quantitatively degraded by alkaline hydrolysis (0.1–0.5 M KOH for 18 h at 37°). This cleavage yields the 2'- and 3' nucleoside monophosphates of the four bases: 2'3'-AMP, 2'3'-GMP, 2'3'-CMP and 2'3'-UMP. This is due to the formation of an intermediate cyclic diester of phosphoric acid (see scheme); it originates from the shift of the ester bond linked to the 5' carbon atom to the OH group in position 2'. In an alkaline milieu it is hydrolyzed to the 2' or the 3' monoester, so that there is always a mixture of the two forms. With DNA a corresponding cyclic diester cannot be formed because there is no OH group in position 2'; DNA is hence stable to alkali. The mononucleotides may be isolated and quantitatively determined by means of paper chromatography (p. 35 ff), thin-layer chromatography (p. 36 f), electrophoresis (p. 379 f) and ion-exchange chroma-

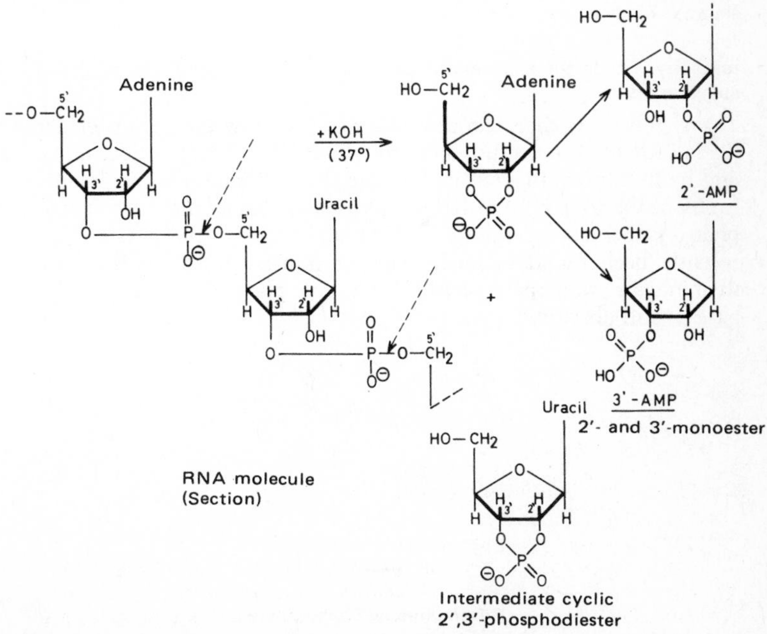

RNA molecule
(Section)

2'-AMP

3'-AMP
2'- and 3'-monoester

Intermediate cyclic
2',3'-phosphodiester

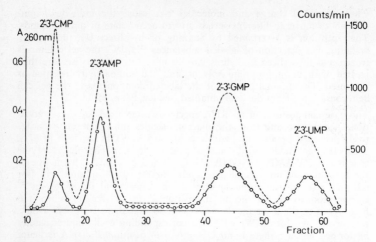

Fig. 92 Ion-exchange chromatography of the cleavage products of ^{32}P-labeled ribosomal RNA from seedlings *(Vicia faba)* after alkaline hydrolysis. For the column an anion exchange resin (Dowex, chloride form) is used. The individual nucleoside monophosphates successively leave the column during elution with increasing concentrations of HCl. By monitoring the ultraviolet absorption (254 nm; o—o) and the ^{32}P-radioactivity (– – – –) the elution diagram shown is obtained.

tography. The last-named procedure is frequently employed; Fig. 92 shows an example.

Transfer RNA is characterized by a relatively low molecular weight (25 000–30 000) corresponding to about 70–80 nucleotide units, and by its content of unusual bases (p. 314). Besides the methylated forms of the four normal bases and pseudouridine, inosine-5′-phosphate, isopentenyladenosine and ribothymidine-5′-phosphate have recently been found in nucleotides of transfer RNA. In this case thymine is a component of RNA! The role played by these unusual bases is still obscure.

Ribothymidine-5′-phosphate

As indicated by the name, molecules of this RNA class transport and transfer amino acids during protein synthesis (for details see p. 386 ff). There is probably at least one specific transfer RNA for each amino acid.

Our knowledge about transfer RNA is based on studies with *E. coli, Euglena,* yeast cells and mammalian liver cells. Although there are species-specific differences in composition and structure, the conditions are clearly the same for all organisms. Therefore, a transfer RNA from another species functions very well in a cell-free system of protein synthesis (p. 383). The same probably holds true for the transfer RNA of plant cells.

The first successful determination of the complete base sequence of a transfer RNA was achieved with the species which specifically transfers the amino acid alanine (Holley* and coworkers). The molecule consists of a chain of 77 nucleotides. According to a recent concept the molecule has a secondary structure with helical regions (= "cloverleaf pattern"). As shown in Fig. 93, the 3' and 5' ends at the open part of the helical region lie close together.

The looped-out portion of unpaired bases on the opposite side contains the *recognition site* (= *anticodon,* cf. p. 390). Two other loops, one with dihydrouridine, the other with a pentanucleotide sequence (-GpTp-pseudo-UpCpG- or GpTp-pseudoUpCpA-), occur to the left and to the right, an arrangement which is probably present in all transfer RNA molecules. A general feature is also the unpaired sequence of the bases cytosine-cytosine-adenine at the 3' end, and the termination of the 5' end in a nucleotide containing guanine. All other sequences – particularly the *recognition site* – differ in the individual transfer RNA species. "Synonymous" transfer RNA species are a certain exception: these are different molecules specific for the same amino acid. Their sequences may differ only in two nucleotide bases as in the case of the two serine-specific transfer RNA molecules from yeast.

In a few cases synonymous transfer RNA species have been found in cells of higher plants; some of these are obviously located in chloroplasts and mitochondria where they are probably active in the protein-synthesizing system of these cell organelles (cf. p. 329).

More recently the sequences of a number of transfer RNA species, particularly from yeast, have been determined. Since several of these have been prepared in crystalline form it has been possible to elucidate their three-dimensional structure (tertiary structure) by means of X-ray diffraction studies. The results obtained confirm the existence of the "cloverleaf structure".

Transfer RNA molecules constitute the "soluble" RNA fraction, a term which indicates that they do not form a sediment even during ultracentrifugation of a cell homogenate at about 100 000 g.

---

* Together with H. G. Khorana and M. W. Nirenberg, R. W. Holley was awarded the Nobel Prize in 1968.

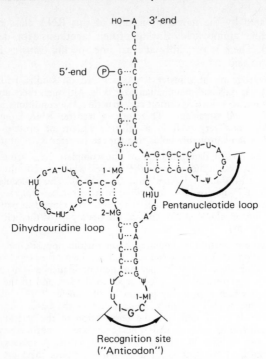

Fig. 93 "Cloverleaf structure" of the transfer RNA species which is specific for the amino acid alanine (p. 355; "t-RNA^Ala"; after Holley). Only the sequence of nucleotide bases is shown; the hydrogen bonding is indicated by dotted lines. Abbreviations of the unusual nucleosides; HU = dihydrouridine; 1-MG = monomethyl guanosine; 2-MG = dimethyl guanosine; Ψ = pseudo-uridine; 1-MI = monomethyl inosine.

Ribosomal RNA, as indicated by its name, is a component of the ribosomes, the organelles of protein synthesis in the cytoplasm. They will be treated in detail when discussing this process (p. 383 ff).

Recent studies indicate that RNA of this type is also present in the cell nucleus (nucleolus), in chloroplasts and probably even in mitochondria as a constituent of ribosome-like particles (cf. p. 385 f).

Ribosomal RNA has a relatively high molecular weight (see below). In plants and in bacteria its base composition seems to be very similar. The high content of guanine and the relatively low content of uracil and cytosine are typical. There is no complementary re-

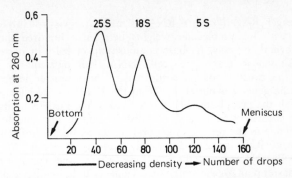

Fig. 94 Isolation of ribosomal RNA components from seedlings *(Vicia faba)* by sucrose gradient centrifugation. In order to locate and identify the components a hole is punched into the bottom of the plastic centrifuge tube; the drops coming out are continually collected in a series of test tubes each receiving two drops. As the number of drops increases (abscissa!) the density of the sucrose solution decreases steadily since the region of highest density (at the bottom!) leaves the tube first. By measuring the absorption at 260 nm in each sample collected, the original location of the RNA components in the sucrose gradient is obtained. The position of the absorption maxima in connection with the sedimentation constants observed in reference preparations permits identification of the RNA components.

lation of bases as in DNA. The molecular size of ribosomal RNA is not uniform; it probably depends on the substructure of the ribosomes (p. 384 f). Two main components can be isolated from the ribosomes by means of centrifugation in the analytical ultra-centrifuge or in a sucrose gradient. In *Chlorella pyrenoidosa* they are referred to as 17 S* and 25 S components in accordance with their different sedimentation rates. Their molecular weights are about 0.6×10^6 and 1.3×10^6, respectively. Similar values – 18 S and 25 S – were found for the major components of ribosomal RNA from seedlings of *Vicia faba* (Fig. 94).

In the blue-green alga *Anacystis nidulans* and in the purple bacterium *Rhodospirillum rubrum,* the values of 16 S and 23 S indicate that the ribosomal components are practically identical with those of bacteria.

Besides these two classes of RNA molecules an additional smaller component of 5 S (see Fig. 94) appears in the sedimentation diagram which is also a native component of ribosomes. Its base sequence was recently determined in preparations from *E. coli* and tumor cells; the 120 nucleotides do not contain any unusual bases.

---

* "S" (= 1 Svedberg; named after the pioneer of analytical ultracentri-fugation) has the value of 10^{-13} sec and is the unit of the sedimentation coefficient. This ("s_{20}") is defined as the speed of sedimentation at 20 °C based on the unit of centrifugal acceleration; its dimension is time. Proteins have values between 1 and 200×10^{-13} (= 1–200 Svedberg units).

Messenger RNA ("matrix RNA"; "m-RNA") is probably also a RNA with high molecular weights between a hundred thousand and several millions. Its base composition, at least in bacteria, is very similar to that of the cell's DNA. This particularity results from the function of these molecules: they transfer information from the genetic material, i.e. from the DNA, to the sites of synthesis in the cytoplasm. The DNA fixed in the nucleus or its equivalent is thereby enabled to mobilize its information and to make it available in the required temporal sequence. Because of this function, this RNA species is termed "messenger" RNA. The term "matrix" RNA refers to the function of these molecules as a matrix or a master plan for the amino-acid sequence to be synthesized.

This new species of RNA was first discovered in phage-infected bacteria, and later also in normal bacteria. The concept of its mode of action as a matrix was developed in phage-infected bacteria, too. It was initiated by the observation that a fraction of rapidly synthesized RNA combines to a large extent with the bacterial ribosomes and directs the synthesis of phage-specific proteins. The involvement of m-RNA in protein synthesis was later confirmed by experiments with non-infected bacteria. In plant and animal cells m-RNA is formed in the nucleus and then probably bound to proteins (p. 348); it is subsequently transferred to the cytoplasm in this form (= "informofer"). Perhaps this m-RNA-protein complex is even present in the polysome (p. 388).

Due to the small amounts in which it occurs and the metabolic lability, i.e. short-lived nature, particularly in bacteria, m-RNA can only be detected in vivo after feeding the cells labeled precursors of RNA synthesis (^{32}P-orthophosphate, ^{14}C- or ^{3}H-nucleosides or -nucleotide bases) in a relatively high concentration for a rather short period of time (= "pulse labeling" experiment); the nucleic acids are extracted and subsequently separated. The rapidly synthesized components among them exhibit radioactivity and high specific activity, respectively.

However, not all rapidly synthesized RNA species are identical with m-RNA. It has been found that transfer RNA and ribosomal RNA are also more or less rapidly synthesized, particularly in organisms with a high growth rate (bacteria, unicellular algae). Apart from its short-lived nature there are other criteria for bacterial m-RNA: a base composition similar to that of DNA, as well as the ability to stimulate protein synthesis in vitro and to form stable complexes, the "DNA-RNA-hybrids", with the one-strand form of the homologous DNA.

The ability of a DNA strand to "hybridize", i.e. to form a double strand structure with a second DNA strand or with an RNA strand, depends on the presence of complementary base sequences in the two strands, either over the entire length or at least over long segments. This provides a technique of determining the complementary sequences in a nucleotide strand of RNA or DNA: the two strands of a DNA molecule are separated by heating thus destroying the hydrogen bonding (= "molecular melting"; p. 326). They are then incubated with the RNA to be tested under appropriate conditions (temperature, salt concentration). If complementary nucleotide sequences are present in one of the two DNA strands and in the RNA strand a double-strand structure, a "DNA-RNA hybrid", is formed. By this method m-RNA molecules can be isolated from a mixture of different RNA species; vice versa it helps to ascertain those segments on a DNA strand which carry information for the sequence of specific RNA molecules (cf. p. 349 f).

RNA Biosynthesis:
the Mechanism of Transcription

Several years after the discovery of m-RNA, enzymes were isolated which catalyze the synthesis of ribopolynucleotides in vitro. One enzyme, extracted and purified from bacteria *(E. coli, Lactobacillus arabinosus, Azotobacter vinelandii),* synthesizes RNA from the four nucleoside triphosphates ATP, GTP, CTP and UTP in the presence of $Mg^{2\oplus}$ and $Mn^{2\oplus}$ with DNA as the sole primer. Equimolar amounts of pyrophosphate are released. Generally, DNA is only active as a primer in the native double-helical structure. Synthesis does not occur in the presence of *DNase* (p. 334 f). Since this enzymatic process is analogous with the one taking place in DNA synthesis the enzyme has been termed *DNA-dependent RNA polymerase* (= *nucleoside triphosphate: RNA nucleotidyl transferase).*

In bacteria another enzyme, *polynucleotide phosphorylase* (= *polynucleotide nucleotidyl transferase*) has been found which catalyzes in vitro the synthesis of ribopolynucleotides from the nucleoside diphosphates ADP, GDP, CDP and UDP, obviously without any DNA dependence. In contrast to *RNA polymerase* polymeric products are formed even when only one, two or three of the four typical nucleosidediphosphates are present in the test mixture. Thus the homopolymers polyadenylic acid ("poly A"), polyguanylic acid ("poly G"), polycytidylic acid ("poly C") and polyuridylic acid ("poly U") may be produced from ADP, GDP, CDP or UDP. Mixtures of nucleoside diphosphates accordingly give rise to formation of heteropolymers containing two, three or all four nucleotide bases (the application of the enzyme for a synthesis of polynucleotides with defined base sequences was an essential prerequisite for the elucidation of the *genetic code;* see p. 389 f). The dependence of this ribopolynucleotide synthesis on the quality as well as the quantity of the substrate does not guarantee the preservation of specificity required of biologically active RNA. Therefore, *polynucleotide phosphorylase* is believed to be an enzyme occurring preferentially in RNA degradation in the cell and not in RNA synthesis. In the presence of inorganic phosphate nucleoside diphosphates are formed from ribopolynucleotid by reversion of the synthesis reaction:

$$\text{polynucleotide} + n\textcircled{P} \rightleftharpoons n[\text{nucleoside} - \textcircled{P} - \textcircled{P}]$$

The enzyme has been partially purified from the blue-green alga *Anacystis nidulans*. It requires $Mg^{2\oplus}$ and $Mn^{2\oplus}$ as cofactors. The biological function of *polynucleotide phosphorylase* as an enzyme of RNA degradation is indicated to a certain extent by its cellular localization: findings in bacteria and blue-green algae suggest that it is bound to ribosomes. In the purified form it cleaves homologous and heterologous ribosomal RNA in the presence of inorganic phosphate.

DNA-dependent RNA-polymerase of the blue-green alga *Anacystis nidulans* was one of the first to be isolated and purified from plant organisms. It has a molecular weight of about 500 000 and consists of four subunits with a molecular weight of 190 000, 145 000, 72 000 and 38 000, respectively; these occur in the stoichiometric proportion 1 : 1 : 1 : 2. Hence a marked similarity exists with the subunit structure of *RNA polymerase* from *E. coli* which has four subunits, too: β', β, σ and α (proportion: 1 : 1 : 1 : 2). In their molecular weight, however, they differ: β' 165 000, β 155 000, σ 95 000, α 39 000. It is not yet clear to what extent these subunits fulfill identical functions in the two enzymes in catalyzing the *transcription,* i.e. the DNA-dependent synthesis of ribopolynucleotide (p. 347 f).

In *RNA polymerase* of *E. coli* the β' subunit is probably implicated in the binding to DNA, while the σ subunit is responsible for initiation of RNA synthesis. The molecular substructure of *RNA polymerase* described here seems to be typical for all prokaryotic organisms – apart from minor differences in the molecular weight of the subunits. In contrast to these enzymes the *RNA polymerases* isolated from higher plants so far are of the eukaryotic type. Here, one has to distinguish between at least two enzymes localized in the cell nucleus: *RNA polymerase "I" ("A")* in the nucleolus, and *RNA polymerase "II" ("B")* in the nucleoplasm. This distribution, first discovered in animal cells *(HeLa,* calf thymus, rat liver) obviously occurs also in plant cells. The two enzymes do not only differ in their chromatographic behavior on DEAE cellulose (p. 381), but also in their sensitivity to α-amanitin, a toxic peptide from the mushroom *Amanita phalloides* (p. 376): *RNA polymerase "II"* from the nucleoplasm is completely inhibited while *"I"* from the nucleolus is not. According to recent findings *RNA polymerase "II"* from cell cultures of *Petroselinum sativum* (parsley) consists of three larger subunits (molecular weight about 200 000, 180 000 and 140 000, respectively) and four smaller ones (molecular weight about 40 000, 26 000, 25 000 and 16 000, respectively). A similar substructure has been demonstrated in *RNA polymerase "II"* from animal cells and from the slime mould *Dictyostelium discoideum.*

As indicated by studies with maize seedlings, chloroplasts obviously have their own *RNA polymerase* the constitution and properties of which are very similar to the enzyme of prokaryotes. It is not yet clear whether a specific *RNA polymerase* occurs in mitochondria.

The activity of highly purified *RNA polymerases* from prokaryotes and eukaryotes depends on the presence particularly of native or denatured DNA in the test system. It does not necessarily have to be homologous, i.e. from the same species; on the contrary, calf-thymus DNA or artificial deoxyribonucleotides such as d(AT) polymer (p. 347) are even more effective. The reason for this relative inefficacy of the homologous DNA is not known.

Methods. Since the amount of RNA enzymatically formed in the test system cannot be measured directly, the labeling technique is employed, just as in measuring the in-vitro synthesis of DNA: the rate of incorporation of ^{14}C- or ^{32}P-(α)-containing nucleoside triphosphates into acid-insoluble material in the presence of native DNA serves as a measure of *RNA polymerase* activity. From the incorporated radioactivity net RNA synthesis can be determined.

If for instance UTP with ^{14}C-labeling of the base moiety serves as substrate together with the three unlabeled nucleoside triphosphates, incorporation of ^{14}C-UMP may occur terminally or non-terminally, i.e. within the chain. The actual position of the unit can be established by examining the cleavage products after alkaline hydrolysis of the synthesized ribopolynucleotide: terminal incorporation gives rise to formation of uridine, released as ^{14}C-labeled terminal nucleoside; non-terminal insertion, however, yields 2', 3'-UMP.

The dependence of RNA biosynthesis on DNA is reflected to a certain extent in the fact that *RNA polymerase* is obviously associated with the same cellular structures as is DNA. In mammalian tissues and plant cells the enzyme is almost certainly bound to the chromatin in the cell nucleus and may be isolated with this. *RNA polymerase* of *E. coli* and of *Anacystis nidulans* as well has been recovered in good yields together with the bulk DNA from homogenates subjected to ultracentrifugation.

RNA Synthesis in Vitro. The ribopolynucleotide formed as reaction product when *RNA polymerase* from bacteria is used, exhibits the same base composition as the DNA added as primer except that uracil in the RNA corresponds to thymine in the DNA. The quotient $\frac{A + U}{G + C}$ has the same value as $\frac{A + T}{G + C}$ of the primer DNA. This obviously DNA-directed arrangement of ribonucleoside triphosphates in chain-like macromolecules (on which the concept of matrix RNA rests, too) corresponds to the mechanism of RNA synthesis in the living cell as indicated by the experimental data obtained so far. We shall later examine the important question whether both or only one DNA strand is required as template (p. 347).

The establishment of the nucleotide sequence of the nascent RNA by DNA very probably takes place according to the principle of base pairing which is employed in DNA biosynthesis, too. The DNA strand provides the *template* or *matrix* for the polynucleotide chain to grow along. Its elongation proceeds as follows: a nucleoside triphosphate is selected which fits the corresponding nucleotide base in the template DNA in a sterically complementary way. For each nucleotide unit added to the new RNA chain one molecule of pyrophosphate is released and a 3′-5′ phosphodiester linkage formed. This process is termed *transcription*. When adenine is the nucleotide base in the DNA, then uracil is the nucleotide base in the growing "m-RNA" chain; in the same way thymine fits with adenine in m-RNA:

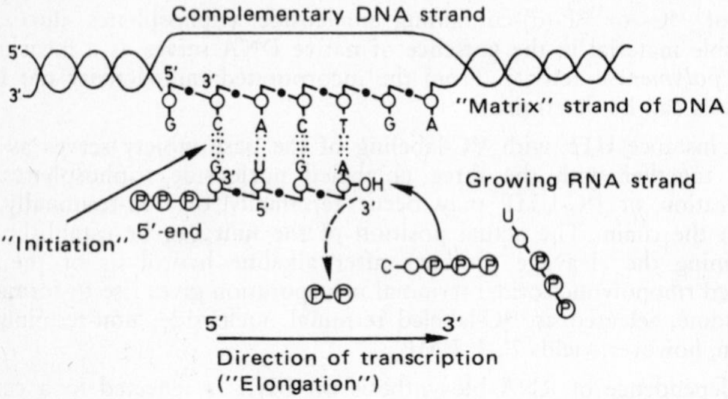

The common fundamental principle underlying DNA and RNA biosynthesis is evident; DNA, in the same way as in its own replication, determines unequivocally the sequence of nucleotides in the new RNA chain which thus has a sequence complementary to that of the DNA strand serving as template. DNA template and the newly formed single-stranded m-RNA combine to a double-helical structure in an intermediate stage of the transcription. Finally, the ribopolynucleotide so formed peels away from the DNA and is released.

The proposed mechanism of RNA synthesis described above has been confirmed by studies in which the DNA template was replaced by synthetic polydeoxyribonucleotides consisting of nucleotides of one or two types only, the test system also containing the four ribonucleoside triphosphates, as well as $Mg^{2\oplus}$ or $Mn^{2\oplus}$. With the polymer dAT (see formula) only ATP and UTP are used to synthesize ribopolynucleotide, which accordingly contains the bases adenine and uracil in alternating and complementary positions in its nucleotides:

Deoxyribose-AT-polymer

Product of synthesis =
Ribose-UA-polymer

In the same way a dGC polymer with CTP and GTP gives rise to formation of a ribopolynucleotide containing the nucleotide bases cytosine and guanine.

In Vivo. Recent findings indicate that in general only one strand of the DNA double helix contains the information required for synthesis of a complementary RNA chain (= "coding strand" or "transcription strand"). This information is obviously not distributed over its entire length, but often restricted to single segments which are moreover separated by inactive sequences (cf. p. 329).*

The second, complementary DNA strand probably stabilizes the DNA structure and is used as template for repairs that might be necessary (see below). The formation of complementary RNA chains along both strands of the primer DNA under in vitro conditions, catalyzed by purified *RNA polymerase,* certainly does not take place either in the bacterial cell or in the cell nucleus. Apparently, only relatively small portions of the DNA are actively involved, while the larger part remains inactive.

As regards precise molecular details of the process of transcription we depend so far on hypothetical models. It probably starts with a local dissociation of the double-stranded DNA in order that the segments of the strand containing information may become accessible for the subsequent formation of the complementary RNA. The chain of the latter is constructed in the $5' \rightarrow 3'$ direction; consequently the RNA strand and the transcription strand run in opposite directions (see scheme, p. 346). It is believed that upon termination of the transcription the DNA segment which has been used as template will be completely degraded. The gap thus formed is probably closed by de-novo synthesis of DNA using the stabilizing second strand as template (the possible involvement of *DNA polymerase I* in this "repair" mechanism has already been mentioned; p. 333).

It may be assumed that in this way a specific RNA is formed for each gene or group of genes which contains the information for the construction of a polypeptide chain (p. 389). This process is only possible if the *RNA polymerase* recognizes and controls the initiation, direction and termination of RNA synthesis when using a specific segment of the DNA

---

* Several bacteriophages, however, are known to contain informational sequences in both strands of their DNA.

which comprises a gene. A master programming system (= *regulation* p. 446) will also have to signal as to when and to what extent transcription is to occur, i.e. when a gene is to be activated and its product is to be formed. First indications regarding the mode of action of this regulatory principle have come from studies on the possible function of the sub-units of bacterial *RNA polymerases* in the process of transcription, par-ticularly with phage DNA as template. A specific nucleotide sequence in the DNA seems to determine the "initiation site" where the first internu-cleotide link in the m-RNA chain is formed; it always has a purine nu-cleotide at the 5′ end followed by UMP: pppGpU···· or pppApU·····. Apparently the conformation of the *RNA polymerase* molecule is changed by the σ subunit in such a way that this specific initiation site is recognized by the enzyme; synthesis of m-RNA ("initiation") starts after binding of the other subunits (= $\beta\beta'\alpha_2$: "holoenzyme" or "core") to the DNA strand. This specific binding site for *RNA polymerase* from which transcription of a gene or a sequence of genes proceeds is termed *promotor*. Once the initial internucleotide bonds have been achieved the factor σ is released; thus presumably permitting the holoenzyme to move along the DNA template and bringing about the subsequent *elongation* of the m-RNA chain. In cells of *E. coli* these newly formed RNA chains attached to DNA can be observed in electron micrographs. Several other protein factors are believed to be involved in transcription, supplementing the action of factor σ. Of these only "rho" (ρ) has been identified so far; this protein (molecular weight 200 000) from *E. coli* is responsible for recognizing specific termination signals by *RNA polymerase* and for the subsequent release of the completed m-RNA chain from the DNA strand.

Details of m-RNA formation in eukaryotic cell nuclei are not yet available. Although it is known that *RNA polymerase II* in the nucleoplasm (p. 344), which catalyzes this process, is composed of several subunits, their func-tions in the process of transcription are still obscure. The RNA molecules, isolated from the nucleus after pulse labeling cells by using suitable precursors of RNA synthesis, are of high molecular weight or high S value (about 10 million; 50 to 70 S). They are heterogeneous in com-position as well as in function. Only a small portion represents native precursors of m-RNA, while the bulk is subject to rapid turnover and does not invade the cytoplasm. The small amount of this *heterogeneous nuclear RNA (hnRNA)* mentioned first may contain nucleotide sequences in its molecules which are complementary to the actual informational sequences of DNA (cf. p. 329). They may well be the result of a maturation process in which hn-RNA is cleaved into smaller pieces and released into the cytoplasm, probably in the form of ribonucleoprotein particles (cf. p. 342).

In fact, such RNA molecules with a molecular weight significantly smaller than that of hnRNA have been found in association with the polysomes (p. 388). A crucial step in the conversion of hnRNA consists in the ad-dition of a tract of polyadenylic acid ("poly A") containing about 200 AMP residues at the 3′-terminus. Due to this poly-A segment hnRNA and m-RNA can easily be distinguished from the other RNA species and can also be selectively isolated, using polyuridylic acid or polydeoxythymidylic acid adsorbed to a carrier (cellulose, cross-linked dextrans) since these form double helical structures with the complementary sequence of the poly A tract.

Our knowledge of the process of transcription in the cell nucleus and of the subsequent formation of the m-RNA molecules is based mainly on findings obtained in specialized animal cells which due to their specific function produce mainly one species of protein and accordingly only one m-RNA species in relatively large amounts, e.g. globin (avian reticulocytes), fibroin (silkworms), histone (synchronized *HeLa* cells, sea urchin) and immunoglobulins. This limitation to only a few experimental objects is due to the fact that in the nuclei of most other cells a large number of m-RNA species are produced simultaneously, with often rapid and drastic changes occurring in quantity and composition, depending on the state of differentiation or function of the cell. It is thus much more difficult to detect and characterize hnRNA and m-RNA in normal cells than in highly specialized ones.

In this connection a rapidly synthesized RNA of high molecular weight from seedling tissue and plant cell cultures is of particular interest; it has a relatively high adenine content ($> 30 \%$) and is hence termed "AMP-rich RNA". Whether it is identical with hnRNA or m-RNA and whether it contains a terminal poly A tract is not yet known.

Ribosomal RNA. Whereas no conclusive evidence has so far been obtained for the existence of a definite precursor and its subsequent conversion to an end product in the formation of m-RNA, a "maturation" process (= *processing*) involving several stages has been elucidated for ribosomal RNA in eukaryotes. This was favored by some unique features of this transcriptional process taking place in the nucleolus which differs from that in the nucleoplasm in various respects: the genes responsible for the specification of ribosomal RNA can be relatively easily identified and isolated since they concentrate in relatively large numbers in the particular region of a single chromosome called the *nucleolar organizer* region. Gene clusters of 300–500 for the 18 S, 25 S and 28 S RNA of the ribosomes are typical for the genomes of many eukaryotes. This "redundancy" of genes is particularly high in certain amphibians where the oocytes contain approximately 1000 nucleoli.

Within the cluster, the DNA regions containing the information for the nucleotide sequences of both the ribosomal RNA species appear to alternate; between these are interspersed equal amounts of a DNA which is not transcribed (= "spacer"). During transcription the first identifiable product is a RNA of high molecular weight; its molecular chain contains the complete sequences of both the ribosomal RNA species separated by short, non-specific nucleotide sequences. The molecular weight of this precursors synthesized in the nucleolus differs from species to species and may vary between 4.8×10^6 and 2.9×10^6 daltons. About 1% of the nucleotides in the molecules carry a methyl group attached mainly to the ribose moiety (2'-methylribose!). Since this methylation appears

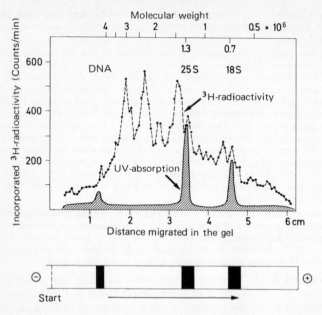

Fig. 95 Electrophoresis in 2% polyacrylamide agarose gel of the total of nucleic acids isolated from a plant cell culture *(Petroselinum sativum)* which had previously been incubated with ³H-uridine for 20 min (for method, see p. 342; after Richter). RNA of low molecular weight (4 S, 5 S) migrates from the gel under these conditions. The absorption profile shows the positions of DNA and the ribosomal 25 S RNA and 18 S RNA. The presence of several rapidly synthesized components (= precursors of ribosomal RNA!) with high specific activities is indicated by the radioactivity profile; the migration in the gel permits an estimation of the molecular weight (see upper scale).

to be typical for all the intermediates and for the final ribosomal RNA species, identification of all the RNA molecules involved in the maturation process is considerably facilitated.

In the nuclei of callus cells *(Petroselinum, Haplopappus)* a precursor of ribosomal RNA of 2.9×10^6 daltons has been demonstrated as well as the intermediates appearing during the "processing" and the end products – 18 S and 25 S ribosomal RNA – (Fig. 95). These findings confirm the results obtained in corresponding studies with tissues of plant seedlings.

Transcription of the 5 S ribosomal RNA seems to take place not in the nucleolus, but elsewhere in the nucleus, and hence irrespective of the "processing" in the former. This RNA species is later inserted into the larger ribosomal subunit (p. 384), a process which most probably takes place in the nucleolus (cf. p. 386).

A very small proportion of the DNA contains sequences with specific information for the molecular structure of the various **transfer RNA** species. These DNA sequences have been demonstrated by the technique of hybrid formation (p. 361). It has been estimated that there are 400 to 1000 genes for transfer RNA molecules per genome in eukaryotes. Like ribosomal RNA, transfer RNA undergoes a process of maturation: a precursor of high molecular weight is initially formed in the cell nucleus and then released into the cytoplasm where it is processed to yield transfer RNA. The most important events of this "trimming" are the enzymatic elimination of non-specific sequences and the methylation of nucleotide bases catalyzed by specific cytoplasmic *transfer RNA methylases*. A similar maturation process has been found for the transfer RNA of prokaryotes *(E. coli)*.

Inhibitors of Transcription. The specific inhibitory effect of the actinomycins and rifamycins, two groups of antibiotics, on DNA-dependent RNA synthesis has been successfully employed for specifying single steps. Particularly actinomycin D (= "actinomycin C_1") has been found to combine with double-helical DNA, probably adjacent to deoxyguanosine residues. The template is thus occupied and the formation of a complementary RNA chain is prevented. While only relatively small amounts of actinomycin are required for this effect, DNA replication is practically uninhibited, even by larger doses. The actinomycins have been found very useful when studying processes in which RNA synthesis dependent on DNA, i.e. formation of m-RNA, is assumed to be involved. Rifamycin and its derivatives, on the other hand, specifically form a stable complex with bacterial *RNA polymerase,* thereby blocking RNA synthesis. This effect has so far been observed in eukaryotic cells only in mitochondria and chloroplasts.

Enzymatic Degradation of Ribonucleic Acids

The specific enzyme of RNA degradation in microorganisms, plant and animal cells is *ribonuclease (RNase)*. It has been isolated from pancreas and purified to a crystalline form. The molecular weight is 13 000. *RNase* was the first protein molecule to have its sequence – 124 amino acids – elucidated (1959). *RNase* was also the first active enzyme to be synthesized (1969).

RNase exhibits a high specificity: it only cleaves phosphodiester bonds in which a pyrimidine riboside participates with its 3'-OH group; the 5' phosphoester bond is hydrolyzed:

Degradation of RNA by
ribonuclease (RNase)

Oligonucleotides with pyrimidine
nucleoside-3'-phosphate

Uridine-   Cytidine-
3'-phosphate   3'-phosphate

RNase degradation yields uridine-3'-phosphate, cytidine-3'-phosphate and oligonucleotides with a terminal pyrimidine nucleoside-3'-phosphate. A cyclic 2'3' phosphodiester is probably formed as an intermediate analogous to that in alkaline hydrolysis of RNA (p. 337 f); in contrast to the latter, it is hydrolyzed specifically to 3' phosphomonoester. This is the reason why *RNase* cannot degrade DNA.

Further cleavage of the oligonucleotides formed is brought about by the non-specific *phosphodiesterases* which have already been discussed (p. 334).

RNase with a limited specificity has been isolated from takadiastase powder, a product from molds (= "T_1"). Its action is exclusively restricted to the internucleotide bonds of RNA between 3' GMP and the 5'-OH group of adjacent nucleotides.

Due to their specificity and different mode of action on the polynucleotide chain RNA-cleaving enzymes are extremely useful aids in determining the sequence of individual RNA species.

References

Allen, R. C., R. Maurer: Electrophoresis and isoelectric focusing in polyacrylamide gel. De Gruyter, Berlin–New York 1974

Darnell, J. E., W. R. Jelinek, G. R. Molloy: Biogenesis of mRNA: Genetic regulation in mammalian cells. Science 181: 1215, 1974.

Geiduschek, E. P., R. Haselkorn: Messenger RNA. Ann. Rev. Biochem. 38: 647, 1969

Harbers, E.: Einführungen zur Molekularbiologie: Nucleinsäuren. Thieme, Stuttgart 1969

Holley, R. W.: The nucleotide sequence of a nucleic acid. Sci. Amer. 214: 30, 1966

Knippers, R.: Einführungen zur Molekular-
biologie Bd. 2: Molekulare Genetik.
Thieme, Stuttgart 1971
Kornberg, A.: The synthesis of DNA. Sci.
Amer. 219: 64, 1968
Loening, U.: RNA structure and meta-
bolism. Ann. Rev. Plant Physiol. 19:
37, 1968
Richardson, C. C.: Enzymes in DNA

metabolism. Ann. Rev. Biochem. 38:
795, 1969
Schötz, F.: Vererbung außerhalb des Zell-
kerns. Umschau 68: 267, 1968
Stewart, P. R., D. S. Letham: The Ribonucleic Acids. Springer, Berlin–Heidelberg–New York 1973
Weinberg, R. A.: Nuclear RNA metabolism. Ann. Rev. Biochem. 42: 329, 1973

Proteins

The second group of cellular macromolecules, the proteins, differ from the nucleic acids in that they do not contain phosphorus, but sulfur. Specificity is as important for proteins as it is for nucleic acids. The information carried by the DNA is expressed in an enormous number of specifically constructed protein molecules. The surprising diversity of proteins in the cell is the result of a vast variability in the molecular constitution of these macromolecules. The cell is able to meet the high demands made on it in this respect mainly because about twenty different amino acids are at its disposal as building blocks. It is obvious that a practically unlimited number of possibilities exist for different amino acid sequences in a protein chain. Moreover, the length of the chain-like molecule is also variable by changes in the number of amino acid units. The conditions for maintenance of high specificity are far more favorable in proteins than in the nucleic acids with their four building blocks; they may be considered nearly ideal in proteins. In the same way as in the macromolecules of nucleic acids, only a strict preservation of the sequence of building blocks guarantees that the protein molecule formed obtains its species-specific structure and is able to carry out its biological function, for example as an enzyme with a high specificity.

There is no doubt that biosynthesis of proteins is one of the most important activities of the cell to which it must devote about 90 % of its energy production. The process is relatively complex since the building blocks must be assembled into molecules in an exact and specific order according to certain "blueprints". This function of a "blueprint" is assumed by molecules of the *messenger RNA* which transfer the information required from the genetic material to the sites of protein synthesis in the cytoplasm. Thus biosynthesis of proteins will only take place in the presence of an active messenger RNA. The nucleotide sequence of its molecules contains the code for the amino acid sequence of each protein (see p. 389 f). We shall later discuss the complex process of protein biosynthesis in detail.

Let us first examine the amino acids, the typical building blocks of proteins. Some of their properties are also characteristic for the macromolecules of proteins which they form.

Amino Acids, the Building Blocks of Proteins

Amino acids represent a further class of compounds containing nitrogen as an additional element, in some instances also sulfur. In the same way as in the purine and pyrimidine components of the nucleotides, nitrogen gives a basic character to the amino acid too, which is here, however, matched by a functional group of acid character: apart from the typical amino group ($-NH_2$), amino acids also contain a carboxyl group ($-COOH$) to which the amino group is always in α-position. Since the substituents of the first two carbon atoms are common to all amino acids a general formula may be obtained by writing "R" for the remaining aliphatic or aromatic residual side chains of the molecules; they may carry additional functional groups:

L-amino acid   D-amino acid   Glycine

The α-carbon atom of all amino acids is asymmetric (provided that "R" is not identical with hydrogen as in glycine!); hence these compounds are optically active. There are two configurations ("enantiomers") of each amino acid (with the single exception of glycine): the one with the amino group on the left is termed "L", the other with the amino group on the right accordingly "D".

Although the two enantiomers do not differ in their chemical and most of their physical properties, they do differ in their ability to rotate the plane of polarized light in solution. The optical rotation, referred to by + (dextro-rotatory) and − (levo-rotatory), is not related to the D- or L-form of a compound*.

---

* A solution containing 1 g/ml of an optically active substance serves as a reference. The rotation of the plane of polarized light is measured on passing through a 10 cm layer of the solution. Since the rotation measured is proportional to the concentration of the substance dissolved, only very small amounts are required. The "specific rotation" is calculated from the values obtained.

With R = H, glycine (formula p. 354) is the smallest amino acid. Glycine lacks an asymmetrically substituted carbon atom, and is hence optically inactive. Neither D-form nor L-form exist. All the other amino acids occurring in proteins are exclusively of the L-form; obviously this form is preferred in most other biologically active amino acids, too. Chemically synthesized amino acids are optically inactive since they are mixtures of the D-form and the L-form. Purified amino acids are colorless and mainly soluble in water; they usually form crystals.

The arrangement of the substituents at the α-carbon atom is shown for the two C_3 amino acids L-alanine and L-serine:

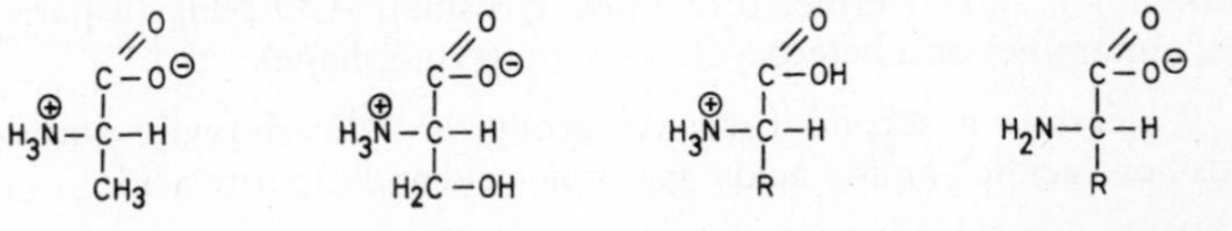

Alanine   Serine   Low pH value   High pH value
                                 (= high H⊕ conc.)   (= low H⊕ conc.)

The two formulae take into account another characteristic feature of amino acids concerning the charge of the amino group and the carboxyl group: in aqueous solution amino acids form "zwitterions" (= dipolar ions) as shown in the diagram, i.e. the two functional groups of the same molecule exhibit opposite charges. The acid carboxyl group loses a H⊕ ion by dissociation while the basic amino group acquires a positive charge by binding a H⊕ ion. By changing the H⊕ ion concentration in the solution it is possible to neutralize one of the two groups: COO⊖ by raising the H⊕ ion concentration (= acidification), NH₃⊕ by lowering it (= alkalization; formulae above).

Therefore, the aqueous solution of an amino acid does not contain any uncharged molecules: it is rather a mixture of zwitterions and positively charged molecules or of zwitterions and negatively charged molecules. The composition of the mixture depends on the dissociation constants of the COOH and NH₂ groups present and on the H⊕ ion concentration in the solution (= pH value). Since the acidic and basic groups react like weak acids and bases, respectively, the law of mass action applies to these, too. Accordingly a dissociation constant (K) can be calculated for each group. Its negative logarithm, the "pK value", is between 2.0 and 2.5 for the carboxyl group, between 9.0 and 9.5 for the amino group.

For each amino acid in solution a state may arise in which most of its molecules are present as zwitterions and there are only very few with a positive or a negative charge. This state is termed the "isoelectric point" because at this pH value the amino acid bears no net charge (positive and negative charges exactly balance one another). The isoelectric point is not only typical for amino acids but also for the macromolecules of proteins consisting of amino acids (cf. p. 379 f).

Classification of Amino Acids. The structures of the individual amino acids (those of significance for protein synthesis are shown on p. 357) illustrate clearly that the side chain "R" may vary greatly according to the attachment of different functional groups several of which may be charged. Accordingly amino acids may be classified into four groups:

1. "R" is an unchanged hydrocarbon chain: glycine, alanine, valine, leucine, isoleucine, proline, phenylalanine.

2. "R" contains non-ionized but polar groups: -SH (cysteine, cystine, methionine), -OH (serine, threonine, tyrosine), $-CO-NH_2$ (asparagine, glutamine) or a heterocyclic system (tryptophane).

3. "R" carries a second carboxyl group = "dicarboxylic amino acids" or "acidic" amino acids: glutamic acid and aspartic acid.

4. "R" has an additional amino group = "diamino monocarboxylic acids" or "basic" amino acids: lysine, arginine, histidine.

Apart from the characteristics shared by an amino acid with other members of the group, individual differences may occur in the structure of the side chain. Let us briefly consider these in a few compounds.

L-valine (L-aminoisovaleric acid), *L-leucine* (L-aminoisocaproic acid) and *isoleucine* (L-amino-β-methylvaleric acid) have a branched hydrocarbon chain. These substances are very similar in structure as well as in their properties. Animal organisms generally lack the ability to synthesize the branched carbon skeleton of these amino acids. Since they must be included in the diet, valine, leucine and isoleucine belong to the amino acids "essential" for animal organisms. On the other hand, the green plant cell and most microorganisms are able to synthesize all amino acids including those with a branched carbon skeleton.

L-proline and *L-hydroxyproline* have a cyclic molecule structure. The N atom in the α-position is incorporated into a ring (pyrrolidine) giving rise to formation of a secondary amine. L-hydroxyproline belongs to Group 2 because of its hydroxyl group. The molecule of *L-phenylalanine* (α-amino-β-phenylpropionic acid) and that of *L-tyrosine* (α-amino-β-hydroxyphenylpropionic acid) contain an aromatic ring. Since the latter cannot be synthesized by higher animals at all, these amino acids are also "essential". The phenolic OH group of tyrosine is acidic by nature: a proton dissociates at low $H^\oplus$ concentration (pH 9.0 and higher). *L-tryptophane,* an amino acid with a N-containing indole ring, is also an "aromatic" amino acid.

Glycine (Glykokoll), L-alanine, L-valine, L-leucine, L-isoleucine, L-proline, L-phenylalanine — Group 1

L-cysteine, L-cystine, L-methionine, L-serine, L-threonine, L-tyrosine, L-asparagine, L-glutamine, L-tryptophane — Group 2

L-aspartic acid, L-glutamic acid — Group 3

L-lysine, L-arginine, L-histidine — Group 4

Each of the three aromatic amino acids exhibits a characteristic ultra-violet absorption, with a maximum near 280 nm. They retain this property when inserted into protein macromolecules. Therefore, the absorption spectrum of a protein solution has a characteristic maximum in the ultra-violet region at 280 nm. A commonly employed method of quantitative protein estimation is based on this feature.

The OH group in *L-serine* (α-amino-β-hydroxypropionic acid) may react with an acid to form an ester. The phosphoric acid ester of serine is an important metabolite in the synthesis of phosphatides (cf. p. 417).

As indicated by the name, *L-threonine* (α-amino-β-hydroxybutyric acid) is related to the C_4 sugar threose. Four diastereoisomers exist of this amino acid since the molecule has two asymmetric centers. Two pairs of compounds are related to each other as an object is to its reflected mirror image: D- and L-threonine, and D- and L-allothreonine, respectively.

$$
\begin{array}{c}
COOH \\
| \\
H_2N-C-H \\
| \\
HO-C-H \\
| \\
CH_3
\end{array}
$$

L-allothreonine

With sulfur a new element enters the side chain of the amino acids *L-cysteine* (α-amino-β-mercaptopropionic acid), *L-cystine* and *L-methionine* (α-amino-β-methylthiolbutyric acid). As a free or bound functional SH group it confers a certain reactivity on these compounds. Thus cysteine readily reacts with a second molecule to form the diamino dicarboxylic acid cystine, in which a disulfide bridge (-S-S-) formed by oxidation links the two residues. Cystine is stable in this form. Cysteine may also form a disulfide bridge when incorporated in a polypeptide chain. In this way two different macromolecules or segments of the same macromolecule are linked by cysteine residues (cf. p. 374).

In *L-asparagine* and *L-glutamine* the additional carboxyl group in the side chain of the corresponding dicarboxylic amino acids, *L-aspartic acid* and *L-glutamic acid*, has been converted to an amide group. This gives a hydrophilic character to the compounds. Hydro-lytic cleavage with alkali or acid releases the dicarboxylic amino acid and NH_3.

L-lysine (α,ε-diaminocaproic acid) is an amino acid with a second amino group in the side chain. At a low $H^\oplus$ concentration (pH 9.0) only this amino group is ionized, while at a higher $H^\oplus$ concentration (pH 5.0) the α-amino group is ionized too:

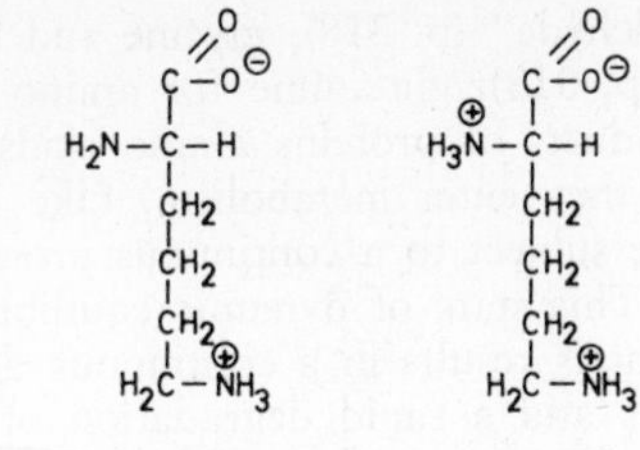

pH 9.0   pH 5.0

L-lysine

L-arginine (α-amino-β-guanidinovaleric acid) also acts as a base because of the guanidino group in the molecule; it marks this amino acid as a derivative of the base guanidine:

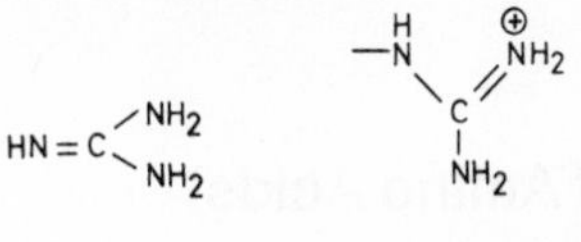

Guanidine   "Guanidino" group

L-histidine (α-amino-β-imidazolylpropionic acid) is endowed with weak basic properties due to the proton attraction by an N atom of the imidazole ring. We have already encountered this hetero-cyclic system as a constituent of the purine bases (p. 313). Histidine is formally derived from alanine the β-carbon atom of which carries the imidazole ring as a substituent. All three "basic" amino acids (lysine, arginine, histidine) happen to contain six carbon atoms each ("hexon bases"!).

The additional acidic and basic groups in the side chains which are not changed when the amino acids are inserted into peptide or protein molecules, cause the "amphoteric" character of these macro-molecules: proteins are also zwitterions and have an "isoelectric point" (cf. p. 379).

Paper chromatography (p. 35 ff), ion-exchange chromatography (p. 36 f) and thin-layer chromatography (p. 337 f) are appropriate methods for the isolation and identification of individual amino acids from a mixture resulting from hydrolytic cleavage of proteins.

Metabolism of Amino Acids. Apart from their function as building blocks of proteins, a number of amino acids fulfill other functions in the cell as well. We have already become acquainted with some of these compounds, e.g. aspartic acid and glycine, which play key roles in the synthesis of the purine and the pyrimidine ring, re-spectively (p. 315 ff); furthermore, several amino acids act as donors of active groups: methionine for "active methyl" (p. 424), serine for "active formaldehyde" (p. 318), glycine and histidine for "active formic acid" (p. 318), glutamine for amino groups (p. 319). As degradation products of proteins amino acids undergo further conversions before they enter metabolism. Like most other compounds proteins are subject to a continuous process of renewal in the cell (p. 309 f). This state of dynamic equilibrium between degradation and synthesis results in a continuous demand for newly formed amino acids and a rapid degradation of the "old" compounds released as cleavage products. The latter may be partially modified to enter other synthetic processes or may be completely degraded to CO_2 and H_2O by the reactions of dissimilation, thus yielding energy. The specific reactions involved in the decomposition of amino acids are discussed in connection with protein degradation (p. 396 ff). Let us now outline the biosynthesis of amino acids.

Biosynthesis of Amino Acids

First evidence for the close coupling of the reactions of photosynthetic CO_2 reduction to the synthesis of some amino acids in the green plant cell has come from short-term pulse-labeling experiments with $^{14}CO_2$ (cf. p. 167). The rapid "flow" of radiocarbon not only into sugar phosphates but also into organic acids and amino acids indicates the close connections between the amino acids and the intermediates of photosynthetic CO_2 reduction cycle, glycolysis and citric acid cycle. Serine, alanine, glutamic acid and aspartic acid which differ from the other amino acids in that they exhibit a preferentially rapid incorporation of radiocarbon are probably formed from the intermediate keto acids. The synthesis of several other amino acids with multiform carbon skeletons which can be only assembled by a sequence of various reactions is less clear. Valine, leucine, isoleucine, phenylalanine, tyrosine and tryptophane belong to this group. Sometimes the rapidly synthesized (= "primary") amino acids serve as precursors in this group of amino acids, sometimes they deliver parts for the construction of the molecules. The scheme below shows the pathway of synthesis of the various amino acids, the origin of their carbon skeletons being used as basis.

The scheme shows clearly that each group of amino acids originates from a typical intermediate of the three most important metabolic cycles in the autotrophic cell. Thus the description of the CO_2 reduction cycle and the citric acid cycle as "emporium" or "reservoir" of metabolic intermediates must be extended to the process of glycolysis which also contributes intermediates for syntheses in this case.

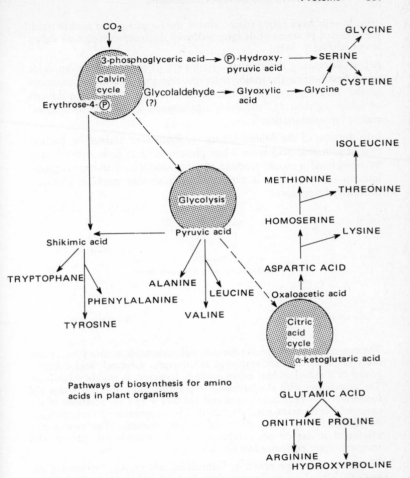

Pathways of biosynthesis for amino acids in plant organisms

The amino acids derived from a common precursor will be treated together. In this way most amino acids (with the exception of histidine and the aromatic amino acids) can be classified as belonging to one of several "families": the "glutamic acid family", the "aspartic acid family", the "pyruvic acid family" and the "serine family".

Some of the reaction sequences described here have so far been studied only in bacteria and fungi; their occurrence in higher plants has not yet been demonstrated. However, a series of results, particularly those obtained from labeling experiments with ^{14}C-precursors and amino acids, indicate that the same reaction pathways are generally followed in plant

cells. In cells from carrot tissue cultures the formation and further transformation of 14 amino acids agree well with the reaction sequences shown
in our scheme (p. 361).

Before discussing the synthetic pathways of individual amino acids
let us first outline the major reactions required to convert ammonia
to the amino group of amino acids. There are three possibilities
of achieving this: 1) "reductive amination"; 2) "amide formation"
and 3) "transamination".

Introduction of the Amino Group. *1) Reductive Amination.* Formation of glutamic acid from α-ketoglutaric acid in higher plants may
be considered a model reaction for introduction of an amino group
into an α-ketoacid; it is catalyzed by a specific enzyme, *glutamic
acid dehydrogenase:*

$$
\begin{array}{c}
COO^{\ominus} \\
| \\
C = O \\
| \\
CH_2 \\
| \\
CH_2 \\
| \\
COOH
\end{array}
+ NH_3
\quad
\xrightarrow[\text{Glutamic acid dehydrogenase}]{\text{NAD-H+H}^{\oplus} \quad \text{NAD}^{\oplus}}
\quad
\begin{array}{c}
COOH \\
| \\
H_2N-C-H \\
| \\
CH_2 \\
| \\
CH_2 \\
| \\
COOH
\end{array}
+ H_2O
$$

α-ketoglutaric acid   Glutamic acid

The enzyme is highly specific for α-ketoglutaric acid: neither pyruvic acid
nor oxaloacetic acid are employed as substrates. *Glutamic acid dehydrogenase* is a mitochondrial enzyme which, however, is readily solubilized.
The association with this organelle is plausible since it is there that the
formation of α-ketoglutaric acid and NAD-H + H$^{\oplus}$ via reactions of the
citric acid cycle takes place (p. 221 f). The occurrence of corresponding
dehydrogenases in plant tissue cells is still disputed. The experiments
conducted so far do not completely exclude possible interference with
transamination processes (see below).

The reversal of the reaction formulated above, i.e. "oxidative deamination", plays an important role in the degradation of amino
acids, particularly in senescent or detached plant organs (cf. p.
398).

Reductive amination of α-ketoglutaric acid does not only bring about the
synthesis of glutamic acid but, as observed in recent studies with yeast
(Torulopsis utilis) and barley roots, obviously plays a key role in NH$_4^{\oplus}$
assimilation of the cell. Therefore, this reaction controls the incorporation of the major part of ammonia nitrogen into organic compounds. A minor part probably comes from glutamine (see below) which
is also formed via glutamic acid (α-ketoglutaric acid → glutamic acid →
glutamine). Thus glutamic acid holds a key position in the metabolism
of free nitrogen compounds.

2) Amide formation. The second important reaction by which ammonia is incorporated directly into amino acids of the plant cell consists in formation of an amide group, characteristic for glutamine and asparagine. Glutamine synthesis is an endergonic process ($\Delta G° = +3500$ cal/mole) and hence only proceeds in the presence of ATP. The specific enzyme involved, *glutamine synthetase,* catalyzes the following reaction:

$$\text{Glutamic acid} + \text{NH}_4^\oplus \underset{\text{"Synthetase"}}{\overset{\text{ATP} \quad \text{ADP} + \textcircled{P}}{\rightleftharpoons}} \text{Glutamine}$$

Glutamine synthetase has been purified from pea seeds. It requires $Mg^{2\oplus}$ or $Mn^{2\oplus}$ as a cofactor. The enzyme also occurs in algae, in many species of gymnosperms and angiosperms and in fungi and bacteria. Association with a cellular organelle has not been demonstrated up to now. It is a most interesting enzyme both in terms of structure and metabolic regulation (see p. 294), the latter reflecting the fact that the reaction product glutamine serves as donor of NH_2 or nitrogen atoms for a number of products including the purine system (p. 319 f).

Synthesis of asparagine from aspartic acid is probably brought about by a similar mechanism. An active *asparagine synthetase* has not yet been demonstrated in plant cells, although formation of asparagine has been observed in cell-free extracts of lupine and wheat seedlings after addition of aspartic acid, $NH_4^\oplus$, ATP and $Mn^{2\oplus}$.

3) Transamination. Most amino acids in the cell are subject to enzymatic conversion to the corresponding ketoacid. Specific enzymes, *transaminases,* cleave off the amino group and attach it temporarily to their prosthetic group, using it later to convert a ketoacid to the corresponding amino acid.

The prosthetic group of the *transaminases* is pyridoxal phosphate, a pyridine derivative which is closely related to the vitamin pyridoxine (formerly called "vitamin B_6" or "adermin"). The compound also acts as prosthetic group in several decarboxylases specific for amino acids (p. 399).

The action of pyridoxal phosphate is probably based in all cases on formation of a reactive intermediate, a "Schiff base", which reacts with the amino acid:

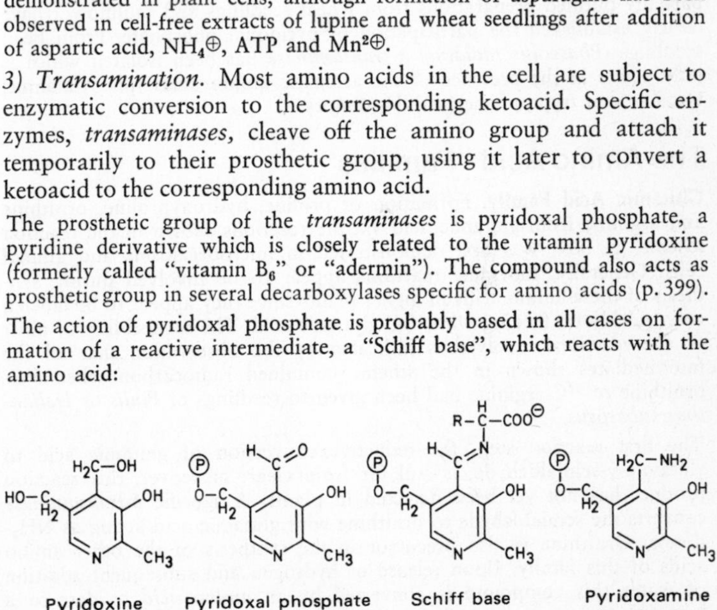

Pyridoxine   Pyridoxal phosphate   Schiff base   Pyridoxamine phosphate

The α-ketoacid is released from the complex via a mesomeric intermediate state while the amino group remains attached to the prosthetic group, converting it to pyridoxamine phosphate.

The α-ketoacid formed enters the "pool" of the citric acid cycle; pyridoxal phosphate is restored by transfer of the amino group to a suitable acceptor. This role is mostly assumed by α-ketoglutaric acid and oxaloacetic acid which are converted to glutamic acid and aspartic acid, respectively (see reaction p. 362). The preferential formation of these two amino acids is obviously due to their key position as precursors in the synthesis of other amino acids, which accordingly belong to the "glutamic acid family" and the "aspartic acid family", respectively (p. 361).

In higher plants transamination probably represents the final reaction step in the biosynthesis of glycine, alanine, valine, leucine, isoleucine, phenylalanine, tyrosine and probably also of serine. Amino acids which are not building blocks of proteins and belong to the "unusual" amino acids in plant organisms may also serve as NH_2 donors, as indicated by recent studies.

Transaminases do not only occur in cells of higher plants but also in algae. In the cell they are mainly located in the mitochondria. Although a plant *transaminase* has not yet been isolated and purified, studies on partially purified preparations from wheat seedlings and cauliflower have clearly established the participation of pyridoxal phosphate. From bean seedlings *(Phaseolus radiatus)* a *transaminase* has been isolated which is only active in the presence of the aromatic amino acids (phenylalanine, tyrosine and tryptophane) as NH_2-donors.

The Amino Acid "Families"

Glutamic Acid Family. Formation of proline, hydroxyproline, ornithine and arginine from glutamic acid via the reactions shown in the reaction scheme (p. 365) was first demonstrated in microorganisms and animal cells. Generally, the same reactions appear to be involved during synthesis of these amino acids in higher plants since they appear to be labeled very quickly on feeding with [14]-C-glutamic acid. The reversibility of most of the reactions involved is indicated by the observation that all the intermediates shown in the scheme contained radiocarbon when ^{14}C-ornithine or ^{14}C-arginine had been given to seedlings of *Pinus* or *Helianthus tuberosus*.

The first reaction step, the reductive conversion of glutamic acid to glutamic-γ-semialdehyde, is still far from clear; moreover, this reaction product has not yet been detected in plants. A specific *δ-transaminase* converts the semialdehyde to ornithine with glutamic acid acting as NH_2-donor. Ornithine is also precursor in the synthesis of the other amino acids of this family. Upon release of hydrogen and subsequent addition of water this compound is converted by an *amino acid oxidase* to a ketoacid, δ-amino-α-ketovaleric acid (p. 365).

This compound spontaneously forms the cyclic structure of pyrroline-5-carboxylic acid. In a similar reaction glutamic-γ-semialdehyde may also form the ring structure of pyrroline-5-carboxylic acid. Both are converted to proline by the action of a *reductase* and in the presence of NAD-H + H$\oplus$ or NADP-H + H$\oplus$. The enzyme is active in extracts of bean and pea seedlings and requires pyrroline-5-carboxylic acid as substrate.

Conversion of proline to hydroxyproline probably proceeds via several intermediates one of which is identical with the threo-γ-hydroxy-L-glutamic acid occurring abundantly in many *Liliaceae* species. Ornithine is the key compound in synthesis of arginine as well; it first reacts with carbamyl phosphate (biosynthesis p. 315) to form citrulline (see reaction scheme on p. 366). The enzyme involved is accordingly called *ornithine transcarbamylase;* it has been found in seedings and algae. Citrulline is a common compound in many plant cells, occurring in relatively

Carbamyl phosphate

L-ornithine   Citrulline   Argininosuccinic acid

Arginine   Fumaric acid

large quantities in several species of the *Cucurbitaceae*. The condensation of citrulline with aspartic acid yields argininosuccinic acid in an endergonic reaction requiring ATP. The elimination of fumaric acid from the latter compound completes the synthesis of arginine. Studies on cell-free extracts of *Chlorella* and of seedlings have confirmed that two enzymes are involved: *argininosuccinic acid synthetase* and *argininosuccinase*. The same reaction principle, whereby aspartic acid contributes an additional N atom for the construction of a new molecule, was observed in biosynthesis of the purine ring (cf. p. 319 f).

Aspartic Acid Family. The importance of aspartic acid as precursor of threonine, isoleucine, methionine and lysine has been shown mainly in studies on bacteria. There are only fragments of information concerning a similar function of this amino acid in higher plants. They indicate, however, that the same pathways are followed as in bacteria. Homoserine exhibits a key position in this connection. This compound is not a protein building block, but occurs freely in seedlings and mature plants. It is formed from aspartic acid via three reaction steps (see scheme). Initially, aspartic acid is converted to the energy-rich compound of aspartyl phosphate in an ATP-dependent reaction catalyzed by *aspartate kinase* (1); hydrogen from NADP-H + H⊕ is enzymatically attached *(aspartate semialdehyde dehydrogenase)*, giving rise to formation of a highly reactive semialdehyde (2); this is reduced to the corresponding alcohol, homoserine, with consumption of an additional equivalent of NADP-H + H⊕ (or NAD-H + H⊕) (3). The specific enzyme, *homoserine dehydro-*

Aspartic acid   β-aspartyl phosphate   Aspartic acid semialdehyde

Homoserine

genase, is activated by both nicotinamide nucleotide coenzymes. The reaction sequence was first elucidated in yeast; however, the enzymes involved occur in pea seedlings as well.

In *E. coli* the reaction catalyzed by *aspartate kinase* occupies a key position as regards the effective metabolic regulation of the synthesis of the three amino acids threonine, methionine and lysine (cf. p. 443). There are three separate and distinct *aspartate kinases;* one of these is specifically inhibited by threonine, one by methionine and one by lysine. Each of these amino acids as an end product controls the specific pathway by exerting influence on the formation of the common intermediate, i.e. aspartyl phosphate.

Homoserine is required for the biosynthesis of methionine, threonine and isoleucine.

The conversion to threonine is performed in two steps during which the hydroxyl group is shifted from the γ- to the β-carbon atom. Homoserine phosphate is formed in an ATP-dependent reaction by *homoserine kinase.* In the terminal step *threonine synthetase,* a pyridoxal phosphate-dependent enzyme, converts homoserine phosphate to threonine. The intermediates and enzymes involved have not yet been found in higher plants. The same holds for the formation of methionine from homoserine via the intermediates cystathionine and homocysteine which has been studied in bacteria. The final reaction consists in a methyl-transfer from N^5-methyltetrahydrofolic acid (cf. p. 318 f) to homocysteine by mediation of a specific *transferase:*

Homocysteine   Methionine

Two separate pathways are employed in the biosynthesis of lysine: one occurs in *E. coli*, algae, fungi and several higher plants, starting with aspartic acid-β-semialdehyde and yielding via several intermediates L,L-α, ε-diaminopimelic acid (formula see below). From this L-lysine is formed by a specific decarboxylase. In *Saccharomyces* and *Neurospora* acetyl-CoA or glyoxylate and α-ketoglutaric acid provide the carbon skeleton of lysine in a complex reaction sequence yielding homocitrate and L-α-aminoadipic acid as intermediates (see below). Reduction by NAD-H + H$^\oplus$ leads to the reactive compound α-aminoadipic-δ-semialdehyde which reacts with glutamic acid to form saccharopine. Cleavage of the latter yields α-ketoglutaric acid and lysine. It is not yet possible to decide which

L,L-α,ε-diamino-pimelic acid	L-lysine	L-α-amino-adipic acid	L-α-aminoadipic-δ-semialdehyde

of the two routes is characteristic for higher plants. While small quantities of L-α-aminoadipic acid occur in extracts of numerous plant species, synthesis of lysine proceeds in maize and pea plants with this precursor as well as with L,L-α,ε-diaminopimelic acid. In wheat seedlings, however, lysine synthesis has only been observed with the latter compound: with ^{14}C in the carboxyl group of L-α,ε-diaminopimelic acid, ^{14}C-carboxyl-labeled lysine was formed.

The initial reaction in isoleucine biosynthesis is catalyzed by *threonine dehydratase* deaminating threonine to α-ketobutyric acid; this reaction has been described for various species of higher plants. The further biosynthetic route has been elucidated by using extracts from bacteria and from *Neurospora*. α-Aceto-α-hydroxybutyric acid is formed by an enzyme-catalyzed condensation of α-ketobutyric acid with hydroxyethyl-TPP (= "active acetaldehyde"). The active coenzyme is thiamine pyrophosphate which is also involved in oxidative decarboxylation of α-ketoacids

α-ketobutyric acid	α-aceto-α-hydroxy-butyric acid	α-keto-β-methyl-isovaleric acid	L-isoleucine

(cf. p. 215 ff). The structural formula on p. 215 shows how the acetaldehyde group is attached to thiamine pyrophosphate.

α-Aceto-α-hydroxybutyric acid is reduced by NADP-H + H⊕ and subsequently converted to the α-ketoacid corresponding to isoleucine (α-keto-β-methylisovaleric acid) by cleaving off water. A transamination completes the synthesis of isoleucine. Several of the enzymes and intermediates involved occur in seedlings of *Phaseolus radiatus*.

Pyruvic Acid Family. The concept that valine and leucine are formed by a reaction sequence very similar to that involved in synthesis of isoleucine has been confirmed to some extent by studies of *Phaseolus radiatus*, mutants of *Neurospora crassa* and of *Salmonella typhimurium*. Pyruvic acid is thought to replace α-ketobutyric acid in the initial condensation reaction in the biosynthesis of valine. The conversion of the α-acetolactic acid formed (see scheme) to α-ketoisovaleric acid is identical in the biosynthetic pathways of both valine and isoleucine. It is probably catalyzed by the same enzymes involved in the formation of α-aceto-α-hydroxybutyric acid in the synthesis of isoleucine (see above). Finally, α-ketoisovaleric acid is converted to valine by transamination. This ketoacid

| Pyruvic acid | α-acetolactic acid | | α-ketoisovaleric acid | L-valine |

also assumes the role of a precursor in the synthesis of leucine proceeding along a separate route. It is condensed with acetyl-CoA yielding β-hydroxy-β-carboxyisocaproic acid which is converted via two intermediates to the immediate precursor of leucine, α-ketoisocaproic acid. This is converted to leucine by transamination.

L-alanine is formed particularly in photosynthetically active cells from pyruvic acid in a direct amination step, glutamic acid acting as NH_2-donor:

| Pyruvic acid | | L-alanine |

Serine Family. Serine and glycine belong to the amino acids in algal cells and leaves which are rapidly labeled during photosynthesis in the presence of $^{14}CO_2$. Their synthesis is closely connected with photorespiration

(p. 159 ff). In this complex reaction sequence glycine is probably first formed via glycolic acid and glyoxylic acid, and subsequently converted to serine by a *serine hydroxymethyl transferase* (cf. p. 161). Another mechanism whereby this amino acid is formed has been discovered in bacteria as well as in higher plants. It proceeds via 3-phosphoglyceric acid, phosphohydroxypyruvic acid and phosphoserine; the amino group is provided by glutamic acid or alanine. In the final step serine is formed by a *phosphatase* cleaving off the phosphoric acid residue. *L-alanine hydroxypyruvate transaminase* has the highest activity in preparations from leaves.

$$
\begin{array}{c}
\text{COOH} \\
| \\
\text{H}-\text{C}-\text{OH} \\
| \\
\text{H}_2\text{C}-\text{O}-\text{P}
\end{array}
\longrightarrow
\begin{array}{c}
\text{COOH} \\
| \\
\text{C}=\text{O} \\
| \\
\text{H}_2\text{C}-\text{O}-\text{P}
\end{array}
\xrightarrow{\text{Transamination}}
\begin{array}{c}
\text{COOH} \\
| \\
\text{H}_2\text{N}-\text{C}-\text{H} \\
| \\
\text{H}_2\text{C}-\text{O}-\text{P}
\end{array}
\xrightarrow[\text{Phosphatase}]{+\text{ H}_2\text{O}}
\begin{array}{c}
\text{COOH} \\
| \\
\text{H}_2\text{N}-\text{C}-\text{H} \\
| \\
\text{H}_2\text{C}-\text{OH}
\end{array}
+\text{ H}_3\text{PO}_4
$$

3-Phosphoglyceric   3-Phosphohydroxy-   Phosphoserine   Serine
acid   pyruvic acid

The antibiotics *azaserine* and *cycloserine* as well as *chloramphenicol* (p. 395) which is closely connected with phenylserine, are derivatives of serine; they are produced by members of the genus *Streptomyces*.

Glycine can be formed in various ways:

1. From serine by cleaving off "active formaldehyde"; tetrahydrofolic acid is involved as coenzyme (p. 318 f).

2. From 3-phosphoglyceric acid via 3-phosphohydroxypyruvic acid which is decarboxylated to phosphoglycolaldehyde and converted to glyoxylic acid via glycolic acid. An amino group is linked to this compound in the final reaction.

3. By cleavage of threonine to acetaldehyde and glycine, catalyzed by a *threonine aldolase*.

4. From a sugar phosphate of the *Calvin Cycle* via *photorespiration* (see above).

The biological significance of reaction 1 lies in the formation of a C_1 fragment, the hydroxymethyl group, which is required as "active formaldehyde" bound to tetrahydrofolic acid in various biosyntheses (p. 318).

Reversal of reaction 1, i.e. formation of serine from glycine and "active formaldehyde" has been observed in *Ricinus* endosperm and in carrots; glycine-2-C^{14} was completely incorporated into the serine molecule.

The transfer of H_2S to serine, first discovered in yeast, may also take place in plant cells and would account for the synthesis of cysteine in these cells. This reaction, which depends on pyridoxal phosphate, has been demonstrated in extracts of spinach leaves. The enzyme involved, *L-serine hydrolyase* or *cysteine synthetase*, is widely distributed. This fact has led to the inference that in many organisms assimilation of inorganic sulfur may proceed via sulfide after an initial reduction of sulfate (cf. p. 303 f).

Another way in which cysteine may be formed consists in a reversible condensation of homocysteine (= methionine without methyl!) with serine and subsequent cleavage of the cystathionine formed (p. 367) to cysteine and homoserine. Reversal of this reaction sequence produces homocysteine and methionine, respectively, from cysteine:

Homocysteine   Serine   Cystathionine   Homoserine   L-cysteine

Aromatic Amino Acids. Unlike animal organisms the autotrophic plant cell can synthesize aromatic compounds, too, i.e. it can assemble the benzene and indole ring systems from carbohydrate building blocks originating in the photosynthetic CO_2 reduction cycle. However, the biosynthetic pathways involved were elucidated not in higher plants but in microorganisms, which also synthesize aromatic compounds from relatively simple building blocks.

Native or experimentally produced biochemical mutants of these microorganisms have proved extremely useful in elucidating metabolic pathways. In each of these mutants the biosynthetic reaction sequence is blocked at another position; accordingly, certain intermediates accumulate. The nature of these intermediates and the ability of other compounds supplied to the cells to overcome the "blocks", permit reconstruction of the reaction sequence of a biosynthetic pathway to a large extent. This important method of *biochemical genetics* has been successfully used many times to elucidate complicated biosyntheses. Unfortunately we cannot go into details here.

Shikimic acid and prephenic acid (structures p. 372) are important intermediates in the biosynthesis of phenylalanine and tyrosine. Shikimic acid is a natural constituent of many plants and obviously occupies a key position in the metabolism of aromatic compounds. That biosynthesis of aromatic amino acids in plant cells is very similar to that in microorganisms is indicated by the formation of ^{14}C-labeled phenylalanine and tyrosine in *Salvia* after feeding with ^{14}C-labeled shikimic acid, and by the appearance of highly radioactive tryptophane when barley shoots are given ^{14}C-anthranilic acid or ^{14}C-shikimic acid.

We are fairly well informed about the synthesis of shikimic acid in plants. It is formed from erythrose-4-phosphate and pyruvic acid via the 5-dehydroquinic acid, a derivative of cyclohexane. When water is cleaved off, 5-dehydroshikimic acid is formed, which is reduced to shikimic acid by NADP-H + $H^{\oplus}$. After phosphorylation of the latter by ATP and condensation with phosphoenolpyruvic acid giving rise to formation of 3-enolpyruvyl shikimic acid-5-phosphate as intermediate, chorismic acid is formed (reaction scheme I). The enzymes involved have been found in a number of higher plants.

At present we know very little about the details of the conversion of chorismic acid to the keto-analogs of phenylalanine and tyrosine. There are probably two ways in which prephenic acid can be formed, leading to

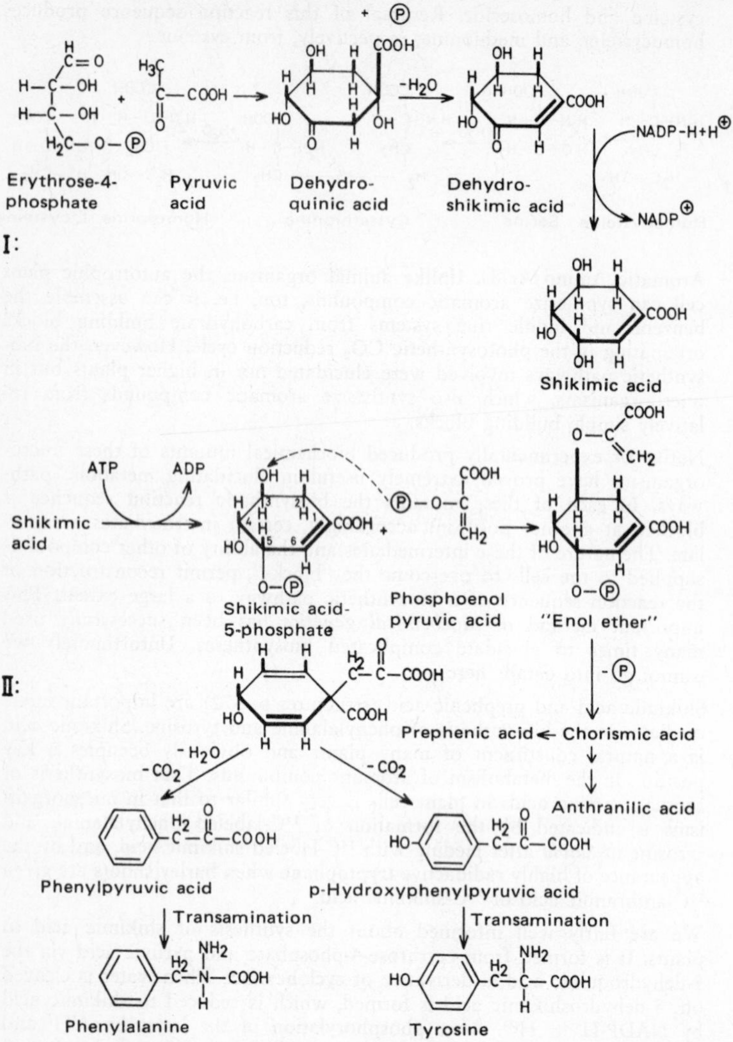

I:

II:

phenylpyruvic acid and p-hydroxyphenylpyruvic acid, the α-keto precursors of phenylalanine and tyrosine (reaction scheme II).

The synthetic pathway of anthranilic acid, the key compound of tryptophane biosynthesis, probably starts with chorismic acid which reacts with glutamine to form anthranilic acid. This compound contributes the aromatic ring and nitrogen to the subsequent construction of the indole ring. The first reaction is a condensation between anthranilic acid and 5-phosphoribosyl-1-pyrophosphate (p. 317) yielding indole-3-glycerolphosphate via two intermediates. In the final reaction the side chain is replaced by serine. This last step is catalyzed by *tryptophane synthetase* which has been found in pea, wheat and maize seedlings. The corresponding enzyme from bacteria also catalyzes a direct condensation of serine and indole.

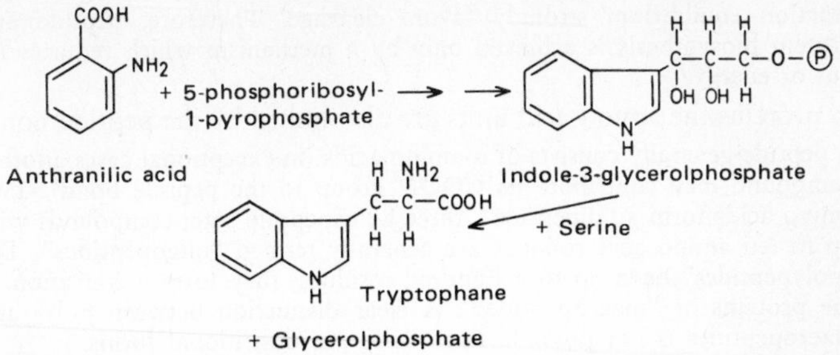

Histidine. The pathway of biosynthesis of histidine in plant cells is still obscure. In bacteria imidazole glycerolphosphate has been identified as an intermediate; it is converted to histidine via histidinol phosphate and histidinol.

Structure of Peptides and Proteins

By analogy with the oligosaccharides which, in terms of molecular structure, may be regarded as a form of transition from mono- and disaccharides to polysaccharides, corresponding compounds exist for the proteins: the *peptides*. As regards molecular size and number of amino acid units, they hold an intermediate position between building blocks and macromolecule. They already exhibit certain typical features of proteins: one very important feature is the nature of bondage between the monomeric units; this bond is commonly termed *peptide bond*.

The Peptide Bond. Linkage of two amino acids leads to a secondary amide bond between the α-carboxyl and α-amino functions of adjacent amino acids, i.e. a condensation reaction with release of water:

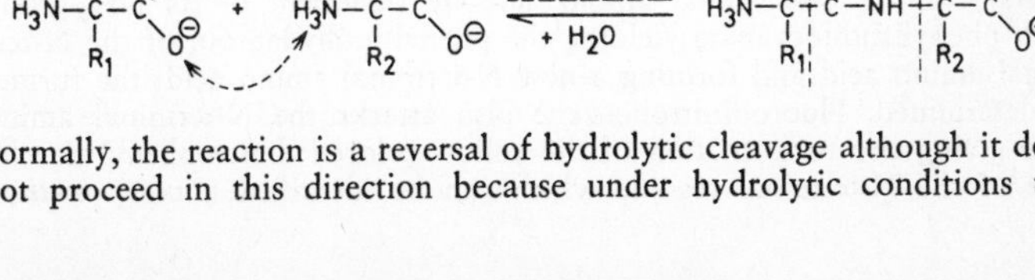

Formally, the reaction is a reversal of hydrolytic cleavage although it does not proceed in this direction because under hydrolytic conditions the reaction equilibrium strongly favors cleavage. Therefore, peptide and protein biosynthesis is achieved only by a mechanism which requires input of energy.

In proteins the amino acid units are also linked by the peptide bond. A peptide generally consists of α-amino acids (in exceptional cases another compound may contribute its COOH group to the peptide bond). Two amino acids form a "dipeptide", three a "tripeptide" etc; compounds with up to ten amino acid residues are generally termed "oligopeptides". The "polypeptides" have up to a hundred residues; they form a transition to the proteins or "macropeptides". A clear distinction between poly- and macropeptides is not possible since there are transitional forms.

An important consequence arises from this type of linkage connecting the amino acids: each peptide and protein chain has a free amino group at the one end and a free carboxyl group at the other. By convention structural formulas of peptides are usually written with the terminal NH_2-group on the left and the terminal carboxyl group on the right (cf. formula). A special kind of abbreviation is often used to designate the linked residues using only the first three letters of each amino acid; the end groups are written "OH" or "H", the free amino groups in the chain "NH_2", and a cystine bond "-S-S-":

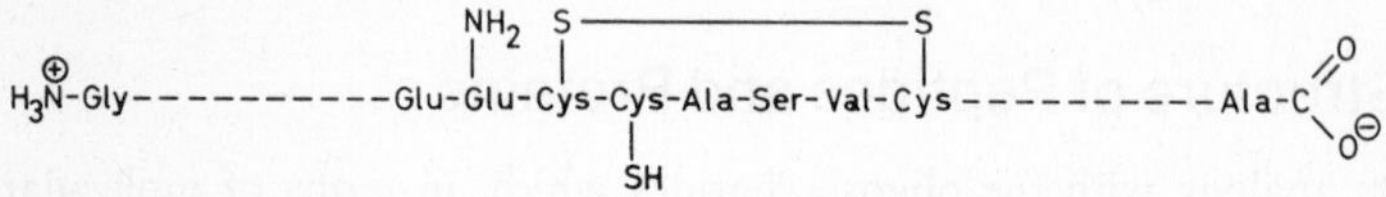

The consequences which result from ionization of the end groups and of other free groups present when a peptide molecule is placed in an aqueous environment will be discussed later (p. 378 f).

In order to characterize a peptide precisely not only the kind and number of the amino acids involved but also their specific order or *sequence* in the chain must be known (e.g. glutathione: glu-cys-gly, not glu-gly-cys! see below). The same applies to the high molecular polypeptides and macropeptides (p. 378).

Methods for elucidation of the primary structure, i.e. an amino acid sequence, consist generally in chemically modifying and cleaving off the terminal amino acid residue which can thus easily be identified (procedures p. 359); then the next amino acid in the chain is treated in the same way, and so on. In the *Edman procedure,* for instance, degradation starts with the N-terminal amino acid by treatment of its NH_2-group with phenylisothiocyanate yielding the phenylthiohydantoin of the N-terminal amino acid and forming a new N-terminal amino acid; the former is determined. Fluorodinitrobenzene also attacks the N-terminal amino acid giving rise to formation of the yellow-colored dinitrophenyl derivative of the N-terminal residue which can be identified (*Sanger method* for determination of end groups). However, in this procedure the peptide has to be completely hydrolyzed; a stepwise degradation as in the Edman procedure is therefore not possible.

Since both procedures can only be used with relatively short peptide chains, longer chains or macropeptides must first be cleaved to fragments of an appropriate length. Protein-splitting enzymes (*proteases*, p. 397) have been successfully used to achieve this. Since each *protease* cleaves a protein molecule in a different way due to its specific mode of action, it is possible to reconstruct the original arrangement of peptide fragments in the protein molecule by comparing the cleavage patterns and the sequences of the peptide cleavage products (= "overlapping cleavage"). The sequences of numerous peptides and proteins have been elucidated by means of this procedure which has been largely automated in the meantime.

Natural Peptides. The most wellknown biologically active peptides have been found in man and in animal organisms: the "peptide hormones" (insulin, glucagon, oxytocin, vasopressin, corticotropin). Their sequences are known in most cases. A number of peptides have also been isolated from higher plants, particularly from storage organs where they constitute a significant proportion of the fraction of soluble nitrogen compounds. These peptides may serve to store nitrogen and sulfur in the form of soluble compounds, thus performing a function very similar to that of glutamine, asparagine, arginine and citrulline. These are the "γ-glutamyl peptides", i.e. derivatives of glutamic acid in which the "γ-carboxyl group" exceptionally participates in peptide bond formation. Alanine, valine, leucine, isoleucine, phenylalanine, tyrosine, methionine and cysteine have been demonstrated as building blocks, together with a few amino acids that are usually not constituents of proteins.

Peptides of low molecular weight have also been found in brown algae; they always contain glutamine. Arginyl glutamine occurs in the green alga *Cladophora*.

Glutathione is a typical tripeptide of many plant and animal cells: "γ-L-glutamyl-cyste(in)yl-glycine"; it has the following structure:

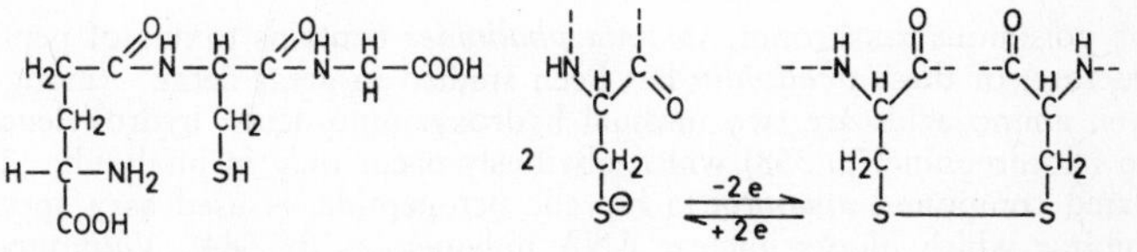

Glutathione

Just as in the case of peptides of plant storage organs mentioned above the γ-carboxyl group is involved in peptide bond formation. The SH-group of the cysteine residue which confers a weakly acidic character to the glutathione molecule, remains reactive even within a peptide structure. It may be induced to release its hydrogen; in this way a disulfide compound is formed from two molecules of glutathione. The process amounts to a removal of electrons (from the ionized form!) and is therefore an oxidation process. Since this reaction is reversible, glutathione has the properties of a redox system (p. 21).

Besides glutathione, derivatives of this tripeptide occur in plant organisms; in these the cysteine residue is replaced by another amino acid.

Protamines belong to the polypeptides. They form a group of compounds with a strongly basic character because of their high arginine content; they do not contain sulfur due to the lack of S-containing amino acids. Their molecular weights are between 1000 and 5000. They are to some extent associated with nucleic acids in the cell, but their biological function is as yet not known.

In terms of chemical structure several antibiotics are peptides, too. *Gramicidin* consists of 10 amino acids, one of which is present in the biologically unusual D-form: D-phenylalanine. They form a cyclic structure with the following sequence:

```
Pro — Val — Orn — Leu — (D)Phe
 |                          |
(D)Phe — Leu — Orn — Val — Pro

        Gramicidin
```

R = benzyl-

Penicillin G-Na

The structure of *penicillin* originates from a dipeptide of valine and cysteine (see formula) to which an organic acid has been added forming a peptide bond via the amino group of the cysteine residue. The residue "R" of the former is variable. The molecule of *actinomycin,* another antibiotic, also contains peptide units (cf. p. 351).

The poisonous mushroom *Amanita phalloides* contains toxins of peptide structure; of these *phalloidin* has been studied in great detail. Among its seven amino acids are two unusual hydroxyamino acids: hydroxyleucine and allothreonine (p. 358) which obviously occur only in phalloidin. The related compound *amanitin,* a bicyclic octopeptide, is used as a specific inhibitor which blocks nuclear RNA polymerases (p. 344). *Valinomycin* is a cyclic peptide compound consisting of six amino acids and six hydroxy acids (lactic acid, hydroxyisovaleric acid) as building blocks, which alternate in the circular sequence. This antibiotic is of great interest as an "uncoupler" of oxidative phosphorylation (p. 241) and as a model "carrier" substance (p. 271).

Biosynthesis of Peptides. Linkage of amino acids to short-chain peptides is probably brought about by a different and less complex mechanism compared to that involved in protein synthesis. In contrast to the latter nucleic acids do not participate; the specificity of the products of synthesis is obviously guaranteed sufficiently by the enzyme. Glutathione offers a good example of peptide biosynthesis. Formation of its peptide bonds is catalyzed by two enzymes: the first is responsible for the linkage of L-glutamic acid and L-cysteine; the second enzyme – *glutathione synthetase* – takes care of the attachment of glycine to the dipeptide formed initially. Since the two reactions are endergonic by nature each is coupled with cleavage of one mole of ATP:

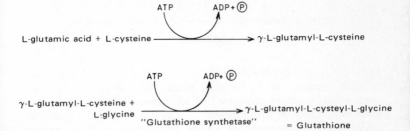

Structure and Properties of Macropeptides

The structure of a protein macromolecule shows the consequent application of the constructional principle which has its origin in the short-chain oligopeptides. In the *macropeptides* or *proteins* it has reached perfection: the insertion of further amino acids and formation of peptide bonds leads, by stepwise elongation of the polypeptide chain, to a long macropeptide chain. The term "macropeptide" comprises all compounds containing from a hundred to several thousand amino acid residues.

The molecules accordingly attain dimensions of 5–100 nm; in solution they are retained by membranes if the pores of these are smaller than 5 nm in diameter, while smaller molecules and ions pass unimpeded. This principle is employed in "dialysis", i.e. the equilibration of a protein solution with a buffer solution across a protein-impermeable membrane.
The molecular weight of proteins is high: between 10 000 and several millions. In determining the molecular weight of proteins, use is made of their sedimentation under the conditions of ultracentrifugation (p. 335 f), apart from measuring the osmotic pressure.

The structural "backbone" of the protein molecule is the regularly repeated sequence –NH–CH–CO– (see molecule segment). Each of these "elementary units" carries a side chain ("R") which is the molecular residue of the amino acid involved (in case of glycine "R" = H!). By analogy with the peptides an NH_2-end and a COOH-end exist in the molecule.

Glycine Tyrosine Cysteine Alanine Valine Asparagine

Protein molecule (segment)

Since the specificity of each protein is encoded in the definite, invariable sequence of its building blocks, knowledge of the "sequence" or "primary structure" of all amino acids involved is of great importance. It is indispensable in exploration of the principle underlying the expression of genetic information in these macromolecules. The amino acid sequence of several biologically important proteins has been determined (for methods, see p. 374 f). Sanger (1954) was the first with insulin, which was shown to be a polypeptide of 51 amino acids. The first macropeptide, *ribonuclease* (p. 351), followed in 1959; its molecule consists of 124 amino acids. Next the sequence of the protein subunit of the tobacco mosaic virus (158 amino acids), that of hemoglobin (574 amino acids) and that of cytochrome c were elucidated.

The functional groups of the side chains or residues "R" including the end groups (–COOH, –NH_2, –OH, –SH) are responsible for a property of proteins which has already been discussed when dealing with the amino acids: in an aqueous environment these groups are usually ionized. Under these conditions a protein molecule acts like a polyvalent ion. We must not forget, however, that in contrast to free amino acids the charges of carboxyl and α-amino group are eliminated when they enter peptide bonds. The only exception are the terminal amino acid residues. Of course the extent of dissociation of the individual groups is determined by the $H^\oplus$ concentration of the solution. For example, under physiological conditions the carboxyl groups of glutamic acid and aspartic acid are negatively charged due to dissociation, whereas the basic groups of lysine and arginine are positively charged due to attachment of protons ($H^\oplus$ ions!).

As described for the functional groups of amino acids, changes in the $H^\oplus$ concentration of the solution also affect the electric charge of the single groups in the protein molecule. Here also, an "isoelectric point" exists, i.e. a pH value at which the number of positive charges equals the number of negative charges. This state is also marked by typical changes in the physical conditions of the protein: at this point it has the lowest solubility and usually precipitates. Because of its electric charges a dissolved protein moves in an external electric field; with a surplus in negative charges it moves towards the anode, with a surplus of positive charges towards the cathode. The sum of the individual charges determines the speed of this movement.

Since the $H^\oplus$ concentration affects the electric charge of individual groups it also exerts its influence on the direction of movement in an electric field: in an acid medium (low pH: surplus of $H^\oplus$ ions) a positively charged protein molecule migrates towards the cathode, in an alkaline medium (high pH; deficiency of $H^\oplus$ ions) as a negatively charged molecule to the anode. Between these two ranges a pH value exists at which the protein does not move at all; it is identical with the isoelectric point.

Electrophoresis takes advantage of charge differences between the proteins at appropriate values of pH. The individual components of a protein mixture move various distances in the electric field because of their different total charges; thus separation is achieved (technique cf. Fig. 96). Electrophoresis is a very convenient technique for analytical and preparative separation not only of proteins but also of mixtures of other biologically important substances such as nucleotides and nucleic acids (p. 350).

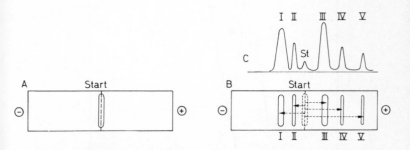

Fig. 96 Electrophoresis of a protein mixture in a buffer-saturated paper strip. A: Application of the sample in a small start zone before the run. B: In the electrical field applied the individual proteins (I–V) have moved various distances away from the starting zone. C: After staining of the bands, quantitative measurement of the proteins separated in the "electropherogram" is possible; the size of the absorption maximum is proportinal to the amount of the particular protein.

Electrophoresis has been successfully employed in the purification of individual proteins which is usually a formidable task. This problem always arises when a biologically active protein – enzyme or hormone – is to be isolated from the natural mixture of cytoplasmic proteins or from a tissue extract. The main difficulty is that most proteins, being closely related compounds, react much the same way when subjected to separation procedures. Moreover, many standard methods of organic chemistry cannot be employed because they would destroy the biological activity of the proteins immediately. Therefore, protein chemistry had to develop its own purification procedures; we shall discuss some of these briefly.

Precipitation. Proteins are precipitated from an aqueous solution without loss of their activity by several agents. Apart from acetone and alcohol, salts are employed, particularly ammonium sulfate which is added either in the form of a highly concentrated solution or as a solid salt.

Column Chromatography. The principles underlying this method have already been discussed (p. 33 f). Column chromatography on ion-exchange resins has recently been used for protein purification: cellulose is modified

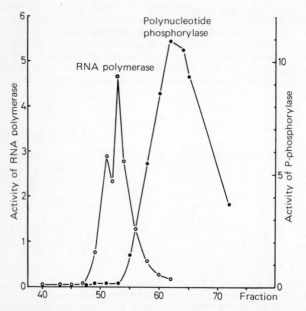

Fig. 97 Separation of two enzymes, *RNA polymerase* and *polynucleotide phosphorylase*, from the blue-green alga *Anacystis nidulans*, by column chromatography on diethylaminoethyl- (= "DEAE") cellulose (after Capesius and Richter).

by linking either anionic or cationic groups (diethylaminoethyl functions = "DEAE" or carboxymethyl functions = "CM") to its hydrophilic macromolecules. The components of the protein mixture adsorbed onto the column are selectively eluted by washing with buffers of increasing ionic strength or by means of salt gradients. Figure 97 shows an example of separation of two enzymes, *RNA polymerase* (p. 343 f) and *polynucleotide phosphorylase* (p. 343 f). Very substantial results in the purification of proteins are also obtained by *gel filtration* which sorts out the molecules according to size (p. 48 f).

Electrophoresis. This purification procedure for proteins based on their difference in charge has already been mentioned as well as its modification using polyacrylamide as a support (p. 336 f).

Another difficult question that arises during the purification of proteins concerns the criterion of purity and the proof of the identity of the preparation. Two methods in particular are employed in this respect: 1) verification of homogeneity by observing the sedimentation in the ultracentrifuge; 2) reaction of the purified protein to specific antibodies in a serological test.

Classification of Proteins

Proteins are rather difficult to classify satisfactorily, not only because of the diversity of their primary structure but also due to the varying spatial structure of their molecules, the "chain conformation" which endows the individual proteins with new properties and functions. Nowadays we favor a classification based on the function of proteins.

1. Scleroproteins and Spheroproteins. The insoluble "scleroproteins", which are fibrous and are hence employed for architectural purposes have their counterparts in the soluble "spheroproteins". The latter are divided into histones, albumins, globulins, glutelins and prolamins. Histones are proteins rich in basic amino acids which occur in the cell nucleus where they are probably combined with nucleic acids. It appears likely that they fulfill a certain function in the regulation of gene activity. Albumins are characterized by a high solubility in water. Precipitation occurs only at relatively high salt concentrations (70–100 %). On the other hand, globulins show only limited solubility in water; it is higher in dilute salt solutions. It is often difficult to make a clear distinction between albumins and globulins.

A great number of enzymes are globulins. *β-Amylase* is probably an albumin, *α-amylase* a globulin. A few of the plant storage proteins are globulins. Some have been isolated and crystallized, e.g. edestin from hemp seeds, zein from maize, arachin and conarachin from peanuts. However, the majority of storage proteins are glutelins and prolamins. These are insoluble in water and dilute salt solutions, but soluble in acids and alkali. Wheat grains contain the prolamin "gliadin" as well as α- and β-glutelin; they constitute the "glutein", the residues of flour after removal of starch. Low-protein flours contain little glutein; in bread-baking a flour with higher protein content is preferred.

2. Proteids ("Conjugated Proteins"). A proteid consists of a protein moiety and a non-protein component, a "prosthetic group". If this is a metal, we have a "metalloproteid", e.g. the Cu-containing plastocyanin (p. 112). In "chromoproteids" the protein is linked to a strongly colored pigment, as in the case of the biliproteids (phycocyanins and phycoerythrins; p. 48 ff). Because of their frequent combination with proteins in the cell, the nucleic acids are termed "nucleoproteids". "Lipoproteids" are major constituents of lamellae and membranes. The prosthetic group of a "glycoproteid" is carbohydrate by nature.

Protein Biosynthesis

The main features of protein synthesis have been elucidated only recently; however, numerous details are still uncertain. In fact, this process is rather complex, not because the chemical mechanism linking the building blocks is difficult to achieve, but because the latter must be inserted into the macromolecules in an exact and specific order or "sequence" depending on laws which were unknown until recently. The biological function of the protein molecule formed is vitally dependent on the exactness of this sequence. An important step in unravelling this complex process was the realization that the "blueprint" or information for the arrangement of the sequence is embodied in the DNA and reaches the site of protein synthesis in a mobile form, the messenger RNA (p. 342 f). This concept of a "flow" of information can be expressed briefly as follows:

$$DNA \rightarrow m\text{-RNA} \rightarrow protein$$

The first reaction step, the *transcription,* i.e. formation of m-RNA with DNA as the template, has already been discussed (p. 345 ff). The second reaction complex has been called *translation* since the information coded in the nucleotide sequence of the m-RNA molecule is now "translated" into the amino acid sequence of the protein molecule being manufactured.

The complex process of protein synthesis comprises three main stages: 1. Conversion of the substrate, i.e. "activation" of the amino acids in the cytoplasm. 2. Formation of peptide bonds between the amino acids on the surface of ribosomes, the specific organelles of protein synthesis. 3. Release of the completed protein macromolecule from the ribosomes and formation of its specific three-dimensional structure.

By combining various isolated cell components from bacteria (*E. coli*) a working "cell-free" system has been established which brings about the linkage of amino acids to peptide chains in the test tube. These experiments have contributed significantly to our present knowledge of protein biosynthesis.

We have already seen that certain biochemical processes like photosynthesis, the citric acid cycle and the respiratory chain take place within specific well-defined cellular organelles. It is hence not surprising that protein synthesis has its own cellular organelle too, the *ribosome,* the structure of which we shall now examine.

Ribosomes. In electron micrographs of many cells roundish submicroscopic particles of about 15 to 25 nm appear which are partly attached to the outer membrane surface of the endoplasmic reticulum, but also occur free in the cytoplasm. The latter form of distribution is typical for the bacterial cell which may contain between 5000 and 50 000 of these particles. Today they are generally termed "ribosomes". They consists of 55–63 % RNA of high molecular weight (p. 340 f); the rest is protein. There are in addition small amounts of substances of low molecular weight.

Since the ribosomes have a rather constant size, they form a homogeneous sediment on differential centrifugation of a cell or tissue homogenate. Bacteria are more suitable for the isolation of ribosomes than animal or plant tissue cells. The degree of purity is higher in bacteria since ribosomes of higher organisms are generally very difficult to free from impurities of low molecular weight, although recently the gel filtration technique (p. 48 f) has been successfully used to remove these impurities.

The sedimentation of isolated ribosomes in the ultracentrifuge has shown that a 70 S particle in bacteria and an 80 S particle in eukaryotic cells are the typical units of this cell organelle. Besides these, aggregates of 110 S (in pea seedlings) and smaller native subunits of 30–50 S may occur. Thus a better definition of the term "ribosome" has been established; originally this term was used to characterize all the ribonucleoproteid particles of the cytoplasm exhibiting sedimentation coefficients from 20–100 S.

Ribosomes specifically dissociate into "subunits" of a definite order of magnitude on removal of magnesium by dialysis or treatment with EDTA ("ethylenediaminetetraacetate cf. p. 75); thus ribosomes from pea seedlings decompose into the following subunits:

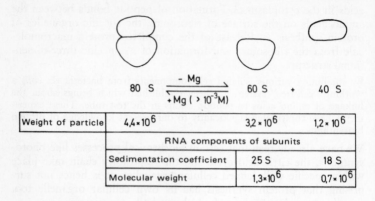

Pea seedling, callus cells:

$$80\,S \xrightleftharpoons[+ Mg\,(\,>10^{-3}M)]{-\,Mg} 60\,S \;+\; 40\,S$$

Weight of particle	4.4×10^{6}	3.2×10^{6}	1.2×10^{6}
RNA components of subunits			
Sedimentation coefficient		25 S	18 S
Molecular weight		1.3×10^{6}	0.7×10^{6}

Fig. 98 shows the sedimentation of the 60 S and 40 S subunits in a sucrose gradient (p. 95 f).

The 70 S ribosome of *E. coli* shows an analogous dissociation into subunits on removal of magnesium:

Escherichia coli:

$$70\,S \xrightleftharpoons[+ Mg]{-\,Mg} 50\,S \;+\; 30\,S$$

Weight of particle	2.6×10^{6}	1.8×10^{6}	0.8×10^{6}
RNA components of subunits			
Sedimentation coefficient		23 S	16 S
Molecular weight		1.1×10^{6}	0.56×10^{6}

Structure of ribosome

Two classes of different size also exist in ribosomal RNA species which are associated with the 60 S and 40 S subunits of the ribosome of eukaryotes as well as with the 50 S and 30 S subunits of the bacterial ribosome and which differ in their sedimentation coefficients. The 28 S RNA (HeLa and animal cells) and the 25 S RNA (higher plants) belong to the 60 S subunit, whereas the 18 S RNA (higher plants, animal cells) belongs to the 40 S subunit (sedimentation diagram for *Vicia* in Fig. 94, p. 341). Values of 16 S and

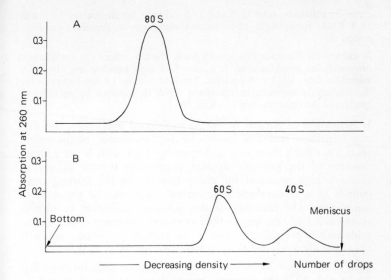

Fig. 98 Centrifugation in a sucrose gradient of 80 S ribosomes (A) and their subunits of 60 S and 40 S (B) formed by removal of magnesium. Preparation from pea seedlings (modified after Bonner and coworkers).

23 S have been found for the RNA of the ribosomal subunits of blue-green algae and bacteria. It is fairly certain that a molecule of the 5 S RNA of low molecular weight (p. 341) is a permanent constituent of the larger 50 S subunit in *E. coli*. The protein moiety of ribosomes has also been analyzed in this organism. The 30 S subunit may contain as many as 21 different proteins and the 50 S subunit as many as 34. The precise functions they fulfill in supporting the ribosomal structure and in the complex reactions of protein synthesis are still largely unknown. Under certain conditions the isolated proteins and the ribosomal RNA species spontaneously re-aggregate into functional ribosomes (= "self assembly") in vitro. Obviously the blueprints for assembling the complex organelle structure are present in the components themselves, i.e. the process does not require any additional genetic information.

It is still an open question whether 90% of the cellular RNA in plant cells is present in the form of ribosomes, as is the case in bacterial and animal cells.

Recent studies have shown that the chloroplasts of higher plants as well as the plastids of the green flagellate *Euglena* contain ribosomes or ribonucleoproteid particles of very similar constitution ranging from 68 S to 73 S. They resemble the ribosomes of prokaryotes (bacteria, blue-green algae). On removal of magnesium isolated ribosomes from chloroplasts generally dissociate into 50 S and 35 S subunits which contain 23 S and 16 S RNA, respectively. Mitochondria also seem to contain own ribosomes.

In eukaryotes the nucleolus is very likely the site where the cytoplasmic ribosomes are manufactured. It is here that their subunits are obviously assembled from RNA and proteins. The information, i.e. the blueprint for the nucleotide sequence of ribosomal RNA is imparted by specific sequences of the chromosomal DNA (p. 349 f).

1. Activation of Amino Acids. As mentioned previously, the first step in protein synthesis consists of the "activation" of the amino acids, i.e. raising them to a higher energy level. This is yet another example of the principle applied successfully before, namely to confer a high potential for group transfer on the starting compounds of a biosynthetic reaction to ensure that it proceeds to completion. The necessity of this substrate activation in protein synthesis is made evident by the fact that the insertion of free amino acids into a polypeptide chain is an endergonic process which is only formally a reversal of hydrolytic cleavage (cf. p. 373).

Each of the about 20 different individual amino acids involved first reacts with ATP, the universal energy donor, to form a mixed anhydride ("aminoacyl adenylate"; see reaction diagram, p. 387). This high-energy derivative is stabilized against hydrolytic cleavage; obviously the reaction product formed remains bound to the protein of the specific enzyme. The latter is termed *aminoacyl-t-RNA synthetase;* it also catalyzes the subsequent transfer of the amino acid from this intermediary product to a molecule of the specific transfer RNA. This reacts via the free 3'-hydroxyl group of the ribose attached to the terminal adenosine moiety of its molecule. The ester bond formed has a relatively high transfer potential as shown by the free energy change of hydrolysis: $\Delta G^{\circ} = -6500$ cal/ mole. Taking alanine as an example, we shall formulate the reactions taking place more precisely (p. 387).

Each amino acid required for protein synthesis is thus attached to its specific transfer RNA by a specific *synthetase*. Specific interactions may exist between the structures of the transfer RNA and the activating enzyme.

Since theoretically one *synthetase* is required for each amino acid at least 20 enzymes of this type must be involved in protein synthesis. In fact, numerous *synthetases* have been demonstrated in microorganisms and animal cells; some have been highly purified. By analogous studies with seedlings *(Phaseolus, Pisum)* as well as with spinach leaves the occurrence of most of these *synthetases* has been verified for heigher plants too. They are found in the particle-free supernatant of a high-speed (100 000 g) centrifuged homogenate of plant tissue, in the "soluble" fraction. Measurement of their activity is hampered by the presence of free amino acids. The specific test gives satisfactory values only when these amino acids have been removed by subsequent purification steps. In the test incorporation of radioactive pyrophosphate into ATP depending on added free amino acids is measured. This exchange which takes place in the absence of transfer RNA corresponds to a reversal of the activation reaction.

Regarding the location of *aminoacyl-t-RNA synthetases* of green plant cells the same holds true as for the transfer RNA species: some of these

Transfer RNA with attached amino acid

enzymes are specific constituents of chloroplasts, as shown with *Phaseolus vulgaris*, *Gossypium hirsutum* and *Euglena*.

Since along with other factors all 20 *aminoacyl-t-RNA synthetases* and the corresponding transfer RNA species must be present in the reaction mixture in order to achieve protein synthesis in vitro (see p. 338 f), the supernatant of high-speed centrifugation mentioned above (= "soluble fraction") or a precipitate formed from it by adjusting the pH value to 5.0 ("pH-5-enzyme") are often used as effective supplement.

When bound to a molecule of its specific transfer RNA each amino acid is ready for being inserted as "activated" building block into the specific sequence of a protein macromolecule. The mode of action of transfer RNA molecules is similar to that of coenzyme A or UDP, "transport metabolites" which are also carriers of "activated" molecules – acetyl residues and sugars, respectively – in metabolism.

2. Processing of a Specific Amino Acid Sequence. While the process of amino acid activation takes place in the cytoplasm the second stage of protein synthesis proceeds at the ribosomes. The blueprint for the amino acid sequence to be constructed is provided by the *messenger RNA* (= *m-RNA*) the function of which as a template was discussed when dealing with RNA biosynthesis (cf. p. 343 ff).

One molecule of m-RNA becomes attached quite firmly to several small and large ribosomal subunits and binds them as complete ribosomes; a cluster is formed resembling a string of pearls (see Fig. 100; p. 394). This combined structure is called a "polyribosome" or "polysome". It may contain from five to a hundred ribosomes depending on the cell type (Fig. 99). Polysomes are unstable structures insofar as the individual ribosomes move along the m-RNA

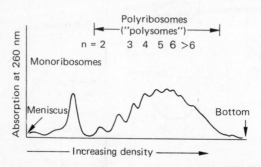

Fig. 99 Distribution of free ribosomes (= monoribosomes) and polyribosomes (= "polysomes") after centrifugation in a sucrose gradient (10—50%). For the method applied, see Fig. 94, p. 341. n = number of bound ribosomes (after Pfisterer).

molecule towards the 3'-end in the course of protein synthesis. This movement occurs stepwise and is characterized by a continuous change of tight and weak bonding between m-RNA and ribosome (see below).

The m-RNA molecule comprises in its nucleotide sequence the encoded information or the *code** as to how the individual amino acids are to be assembled in a specific sequence. This nucleotide "shorthand" is read off and "translated" into that of the amino acids (cf. p. 378 f).

How these four nucleotide bases – in shorthand: A, G, U and C – are able by virtue of their arrangement in the m-RNA molecule to assign each of the about 20 amino acids involved its proper place in the growing chain of the protein macromolecule, i.e. the decoding of the four-letter language of m-RNA, represents the problem of the *genetic code*. Its elucidation, almost completed now, is one of the most impressive accomplishments of modern biology. For details

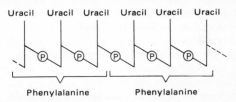

the reader is referred to modern textbooks of genetics. We shall outline briefly the major elements of this "code" which are indispensable for understanding the processes of protein synthesis on the ribosome. It has been found that the base sequence of three consecutive nucleotide residues (= "triplet") in the m-RNA represents the symbol or "codon" for an individual amino acid; e.g. UUU for phenylalanine. A number of these codons of m-RNA and the amino acids they signify are shown in Table 1.

Accordingly the nucleotide sequence of an m-RNA molecule consists of a consecutive sequence of such base triplets which signifies a definite sequence of amino acids. The former is recognized and translated continuously and without interruption; the codons certainly do not overlap. The problem in assembling the amino acid sequence is to transfer the individual amino acids to the ribosome in such a way that they can be inserted into the growing peptide chain in the correct position at the right moment. The ribosome helps to meet this requirement in that it offers at least two specific

---

* A "code" is a system of signals corresponding to units of information.

Table 1

1st and 2nd nucleotide base	3rd nucleotide base	Amino acid
U U	C; U	Phenylalanine
U C	C; U; A; G	Serine
C A	C; U	Histidine
G A	C; U	Aspartic acid
G A	A; G	Glutamic acid
G U	C; U; A; G	Valine
A A	A; G	Lysine
C C	C; U; A; G	Proline

reaction and binding sites for linking amino acids. These are obviously arranged side by side so that each may bind a codon of the m-RNA attached; a total of two consecutive triplets of nucleotide bases thus occupy this region on the ribosome (cf. scheme, p. 391). This construction enables a molecule of transfer RNA to react by virtue of its chemical structure, i.e. through its "recognition site" or "anticodon" which we have already discussed (p. 339 f). This segment of the sequence comprises one triplet and has a specific base sequence in each species of transfer RNA: it is complementary to the sequence of the corresponding amino acid codon on the m-RNA strand*. For the transport and particularly for the precise attachment of aminoacyl transfer RNA to the two reaction or binding sites, the same mechanism of base pairing is put into operation which we have already recognized as being typical for the biosynthesis of DNA and RNA (cf. p. 329 ff).

The elucidation of protein biosynthesis is most advanced in prokaryotes, particularly in E. coli. Many findings indicate, however, that the rules adhered to in this organism apply in principle to eukaryotes as well.

In E. coli an *initiation complex* is first formed by attachment of a 30 S ribosomal subunit and a molecule of transfer RNA carrying formylmethionine ("N-formyl-met-t RNA"; scheme p. 391) to the 5'-end of a m-RNA molecule. This process only takes place when GTP, $Mg^{2\oplus}$ as well as specific proteins or *initiation factors* (F_1, F_2, F_3) are present; the

---

* A codon on the m-RNA is hence written in the $5' \to 3'$ direction, the complementary anticodon on the transfer RNA, however, in the $3' \to 5'$ direction in order to document the antiparallelism of base pairing (cf. scheme p. 392).

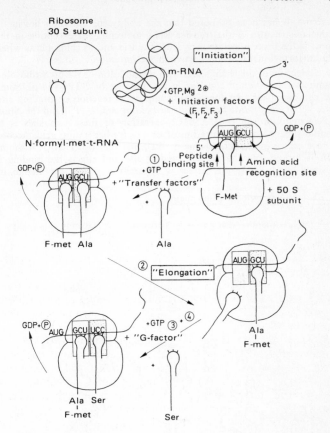

function of the latter is still obscure. Obviously the triplet AUG signifies the attachment of N-formyl-met-t- RNA. It is often located at the 5′-end of m-RNA molecules; for this reason it is generally accepted today that this base sequence marks the starting point for the peptide chain to be synthesized which consequently begins with methionine as the first building block. Since the NH_2-group of this substance, due to formylation, cannot form a peptide bond, this can be established only via the carboxyl group. In this way the first amino acid determines the direction of growth of the peptide chain, i.e. from the amino end to the carboxyl end. In a number of proteins from *E. coli*, however, neither formyl methionine nor methionine form the amino end; this is because after completion of the protein chain they lost either the formyl group or methionine due to a subsequent enzymatic cleavage.

In eukaryotes a methionine-charged transfer RNA is also involved in the *initiation* though the amino acid lacks the formyl group in this case.

Therefore, it has been proposed that the ability for *initiation* lies in the particular structure of the transfer RNA molecule. An exception in this respect is the protein synthesis in mitochondria and in chloroplasts where again N-formyl-met-t RNA is active in the *initiation*.

The *elongation* of the chain starts with the addition of the 50 S ribosomal subunit to the *initiation complex* in the presence of GTP. The ribosome, now complete, has two functionally important reaction or binding sites (cf. p. 389 f); the *peptide-binding site* or *holding site* (="P") and the *amino acid recognition site* (= "A"). Co-operation of these two sites is an essential precondition for attachment of any further amino acid. According to a recent model (cf. scheme p. 391), first the triplet AUG of the m-RNA is locked into the *peptide binding site* and effects the binding of N-formyl-met-t RNA; this triplet is apparently specific for methionine

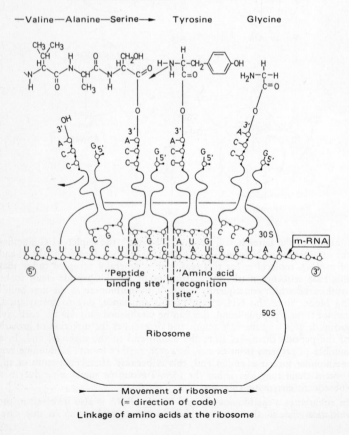

Linkage of amino acids at the ribosome

as starting amino acid (see above). At the same time the *amino acid recognition site* is occupied by the aminoacyl-t-RNA specified by the codon next to AUG, probably involving specific proteins *(transfer* or *elongation factors)* and GTP ①. Then the peptide bond is formed by *peptide synthetase (peptidyl transferase),* a protein of the 50 S subunit. Hereby, formyl methionine is separated from its transfer RNA and attached via its carboxyl group to the amino acid which, together with its transfer RNA, occupies the *amino acid recognition site* ②. Thus the dipeptide formed is not only bound to the ribosome by this transfer RNA but also moves in this form to the *peptide binding site* ③ which has become vacant in the meantime. It is possible to isolate such complexes of ribosomes with attached peptide chains. During this *translocation* the ribosome moves the length of one triplet along the m-RNA in the $5' \rightarrow 3'$ direction: thus the next triplet enters the *amino acid recognition site* ④. GTP again supplies the energy required; furthermore, another ribosomal protein, *translocase,* is involved. The molecular basis of the translocation process is unknown, but it is assumed that conformational changes of the ribosomal subunits take place. With the attachment of a new transfer RNA charged with amino acid at the *amino acid recognition site* the process of peptide-bond formation and translocation (①–④) starts to be repeated. It is completed when the peptide chain has been elongated by a further amino acid and is attached to the *peptide binding site* by the corresponding transfer RNA. This reaction sequence is repeated with yet another of the 20 amino acids serving as building blocks until the chain has reached its final length (i.e. 600 times for a protein molecule consisting of 600 amino acid residues!).

As already mentioned, several ribosomes are usually attached to the m-RNA molecule at the same time. Since each ribosome has at least two reaction or binding sites, it is possible that several molecules of the same protein may be formed on the one m-RNA molecule. The possible mechanism of this process is shown schematically in Fig. 100. When the m-RNA chain has "passed" all the ribosomes (1–5), five identical protein molecules have been produced in one operation. The economic significance for the cell is obvious. Movement of the individual ribosomes along the m-RNA chain is a prerequisite for producing several proteins simultaneously. Hence a stable bond between the two structures need exist only for the short interval of *recognition* and formation of a peptide bond.

Measurements of the rate of synthesis in vivo with *E. coli* indicate that the individual steps proceed very fast: about 20 amino acids per second can be linked. In eukaryotic cells this rate is significantly lower.

A repeated employment of the same m-RNA molecule as template is also possible though this possibility is limited by the relatively short life-time of m-RNA in bacteria (only a few minutes; cf. p. 342); thus an average of only 10–20 copies of the one protein molecule are produced. Rapid degradation of m-RNA is an important principle in the regulation of protein synthesis (cf. p. 446).

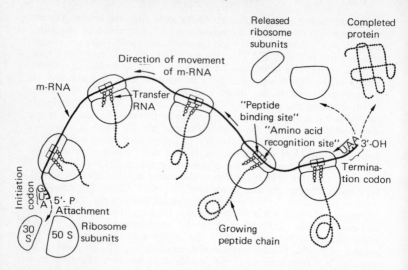

Fig. 100 Simultaneous formation of identical proteins on a polysome. Five ribosomes are linked by a molecule of messenger RNA to a functional unit. For details see text.

3. Release of the Completed Protein Molecule. When the last codon of the m-RNA molecule has passed the two reaction sites on the ribosome, and the corresponding amino acid has been added as terminal building block to the poly- or macropeptide chain, release of the macromolecule from the ribosome takes place, the last stage, in protein synthesis.

Several lines of evidence have established that in this process of *termination* there exists, in analogy to the initiation codon, a corresponding *termination codon* or *stop codon* – usually at the 3'-end of the m-RNA. This terminates the linkage of amino acids and signals the end of the chain. The triplets UAA, UAG or UGA (= "nonsense triplets") are being discussed as a possible base sequence for this stop codon. The actual release of the peptide chain from the last transfer RNA and thus from the ribosome is obviously brought about by proteins which bind specifically to the ribosomes in response to the specific stop codon: *R-factors* or *release factors*.

Termination codons are particularly important if an m-RNA molecule contains the information for several different kinds of proteins. Arrangements must thus be made in the polynucleotide chain to ensure that only the region specific for one protein is translated at a time and that the product formed is released from the ribosome at the right moment. Only hypothetic models exist at present regarding subsequent events during which the "primary structure" of the protein chain is folded into its characteristic three-dimensional conformation (= "secondary" or "tertiary structure").

The completed and biologically active protein finally appears in the cytoplasm or is incorporated into defined cellular structures or organelles, e.g. as a specific enzyme into a mitochondrion or a chloroplast.

The ribosomes, having accomplished their function, separate from the m-RNA, decompose into their subunits and are free to repeat the process (= "ribosome cycle"). This explains why besides polysomes mainly the subunits of "free" ribosomes are observed in the bacterial cell.

Finally, we shall briefly summarize the consequences which result from the genetic code for transcription and translation. Given a nucleotide base sequence in the DNA template of -AAACAATATTTT- the m-RNA molecule complementary to this region will have the sequence -UUU GUU AUA AAA-, comprising four codons as indicated. Each triplet signifies one amino acid; accordingly, this region of m-RNA contains the information for the following amino acid sequence: -Phe-Val-Ile-Lys-. The corresponding transfer RNA species are characterized by the following anticodons or "recognition sites": AAA, CAA, UAU and UUU, respectively. The DNA and the molecules of the transfer RNA exhibit base sequences which are complementary to the corresponding codons of the m-RNA (with uracil replacing thymine in the transfer RNA).

Effect of Antibiotics. A number of antibiotics have turned out to be very useful in the elucidation of protein biosynthesis since they rather specifically interfere with certain reaction steps. Furthermore, new insights into their mode of action in vivo have been obtained. *Puromycin*, for instance, prematurely terminates the growth of the peptide chain and causes its release, because it displaces the structurally very similar aminoacyl-t-RNA species from the *peptide binding site* on the ribosome; in this position the antibiotic can form a peptide bond with the previously attached amino acid but not with the one arriving subsequently. *Streptomycin* binds to a protein of the 30 S subunit thereby causing errors in the reading of the code. *Chloramphenicol* selectively blocks protein synthesis in bacteria, mitochondria and chloroplasts by binding to the 50 S subunit, possibly at the *peptide binding site;* this reaction, however, does not take place with the 60 S subunit of eukaryotic ribosomes. The latter are affected especially by attachment of *cycloheximide (actidione),* which probably interferes with *initiation* and *translocation* processes leading to an inhibition of protein synthesis.

Protein Synthesis in Plant Cells. Whether protein synthesis in lower and higher plants is brought about by the same reaction mechanism as in bacteria is still an open question since only few experimental data are available. Corresponding in-vitro systems have been assembled from plant tissues (maize endosperm and wheat germ) and algal cells *(Chlorella, Anacystis)*. Apart from partly purified ribosomes and a "soluble fraction" (activating enzymes + transfer RNA) these systems contain $MgCl_2$, ATP and GTP, as well as phosphoenolpyruvic acid and *pyruvate kinase* as energy-regenerating system (reaction mechanism, p. 213).

The presence of DNA, ribosome-like particles, transfer RNA species and *aminoacyl-t-RNA synthetases* in chloroplasts gives rise to the question whether these organelles accomplish a specific protein synthesis independent of nuclear genes. Recent findings appear to confirm a limited formation of proteins in this way; ribosomal proteins of the chloroplast and several structural proteins of the thylakoids may thus be formed under the control of the chloroplast DNA. The information furnished by the latter, however, represents only a very small portion of its total informational content; the function of the remaining 80–90 % is unknown. On the other hand, one has to consider that evidently numerous proteins of the chloroplast are synthesized outside the organelle, i.e. on cytoplasmic ribosomes, and are subsequently transferred into the chloroplast. Their blueprints are certainly provided by the nuclear DNA. In this respect, *carboxydismutase* (p. 155) is of utmost interest since recent results indicate that one of its subunits is coded by nuclear DNA, while the other one is coded by chloroplast DNA. Accordingly, the smaller polypeptide chain is synthesized on cytoplasmic ribosomes, the larger one on chloroplast ribosomes. Only after their union in the chloroplast the active enzyme will have been formed.

Protein Degradation and Utilization of the Products

The state of "dynamic equilibrium" which is present for many substances of the living cell and which brings about their permanent renewal also applies to the protein macromolecules (cf. p. 309 f). Simultaneously with the biosynthesis of new molecules, a degradation of "aged" molecules takes place. In unicellular organisms and rapidly dividing meristem cells this *proteolysis* is usually masked by a vigorous synthesis of new protein molecules. Vice versa degradation processes predominate when reserve proteins are being mobilized in plant storage organs. In living matter, degradation of proteins is mainly achieved by microorganisms.

Protein degradation is an exergonic process; the equilibrium of the hydrolysis involved favors cleavage. A protein molecule is enzymatically split either into fragments or directly into amino acids by an attack from both ends of the chain or fragment. The amino acids are utilized for synthesis of new proteins or, as in the case of mobilized reserve proteins, are fed into the "operating" metabolism or "building" metabolism after previous removal of the amino group (= *deamination*, p. 398). In the form of keto acids they may undergo degradation in the citric acid cycle or be used for synthesis of carbohydrate and fatty acids (cf. p. 231 f). Mature higher plants, in contrast to man and animals, do not make use of the first possibility under normal conditions. A dissimilatory degradation of the carbon skeleton of amino acids to obtain energy (= "dissimilation of protein") is only observed in plants suffering from malnutrition.

We shall begin the treatment of the degradation processes with a brief compilation of enzymes which specifically cleave the peptide bond, and are generally termed *proteolytic enzymes* or *proteases*.

The enzymes of protein degradation are generally classified as *proteinases* and *peptidases* according to their substrate specificity. However, it has been shown that *proteinases* by way of exception hydrolyze some peptides too, and that many *peptidases* may also cleave proteins. Today, the term *protease* is hence preferred for all those enzymes which split peptide bonds. According to their mode of action we distinguish between *endopeptidases* and *exopeptidases*. Enzymes of the first group prefer proteins and higher polypeptides as substrate (for this reason they are also called *proteinases*); they degrade the molecules by cleaving bonds within the peptide chain yielding peptide fragments. Each of these enzymes exhibits a specificity for the type of peptide bond which it attacks in the peptide chain. Their action is accordingly restricted not to a specific protein, but to specific structural properties of the macromolecule. Therefore, all proteins are practically subject to cleavage by a few *endopeptidases* though the cleavage products formed are of various size. These enzymes cannot degrade starting from the molecule ends. The employment of *proteases* to elucidate the sequence of proteins has already been discussed (p. 375).

Among the wellknown and intensively studied *endopeptidases* are the digestive enzymes *pepsin, trypsin and chymotrypsin,* as well as the plant *proteases ficin* in latex of *Ficus* species, *papain* (melon tree, *Carica papaya*) and *bromelin* (pineapple fruit). Little is known about the *endopeptidases* active in animal and plant cells.

Degradation of proteins or peptides exclusively from the molecule ends is restricted to the *exopeptidases,* as indicated by the name. These enzymes require either the presence of a free α-amino function, hence splitting only the N-terminus off the polypeptide chain (= *aminopeptidases*) or that of a free α-carboxyl function, hence liberating the C-terminus (= *carboxypeptidases*). By continuous repetition of the cleaving reaction the protein or peptide is in the end converted into a mixture of its amino acid building blocks.

Aminopeptidases have been demonstrated in higher plants though details as to their properties are yet unknown. However, we have some knowledge about the *exopeptidases* in the gastrointestinal tract of man and animals.

Among the enzymes which cleave mainly lower peptides or fragments of proteins the *dipeptidases* of animal organisms have been studied in detail. As indicated by their name, they degrade exclusively dipeptides into their two amino acid building blocks. *Glycyl-glycine dipeptidase, prolinase* (a prolyl-peptide splitting enzyme) and *prolidase,* an aminoacylproline splitting enzyme, belong to these specific *peptidases*.

Metabolism of Free Amino Acids. The free amino acids formed by proteolysis are partly converted to the corresponding α-ketoacids by transamination (p. 363 f), and in part utilized as building blocks in the synthesis of new protein molecules. When a surplus arises, e.g. in seedlings during mobilization of reserve proteins in the seed, the amino acids may enter various metabolic pathways. In the beginning a characteristic conversion generally takes place by which the molecule is "fitted" for the particular reaction pathway.

Of particular importance among these initial reactions is the "oxidative deamination" of the amino acid to an α-ketoacid. It is the reversal of reductive amination, which we have encountered in the formation of glutamic acid from α-ketoglutaric acid catalyzed by *L-glutamic acid dehydrogenase* and NAD-H + H$^\oplus$ (cf. p. 362). This enzyme also catalyzes the reverse process:

$$H-\underset{\underset{COOH}{|}}{\overset{\overset{R}{|}}{C}}-N\overset{H}{\underset{H}{\diagdown}}H \xrightarrow[\text{NAD}^\oplus \quad \text{NAD-H+H}^\oplus]{} \underset{\underset{COOH}{|}}{\overset{\overset{R}{|}}{C}}=NH \xrightarrow{+ H_2O} \underset{\underset{COOH}{|}}{\overset{\overset{R}{|}}{C}}=O \ + \ NH_3$$

At least nine of the common amino acids may enter the citric acid cycle in the form of a dicarboxylic acid, particularly in microorganisms and animal cells. For histidine, proline, ornithine, arginine and lysine it is α-ketoglutaric acid which originates from the common intermediate glutamic acid. Another intermediate of the citric acid cycle, fumaric acid, is formed from the aromatic amino acids. Oxidative deamination of aspartic acid yields oxaloacetic acid. Acetyl-CoA is formed via pyruvic acid from alanine and serine; the function of acetyl-CoA as a building block in the biosynthesis of fatty acids and isoprenoids will be treated later (p. 405 ff and p. 430 ff).

Apart from these conversions glycine, serine as well as histidine are of great importance in the "C_1-fragment metabolism" (p. 318). Serine provides "active formaldehyde", while "activated formic acid" originates from glycine and histidine. Participation of these C_1-fragments in the formation of complex cellular compounds is apparent in the synthesis of pyrimidine and purine nucleotides (cf. p. 317 f and p. 319 f).

The α-ketoacids produced by transamination or oxidative deamination and transferred to the pool of the citric acid cycle either undergo complete degradation to CO_2 and H_2O, thus contributing to the energy supply of the cell, or serve as starting substances in the synthesis of various cellular compounds. Conversion to carbohydrates is brought about by the mechanism of "gluconeogenesis" (cf. p. 229 f), i.e. by a partial reversal of aerobic glucose degradation, with oxaloacetic acid as key compound. This is particularly important for the cells of plant storage organs and tissues. All the amino acids which are suitable precursors for this pathway are termed "glucoplastic" (p. 230). Starting from the metabolic "junction" of acetyl-CoA several amino acids may thus enter fatty acid synthesis after having been converted to an α-ketoacid.

Conversion of alanine, valine, leucine and isoleucine into "activated fatty acids" by transamination reactions and subsequent oxidative decarboxylation of the ketoacids formed must also be mentioned in this context. We have already treated the mechanism of the latter reaction with pyruvic acid serving as example (p. 215 ff). The carbon skeletons of valine, leucine and isoleucine, shortened by one carbon atom and bound to CoA, may enter directly into fatty acid metabolism. Isoleucine may also become an intermediate of the citric acid cycle or a precursor of porphyrin synthesis (p. 418 ff) after having been converted to succinyl-CoA via propionyl-CoA.

A further reaction for degradation of amino acids, employed mainly in bacteria and fungi, is direct decarboxylation. It is catalyzed by *amino acid decarboxylases* and gives rise to formation of primary amines; the active coenzyme is pyridoxal phosphate (p. 363).

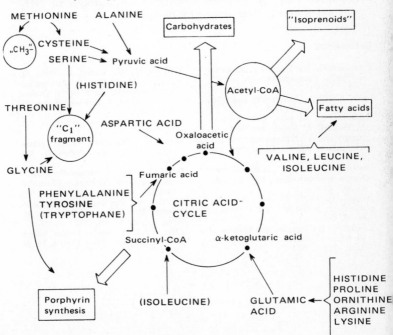

The reaction products, the "biogenic amines", are either important precursors in the biosynthesis of coenzymes and biologically active compounds or play an important role as active substances in pharmacology.

The scheme on p. 399 summarizes the most important conversions of amino acids and their interactions with the various biosynthetic processes.

NH₃-Detoxication. As a consequence of protein degradation, particularly during oxidative deamination of the released amino acids, NH_3 should accumulate in the cell and would gradually poison it. This does not occur because the plant has various mechanisms at its disposal to bind free NH_3. These differ from those of animal organisms insofar as nitrogen is not excreted by plants due to economic reasons. The NH_3 is rather converted into a chemical form from which nitrogen may easily be mobilized and provided for syntheses. Thus NH_3 may be bound to aspartic acid or glutamic acid giving rise to formation of the amide compounds asparagine and glutamine, respectively. More complex substances such as allantoin or allantoic acid, present in roots, are also effective in detoxication and simultaneous binding of the nitrogen of NH_3. Formation of urea is characteristic for fungi.

References

Braunitzer, G.: Die Primärstruktur der Eiweißstoffe. Naturwissenschaften 54: 407, 1967

Crick, F. H. C.: The genetic code. Sci. Amer. 215 no. 4: 55, 1966

Hartmann, G., W. Behr, K. A. Beissner, K. Honikel, A. Sippel: Antibiotika als Hemmstoffe der Nucleinsäure- und Proteinsynthese. Angew. Chemie 80: 710, 1968

Haselkorn, R., L. B. Rothman-Denes: Protein synthesis. Ann. Rev. Biochem. 42: 397, 1973

Jockusch, H., H. G. Wittmann: Entschlüsselung des genetischen Codes. Umschau 66: 49, 1966

Kurland, C. G.: Structure and function of the bacterial ribosome. Ann. Rev. Biochem. 41: 377, 1972

Mathaei, H., G. Sander, D. Swan, T. Kreuzer, H. Caffier, A. Parmeggiani: Reaktionsschritte der Polypeptidsynthese an Ribosomen. Naturwissenschaften 55: 281, 1968

Nomura, M.: Ribosomes. Sci. Amer. 221: 28, 1969

Ochoa, S.: Translation of the genetic message. Naturwissenschaften 55: 505, 1968

Ochoa, S.: Die Rolle ribosomaler Fraktionen beim Beginn der Polypeptidkettenbildung. Naturwiss. Rundschau 23: 1, 1970

Schreiber, G.: Die Translation der genetischen Information am Ribosom. Angew. Chemie 83: 645, 1971

Zalik, P., B. L. Jones: Protein biosynthesis. Ann. Rev. Plant Physiol. 24: 47, 1973

Lipids

The lipids characteristic for higher plants are made up of various groups: neutral fats, waxes, glycerol phosphatides, sphingolipids and glycolipids. A common property is the complete insolubility in water, but good solubility in organic solvents. Thus the overall group of *lipids* also includes lipid-like substances *(lipoids)*.

Neutral Fats

Chemical Constitution. Neutral fats are esters of the trihydric alcohol glycerol with one or more unbranched monocarboxylic acids, the fatty acids. If only one of the hydroxyl groups in the glycerol molecule is esterified with a fatty acid the compound is a monoglyceride; if two or three are esterified, a di- or triglyceride is formed (new nomenclature: "monoacyl glycerol", "diacyl glycerol" and "triacyl glycerol"). In the oil and fat-containing seeds of higher plants, the triglycerides predominate over the di- and monoglycerides.

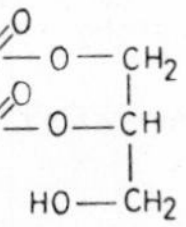

Glycerol

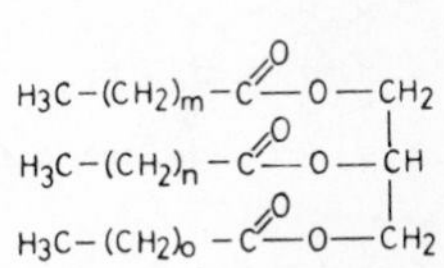

Monoglyceride

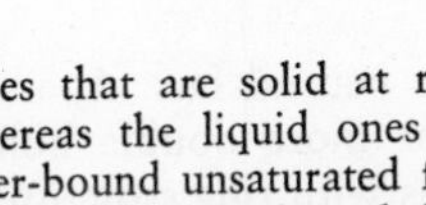

Diglyceride

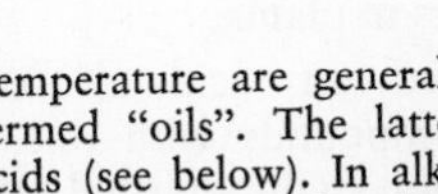

Triglyceride

Triglycerides that are solid at room temperature are generally termed "fats", whereas the liquid ones are termed "oils". The latter contain mainly ester-bound unsaturated fatty acids (see below). In alkaline conditions triglycerides are degraded to free glycerol and the alkali salts of the fatty acids (= "soaps"). This process which has been used for a long time in the production of soap is termed "saponifying". In modern organic chemistry this term is generally applied to hydrolytic cleavage of various compounds.

Triglyceride   Glycerol   Alkali salt ("soap")

Saponification of a fat

Fatty Acids

It is the straight-chain hydrocarbon fatty acids containing an even number of carbon atoms that mainly account for the unusual chemical and physical properties of lipids. Their special structure ensues from C_2-units, i.e. acetyl residues which are utilized as building blocks in their biosynthesis (p. 405). Branched-chain fatty acids do not seem to occur in higher plants.

Particularly prevalent are saturated fatty acids with 16 and 18 carbon atoms: palmitic acid ($C_{16}H_{32}O_2$) and stearic acid ($C_{18}H_{36}O_2$). Their chains exhibit a typical zig-zag configuration as shown schematically in the following formulas:

Palmitic acid   ($C_{16}H_{32}O_2$)

Stearic acid ($C_{18}H_{36}O_2$)

Among the saturated fatty acids, lauric acid (C_{12}), myristic acid (C_{14}) and arachidic acid (C_{20}) represent further constituents of neutral fats in plants.

Unsaturated fatty acids also occur in neutral fats. Apart from the typical compounds with one, two or more double bonds in the hydrocarbon chain, unusual fatty acids with ring structures or polar functions in the molecule have been identified in plant neutral fats. The C_{18} compounds of oleic acid ($C_{18}H_{34}O_2$), linoleic acid ($C_{18}H_{32}O_2$), linolenic acid ($C_{18}H_{30}O_2$) and the C_{20} compound of arachidonic acid ($C_{20}H_{32}O_2$) belong to the first group:

Oleic acid ($C_{18}H_{34}O_2$)

Linoleic acid ($C_{18}H_{32}O_2$)

Linolenic acid ($C_{18}H_{30}O_2$)

Arachidonic acid ($C_{20}H_{32}O_2$)

Linoleic acid and linolenic acid are present in large quantities in linseed oil (name!). These two substances are essential in human and animal nutrition, i.e. they cannot be synthesized by these organisms and must be taken up with the nutritional plant material. While oleic acid and palmitic acid are widely distributed in the reserve fats of plants, other fatty acids accumulate in cellular organelles, e.g. linolenic acid in the chloroplasts. Other unsaturated fatty acids are found in varying distribution and quantity in the fats of certain plant species or families; up to six double bonds may be present in the molecules.

The presence of a double bond in the molecule means that the substituents at the carbon atoms involved may be positioned either *cis* or *trans*. Oleic acid in its natural form has the *cis* configuration; the *trans* compound termed elaidic acid can only be synthesized chemically.

The double bond in the center of the carbon chain affects the shape of the molecule in that it tends to achieve the most stable form. This probably leads to a slightly bent structure:

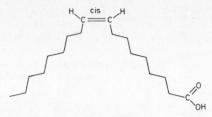

Since the position of the double bond may vary, isomeric compounds result; a few of these have been demonstrated in higher plants. Recently several complex unsaturated fatty acids have been found in plant fats which contain double bonds with *trans* configuration, e.g. α-eleostearic acid.

When more than one double bond is present in the hydrocarbon chain, it is separated from the neighboring double bond either by a methylene group or by two single bonds. Thus their arrangement differs markedly from a conjugate system (cf. p. 37 ff) of the kind typical for carotinoids. We speak therefore of "isolated double bonds" or a "nonconjugate system".

Metabolism. Glycerol and fatty acids, the molecular building blocks of fats, are produced by separate biosynthetic pathways. They form a fat molecule only in a final reaction. The ability of human and animal organisms to convert carbohydrates and proteins into fats is wellknown. Similar conversions occur in plant organisms, particularly in those which store fat as reserve substance, e.g. in their seeds. Carbohydrates probably serve as the main substrate. First they have to be degraded to acetyl-CoA via hexose and pyruvic acid. The conversions required are brought about by glycolysis and by decarboxylation of the C_3 fragment (p. 208 ff, 215 ff).

As pointed out previously acetyl-CoA holds a key position in metabolism in that its "pool" provides substrates for various coupled processes of synthesis (cf. scheme p. 232). One of these is the biosynthesis of fatty acids which utilizes activated C_2 fragments to construct long-chain molecules. Thus an important communication exists between carbohydrate metabolism and fat metabolism which may also work in the opposite direction, i.e. mobilization and conversion of fat to carbohydrates (p. 226 ff), e.g. in the germination of seeds.

The precursor of the second building block of fat, glycerol, is also supplied by glycolysis in the form of dihydroxyacetone phosphate which yields L-α-glycerophosphate after reduction; this is ready for insertion into a fat molecule (p. 410 f). Let us consider first the formation of fatty acids.

Biosynthesis of Fatty Acids

Two synthetic pathways for long-chain saturated fatty acids have been discovered in microorganisms and animals: 1) de-novo synthesis from C_2 building blocks in the cytoplasm, and 2) elongation of preformed fatty acids in cell organelles. We shall treat first the de-novo synthesis. Initially CO_2 is bound to acetyl-CoA with consumption of ATP giving rise to formation of malonyl-CoA:

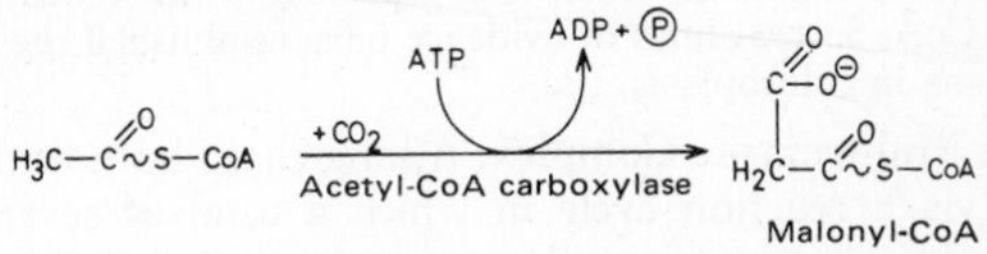

Biotin serves as coenzyme of the *acetyl-CoA carboxylase* involved which may be covalently bound to the enzyme protein. The first reaction step consists in an "activation" of carbonic acid which is bound to biotin with concomitant cleavage of ATP; it is hence raised to a higher energy level and thus able to react as an "active C_1 fragment" with acetyl-CoA:

$$\text{Biotin carboxylase} + HCO_3^{\ominus} + ATP \xrightarrow{Mg^{2\oplus}} CO_2\text{-biotin carboxylase} + ADP + \textcircled{P}$$
$$CO_2\text{-biotin carboxylase} + \text{acetyl-CoA} \longrightarrow \text{malonyl-CoA} + \text{biotin carboxylase}$$

Biotin is another coenzyme of C_1 metabolism (cf. p. 398). It serves as an intermediate carrier for the carboxyl group being transferred. Biotin was first demonstrated in liver extracts ("vitamin H") and in yeast as a growth factor. It has been isolated from egg-yolk. It may form a stable complex with avidin, a constituent of the egg-white, thereby losing its specific function. If biotin is fed in this form to animals, typical symptoms of avitaminosis appear. Biotin is a cyclic derivative of urea to which a sulfur-containing ring has been added. The complex with the

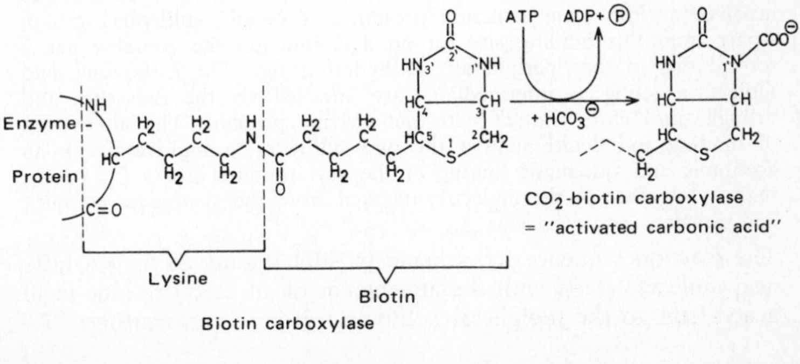

CO$_2$-biotin carboxylase = "activated carbonic acid"

enzyme is achieved through an amide linkage between the ε-amino group of a lysine residue of the enzyme and the carboxyl group of biotin.

The actual activation reaction consists in the binding of carbonic acid to a nitrogen atom of biotin in an endergonic reaction, the energy requirement being met by ATP, with formation of biotinyl-AMP as intermediate (see above). Bound as carboxybiotin, the carbonic acid is able to enter a carboxylation reaction. Free carboxybiotin is an extremely labile molecule which may be stabilized by conversion to a dimethyl ester.

Purified *acetyl-CoA carboxylase* from wheat seedlings appears to consist of a biotin-containing component and a protein which binds acetyl-CoA or malonyl-CoA. Several lines of evidence have established the occurrence of this enzyme in chloroplasts, too.

The Active Multienzyme Complex. A long-chain fatty acid molecule is formed via a reaction cycle in which a total of seven enzymes participate. In yeast and in cells of certain animal tissues these enzymes are assembled in a functional unit, a "multienzyme complex" which is termed *fatty acid synthetase;* it has been isolated and purified from these cells.

The particles obtained from yeast move in the gravity field of the ultracentrifuge (p. 335 f) as a single component with a molecular weight of 2.3×10^6. As revealed by electron microscopy the complex is a hollow, oval particle with an equatorial ring; it is composed of subunits.

Studies on the mode of action of *fatty acid synthetase* have shown that a close relationship exists between the spatial arrangement of the individual enzyme proteins and the reaction steps catalyzed by these. Free intermediates are obviously not released, but directly passed on to the next enzyme and tightly bound again by means of appropriate binding sites. This system exhibits a number of advantages in terms of reaction kinetics as far as the repetition of the seven reactions required for stepwise elongation of the carbon chain is concerned.

Certain models concerning the structure of the multienzyme complex have been proposed recently, taking into account the typical reaction sequence. As indicated in the reaction scheme (p. 407) the seven enzymes involved form a ring structure around a "central" sulfhydryl group. Apart from this binding site for an acyl function the complex has a second one in the "peripheral" sulfhydryl group. The carboxylic acid molecules acting as intermediates are attached via the two sites and brought into close contact with the enzyme proteins. The alternation of binding and detachment at the two sulfhydryl groups results in an economic and systematic linking of the acyl residues into a C_{16} or C_{18} chain. Only then is the molecule released from the *synthetase* complex and transferred to coenzyme A.

The reaction sequence (cf. scheme p. 407) leading to a new fatty acid molecule starts with the attachment of an acetyl residue from acetyl-CoA to the peripheral sulfhydryl group *(start-transfer;* "R"

Mode of action of the multienzyme complex
in fatty-acid biosynthesis

is in this case = H!). Subsequently Malonyl-CoA reacts with the
central sulfhydryl group of the enzyme complex *(malonyl-transfer)*.
In the acetyl-malonyl enzyme formed the C_2 and C_3 residues bound
to the two functional groups are ready for the next step, the *con-*

densation. The acetyl group reacts with malonic acid leading to formation of acetoacetyl enzyme accompanied by decarboxylation. The C_4 product remains attached to the central SH-group during the following steps of reduction, removal of water and subsequent reduction. The ketoacid is first reduced by NADP-H + H$^\oplus$, the specific transport metabolite for hydrogen in reductive syntheses *(1st reduction)*. The reaction product, β-hydroxybutyric acid, is converted to the unsaturated C_4 compound, crotonic acid, by removal of water.

This forms the substrate for the *2nd reduction* with NADP-H + H$^\oplus$ as hydrogen donor. It does not, however, reduce the molecule directly but with mediation of flavin mononucleotide (FMN, p. 17 f) as the active group specific for the ethylene bond. FMN is firmly bound to the specific structural protein, too. The saturated carboxylic acid is now transferred to the peripheral SH-group *(acyl transfer)* while a malonyl residue supplied by malonyl-CoA is again attached to the central SH-group. The reaction cycle just described is now repeated with butyryl-malonyl enzyme as starting complex giving rise to a chain elongated by two carbon atoms. It is transferred to the peripheral SH-group. The cycle is repeated with additional molecules of malonyl-CoA until a C_{16} or C_{18} fatty acid has been produced by successive chain elongation. Now the terminal reaction product is released from the *synthetase* complex, the palmityl or stearyl residue being transferred to coenzyme A *(palmityl transfer;* below).

The two functional SH-groups having been restored, the *synthetase* complex is now capable of synthesizing a new fatty acid molecule. The cycle is initiated by the attachment of acetyl-CoA to the peripheral SH-group as described above. It is unknown which factors determine the chain length at which synthesis ceases.

Palmityl - CoA

A similar multienzyme complex is active in fatty acid synthesis of bacteria. It has been obtained in pure form. Since it is more easily fractionated into its subunits it is convenient to use for isolation and characterization of these subunits. Obviously the central SH-group is part of an independent non-enzymatic protein *(acyl carrier protein)* the active group of which is identical with pantetheine (p. 218). It is attached to a serine residue of the enzyme protein by a phosphodiester bond:

4'-phosphopantetheine, bound to serine

According to a current model the central SH-group in the molecular chain of pantetheine has a long mobile arm by which the acyl function bound at its end can be brought rapidly and specifically into contact with the active sites of the enzymes arranged in a ring. The pheripheral SH-group belongs to a cysteine residue which is probably a building block of the enzyme involved.

Arrangement of the enzymes of a common synthetic pathway in a functional complex is advantageous in two ways: diffusion of intermediates is largely restricted, and interference by enzymes involved in other reaction sequences is avoided, particularly interference by the enzymes of fatty acid degradation which, in contrast to the biosynthesis, takes place in the mitochondria.

By assigning separate compartments to the two reaction pathways of synthesis and degradation, the cell is able to operate the two processes simultaneously, while regulating them independently (cf. p. 178).

Cell-free systems which synthesize long-chain saturated fatty acids have been obtained from tissues of higher plants, too. However, it is still uncertain whether *fatty acid synthetase* is also present in the cells as a multienzyme complex similar in structure and function to that of yeast and bacteria. *Acyl-carrier protein* (above) isolated from avocado fruit and spinach leaves differs from the corresponding compound in bacteria by its higher lysine content. Cells of *Euglena* also contain active *fatty acid synthetase,* its formation depending on growth conditions. Two forms of the enzyme have been found during illumination: one in the chloroplasts which is activated by *acetyl-CoA carboxylase,* and a second one which is independent of the former and is probably localized in the cytoplasm. In dark-grown cells of *Euglena* only the second form seems to be present. According to these findings the system of fatty acid biosynthesis in higher plants appears to resemble that in bacteria, yeast and animal organisms to a large extent.

Synthesis of long-chain fatty acids by a mechanism of chain elongation is probably restricted to mitochondria and microsomes since the enzymes involved are bound to their membrane structures.

It has been proposed that in mitochondria palmityl-CoA is converted to stearyl-CoA by means of acetyl-CoA and in the presence of coenzyme-bound hydrogen. The same reaction proceeds in microsomes, except that malonyl-CoA is the unit being attached; accordingly CO_2 is released. The enzymes responsible for elongation of preformed fatty acid molecules also occur in cells of higher plants, particularly in those in which waxes (cf. p. 416) with extreme long-chain fatty acids (C_{20} to C_{28}) are synthesized. Microsomes are probably the reaction sites.

In the biosynthesis of mono- and polyunsaturated fatty acids the principle of "desaturation" is obviously applied: in the presence of molecular oxygen and NADP-H + H$\oplus$ or NAD-H + H$\oplus$ double bonds are introduced into previously saturated fatty acid molecules by specific enzymes (= *desaturases*). In this case the most important precursors are stearyl-CoA and palmityl-CoA. Labeling experiments with appropriate precursors (^{14}C-labeled saturated fatty acids from C_{14} to C_{19}; ^{14}C-acetate) have confirmed that this reaction mechanism operates in tissue sections, cell organelles and subcellular particle fractions from higher organisms. The isolation and characterization of the enzymes involved, however, has only just started. A particle-bound *acyl-CoA desaturase* was recently isolated from baker's yeast, and a *stearyl-CoA desaturase* system from *Euglena;* components of the latter were *NADP-H oxidase, desaturase* and ferredoxin.

The conversion of oleic acid to linoleic acid and linolenic acid in greening cells and in several seeds at the stage of their CoA compounds is a particularly significant example of the formation of polyunsaturated fatty acids. They are intensively synthesized during chloroplast differentiation as shown by labeling experiments with ^{14}C-acetate.

Mono- and polyunsaturated fatty acids may undergo numerous secondary conversions in various plant species or families, e.g. additional chain elongation, incorporation of oxygen (= epoxy-fatty acids) and conversion of double bonds to triple bonds (= alkene acids).

Glycerol enters the final ester formation with the fatty acids in the form of glycerolphosphate ("L-α-glycero-phosphate"). This phosphorylated derivative of the trihydric alcohol is formed mainly by reduction of dihydroxyacetone phosphate with NAD-H + H$\oplus$ catalyzed by *glycerol-1-phosphate dehydrogenase*:

Dihydroxyacetone phosphate   L-α-glycerophosphate

This enzyme, however, has not yet been demonstrated in higher plant cells.

The esterification leading to formation of neutral fats proceeds in two reaction steps. In the first step, catalyzed by a specific enzyme, two molecules of "activated" fatty acids, i.e. bound to coenzyme A (C_{16}, C_{18}) react with glycerolphosphate yielding a diglyceride phosphate; two molecules of CoA-SH are released (see below). The reaction product is termed "phosphatidic acid"; compounds of this class play an important role as intermediates in the biosynthesis of phospholipids (p. 417). Upon restoration of the third hydroxyl group by dephosphorylation, catalyzed by a *phosphatase,* the second esterification can take place. A triglyceride is formed during the subsequent acylation with another activated fatty acid molecule; the reaction product is at the disposal of the plant cell as fat reserve.

$$
\begin{array}{l}
H_2C-OH \\
HO-C-H \\
H_2C-O-\textcircled{P}
\end{array}
+\ 2\ R-C\overset{O}{\underset{}{\diagup}}\!\!\sim\!S-CoA
\longrightarrow
R-C\overset{O}{\underset{}{\diagup}}
\begin{array}{l}
H_2C-O-\overset{O}{\overset{\|}{C}}-R \\
O-C-H \\
H_2C-O-\textcircled{P}
\end{array}
+\ 2\ HS-CoA
$$

L-α-glycerophosphate   Diglyceride phosphate
                                        = phosphatidic acid

Degradation of Fats

Before the accumulated reserve fats can be utilized in metabolism their molecules have to be decomposed into appropriate fragments. During germination of fat-containing seeds the lipid content decreases rather rapidly, for the benefit of soluble compounds. At the same time the enzymes specifically catalyzing fat cleavage, the *lipases,* reach their highest activity; according to their mode of action they are classified as *hydrolases* belonging to the group of *esterases.* They cleave the ester bonds in fat molecules concomitant with uptake of water. With diglycerides and monoglycerides as intermediates free fatty acids and glycerol are eventually formed provided the exposure to the enzyme has been long enough:

$$
\begin{array}{l}
H_3C-(CH_2)_{14}-C\overset{O}{\overset{\diagup}{}}O-CH_2 \\
H_3C-(CH_2)_{14}-C\overset{O}{\overset{\diagup}{}}O-CH \\
H_3C-(CH_2)_{14}-C\overset{O}{\overset{\diagup}{}}O-CH_2
\end{array}
\xrightarrow[\text{Lipase}]{+H_2O\quad +H_2O\quad +H_2O}
\begin{array}{l}
H_3C-(CH_2)_{14}-COOH \\
+ \\
H_3C-(CH_2)_{14}-COOH \\
+ \\
H_3C-(CH_2)_{14}-COOH
\end{array}
+
\begin{array}{l}
HO-CH_2 \\
HO-C-H \\
HO-CH_2
\end{array}
$$

The two cleavage products pass on to various metabolic pathways. Glycerol preferentially enters the carbohydrate metabolism. In fact, it has been demonstrated that added glycerol-1,3-C^{14} is converted to ^{14}C-sucrose in etiolated cotyledons, in cell-free extracts obtained from these as well as in tissue slices from higher plants. Concomitant release of $^{14}CO_2$ has also been observed. These findings indicate that glycerol is converted to sucrose via triose and hexose phosphates. Glycerol may also be degraded to dihydroxyacetone phosphate and pyruvic acid by reactions identical to those in glucose dissimilation, comprising the reaction sequences of glycolysis and the citric acid cycle.

Long-chain fatty acids, on the other hand, are cleaved mainly by a special enzyme-catalyzed mechanism termed *β-oxidation** yielding C_2 fragments which enter the metabolism as acetyl-CoA (*a-oxidation* is of less importance; p. 416). The activated C_2 fragment serves in particular as substrate of the citric acid cycle; it is used either in the energy production connected with the respiratory chain or in various synthetic processes via the intermediates of the citric acid cycle (cf. diagram p. 232). The entry of acetyl-CoA into the glyoxylic acid cycle whose reactions are characteristic for fat-storing tissues of higher plants (see p. 226 ff) guarantees a rapid synthesis of carbohydrates, particularly in the transport form of sucrose. This process of "gluconeogenesis" (p. 229 f) forms a link between lipid metabolism and carbohydrate metabolism in the cell. Therefore, fat as a storage substance in plant cells and tissues serves firstly as substrate in the production of energy, and secondly as starting material in de-novo synthesis of cellular compounds.

β-Oxidation of Fatty Acids and Energy Production

This complex process involving several enzymes and coenzymes appears to be identical in plants and animals. The initial reaction is an "activation" of the chemically rather inaccessible fatty acid molecule which is converted to the high-energy acyl derivative of

---

* Since the essential feature of this mechanism leading to the release of a terminal C_2 fragment is the involvement of the β-carbon atom in the reaction concerned, it has been termed "β-oxidation". It was discovered by F. Knoop in 1905, but the details have only recently been elucidated, especially through the studies of F. Lipmann, F. Lynen, D. E. Green and others.

CoA, with consumption of ATP. Acyl adenylate is the intermediate; the reaction is catalyzed by *acyl thiokinase:*

$$ATP \qquad AMP + \textcircled{P}-\textcircled{P}$$

$$CH_3-(CH_2)_n-COOH + HS\text{-}CoA \xrightarrow[\text{Acyl thiokinase}]{} CH_3-(CH_2)_n-C\overset{\displaystyle\parallel}{}\sim S-CoA$$

This process may be compared with the conversion of the acetyl residue arising from pyruvic acid to the reactive compound of acetyl-CoA (p. 215 ff).

The fatty acyl residue is now transferred from CoA to a specific carrier substance which transports it across the natural barrier of the inner mitochondrial membrane to the actual site of degradation. Carnitine ester functions as an acyl-group carrier; its group transfer potential roughly equals that of the acyl-CoA compound, so that the transfer can take place without additional input of energy:

$$CH_3-(CH_2)_n-C\overset{\displaystyle\parallel O}{}\sim S-CoA$$

Fatty acid CoA
+

$$HO-CH-CH_2-\overset{\oplus}{N}\begin{smallmatrix}CH_3\\CH_3\\CH_3\end{smallmatrix}$$
$$H_2C-COO^{\ominus}$$

Carnitine

$$\rightleftharpoons \quad CH_3-(CH_2)_n-C\overset{\displaystyle\parallel O}{}\sim O-CH-CH_2-\overset{\oplus}{N}\begin{smallmatrix}CH_3\\CH_3\\CH_3\end{smallmatrix}$$
$$H_2C-COO^{\ominus}$$

Acyl carnitine
+
HS-CoA

Since fatty acid degradation occurs in the matrix space of the mitochondrion (cf. p. 203) and requires acyl-CoA compounds as substrate, the fatty acid residue must be transferred to CoA again in reversion of the reaction described above.

Hydrogen is removed in the first reaction giving rise to formation of the $\alpha\beta$-unsaturated acyl derivative of CoA (see scheme of pathway, p. 414).

The specific enzyme, *acyl-CoA dehydrogenase*, is a flavoprotein belonging to a group of closely related enzymes in mitochondria which all contain FAD (p. 17 f) as prosthetic group. Slight differences exist in connection with substrate specificity, particularly regarding the length of the chain of fatty acids. This may be due to the mechanism of fatty acid degradation which like that of biosynthesis, proceeds by repetition of a cyclic reaction sequence. During this process the carbon chain of the fatty acid residue is shortened by a C_2 fragment each time. The individual reaction steps are summarized in the reaction scheme below. There is a specific *dehydrogenase* for each of the different chain lengths. The flavoproteins transfer the hydrogen to the respiratory chain (p. 235 ff) and, having restored their oxidized state, may engage in fatty acid degradation again.

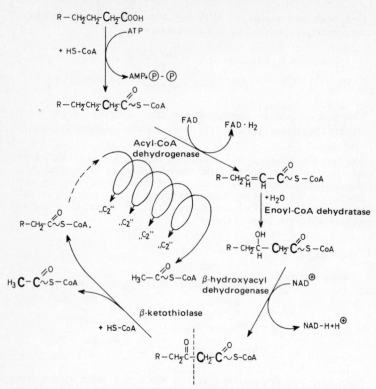

Reaction sequence in β-oxidation of a fatty acid

Water is added to the unsaturated fatty acyl derivative of CoA formed giving rise to a β-hydroxy fatty acid; the reaction is catalyzed by *enoyl-CoA hydratase*. The secondary alcohol group formed undergoes a second oxidation catalyzed by *β-hydroxyacyl-CoA dehydrogenase* with $NAD^{\oplus}$ as coenzyme. The latter transfers the hydrogen to the respiratory chain.

This second dehydrogenation yields a β-ketoacid which, being a thioester, is unstable and readily releases a C_2 fragment as acetyl-CoA while the molecule residue reacts with a free molecule of CoA-SH; *β-ketothiolase* acts as specific enzyme. This "thiolytic cleavage" is the key reaction step in the process of degradation since the -C-C-bond in the molecular chain is split. It is also important in terms of energy because the released C_2 fragment remains bound to CoA (= "activated acetic acid") and the molecule residue leaves the cleavage reaction as an acyl-CoA compound with a high transfer potential. Thus the free energy of this reaction is largely conserved in chemical bonding; no further "activation" is hence required for the next reaction step in degradation. Activation is thus restricted to a unique "starting reaction" (see above) in which a fatty acid molecule is bound to CoA independent of its chain length.

The molecule residue, shortened by a C_2 fragment and bound to CoA, again enters the reaction cycle described above and loses a further C_2 fragment in the form of acetyl-CoA via the reaction steps of dehydrogenation, hydration and second dehydrogenation. The process is repeated until the carbon chain has been completely split, yielding acetyl-CoA; then another fatty acid molecule is fed into the degradation machinery.

The balance of the overall process is as follows: there is no direct gain of energy in the form of ATP; on the other hand, hydrogen bound to specific transport metabolites is provided in relatively large amounts for oxidation by the respiratory chain, thus yielding energy. Each C_2 fragment gives rise to 2 (H) bound to NAD and 2 (H) bound to the FAD system. Since oxidation of 1 mole of NAD-H $+ H^\oplus$ via the respiratory chain yields 3 moles of ATP under optimal conditions, and that of 1 mole of $FADH_2$ yields 2 moles of ATP, a total of 5 moles of ATP results from each acetyl residue. For palmitic acid seven cleavage steps have to take place; these yield $7 \times 5 = 35$ moles of ATP. Since 1 mole of ATP was required for the initial activation, a net yield of 34 moles of ATP results from the degradation of 1 mole of palmitic acid to 8 acetyl-CoA fragments. Complete oxidation of the latter via citric acid cycle and respiratory chain yields an additional $8 \times 12 = 96$ moles of ATP. The production of ATP on complete oxidation of 1 mole of palmitic acid to CO_2 and H_2O thus amounts to $96 + 34 \doteq 130$ moles of ATP.

The efficiency of this process is easily calculated since we know both the free energy change ($\Delta G^\circ = -2\,300\,000$ cal/mole) and the free energy of hydrolysis of ATP under standard conditions ($\Delta G^\circ = -7000$ cal/mole):

$$\frac{130 \times 7000}{2\,300\,000} \times 100 = 40\,\%$$

This value corresponds to that obtained for aerobic oxidation of glucose to CO_2 and H_2O (p. 246).

The functional capacity of the fat-degrading system and the gain of chemically bound energy resulting from it do not only depend on rapid re-activation, i.e. oxidation, of the reduced coenzymes. Apart from this, the acetyl-CoA produced must be metabolized at a high rate in order to avoid a spilling over of its "pool" and gradual blocking of the process of fatty acid degradation. Figuratively speaking, the end products must be continuously removed from the assembly line to prevent them from piling up. The bulk of acetyl-CoA enters the citric acid cycle and is degraded to CO_2 and coenzyme-bound hydrogen. The remainder serve as precursors in biosynthetic processes, e.g. via the glyoxylic acid cycle. Supply of C_2 fragments for the latter and an adequate gain of energy are only guaranteed by a harmonious cooperation of β-oxidation, citric acid cycle and respiratory chain. It is hence not by mere chance that the three processes are localized in close spatial contact in the mitochondrion. However, recent studies indicate that in fat-containing seeds of some plant species, e.g. *Ricinus,* β-oxidation may take place outside the mitochondria. The specific enzymes appear to be bound to glyoxysomes (p. 227 f). In this case, a close functional relationship would exist between the degradation of fatty acids and the glyoxylic acid cycle (p. 226 ff); the acetyl-CoA formed would rapidly and preferentially be converted first to malate and then utilized as precursor for sucrose synthesis (cf. p. 228).

On the other hand, the complete reaction sequence of β-oxidation has been demonstrated in mitochondria of peanut cotyledons (Stumpf et al.).

α-Oxidation of Fatty Acids

This less important mechanism of fatty acid degradation has been observed in cotyledons and young leaves. It involves the following reaction steps: elimination of the carboxyl group by oxidation with H_2O_2 catalyzed by a *fatty acid peroxidase* and subsequent oxidation of the aldehyde compound formed (shortened by one carbon atom!) by a *dehydrogenase*.

$$R-CH_2-CH_2-COOH \xrightarrow{+H_2O_2} R-CH_2-CHO + 3H_2O + CO_2$$

$$R-CH_2-CHO \xrightarrow{+H_2O, NAD^{\oplus}} R-CH_2-COOH + NAD-H + H^{\oplus}$$

Molecular oxygen may also be effective in the first step as indicated by studies of a reaction system in young leaves.

Waxes

Typical compounds of this class of substances structurally resemble fats in that they are also esters of higher fatty acids with an alcohol. The waxes, however, contain a univalent higher aliphatic alcohol instead of glycerol. Recent studies indicate that natural waxes contain other constituents besides with different chemical structures: saturated hydrocarbons with C_{21} to C_{37} carbon chains, primary alcohols with an even number of carbon atoms (C_{22} to C_{32}) and long-chain free fatty acids (C_{14} to C_{34}) or hydroxy fatty acids. Complicated mixtures result which are difficult to analyze.

Glycerophosphatides and Phytosphingolipids

It is only recently that the isolation and identification of characteristic glycerophosphatides from plant organisms has been achieved. L-α-glycerolphosphate, formed by reduction of dihydroxyacetone phosphate, acts as the elementary building block. Its two free OH-groups are esterified with two "activated" fatty acid molecules bound to CoA: a "phosphatidic acid" (p. 411) is formed.

Phosphatidylcholine from the group of the "lecithins" is a typical plant phospholipid. Ethanolamine or inositol replace choline in other compounds. The corresponding compounds, phosphatidylethanolamine and phosphatidylinositol, occur in seeds, just like phosphatidylcholine. Phosphatidylglycerol appears to be the typical phosphatide in leaves. The sphingolipids are characterized by the presence of the long-chain amino alcohol sphingosine, which takes the place of glycerol in the molecule. Closely related substances are derived from dihydroxysphingosine. Phytosphingosine is a 4-hydroxy derivative of dihydroxysphingosine.

Glycolipids

In compounds of this group of lipids glycerol is esterified with two fatty acid molecules as well as with one or more sugar units, particularly with galactose, by glycosidic bondage via the third OH-group. The fatty acid component is mainly linolenic acid (p. 402 f). Substances of this group constitute the major part of the lipids of green leaves. In quantity they exceed the phosphatides by a factor of five. The most important compounds are monogalactosyldiglyceride and digalactosyldiglyceride.

Monogalactosyldiglyceride   Digalactosyldiglyceride

References

Goldfine, H.: Lipid chemistry and metabolism. Ann. Rev. Biochem. 37: 303, 1968

Kull, U.: Die pflanzlichen Glykolipide. Naturwiss. Rundschau 19: 109, 1966

Lynen, F.: Aufbau der Fettsäuren in der Zelle. Naturwiss. Rundschau 20: 231, 1967.

Mudd, J. B.: Fat metabolism in plants. Ann. Rev. Plant Physiol. 18: 229, 1967.

Mazliak, P.: Lipid metabolism in plants. Ann. Rev. Plant Physiol. 24: 287, 1973

Stumpf, P. K.: Metabolism of fatty acids. Ann. Rev. Biochem. 38: 159, 1969

Chlorophylls and Cell Hemins

Several lines of evidence indicate that the porphyrin system (p. 38 f), the common structure of chlorophylls and heme compounds is synthesized by the same reaction mechanism in plant and animal cells. It is only relatively late, at the stage of protoporphyrin, that the switches are thrown for different pathways the end products of which are chlorophylls in plant cells, hemoglobins in animals, and cytochromes, catalases and peroxidases in both.

Catalases and *peroxidases* are iron-containing porphyrin proteins which, like cytochromes, belong to the group of heme enzymes. In contrast to the latter the iron in the active group does not change its valency; the prosthetic group is normally present as $Fe^{3\oplus}$-protoporphyrin. Both types of enzyme may use hydroperoxide (H_2O_2) as substrate. The *catalases* decompose this compound to H_2O and O_2; they exhibit a particularly high activity. The *peroxidases* catalyze oxidations utilizing H_2O_2. While the plant enzymes of this group generally contain $Fe^{3\oplus}$ protoheme or related compounds as active group, "green" heme compounds of unknown structure (= "*verdoperoxidases*") may occur in addition among the animal enzymes. A pure crystalline *peroxidase* has been isolated from horseradish; it contains one protoheme group in the molecule.

Two methods have been particularly useful in elucidating the principal features of biosynthesis: 1) Feeding of radioactively labeled precursors to organisms, especially to unicellular ones, and 2) analysis of single reactions in vitro by means of specific enzymes from erythrocytes, higher plants and bacteria. Application of ^{15}N- or ^{14}C-labeled glycine led to the elucidation of heme biosynthesis, one of the first great successes of the isotope-tracer method just developed at that time (Shemin and coworkers, 1950). Another technique, as original as it is effective, has been employed to study chlorophyll synthesis in some mutant strains of *Chlorella pyrenoidosa;* this will be discussed presently (p. 423).

The Establishment of the Porphyrin Structure

In analyzing the biosynthesis of these complex compounds which exhibit such diverse functions it is reasonable to begin with the reactions connecting it with the substrate-supplying pathways of metabolism. As can be seen from the schemes on p. 232 and p. 399, the precursors required are provided from the "pool" of the citric acid cycle and that of the amino acids in the form of succinyl-CoA and glycine, respectively. They are fed into the machinery of porphyrin synthesis through a special "outlet". The close cooperation between "operating metabolism" and "building metabolism" is apparent here once again.

First in a condensation reaction the two compounds form α-amino-β-ketoadipic acid which is unstable and is immediately converted to δ-aminolevulinic acid with release of CO_2. The enzyme catalyzing this reaction is *δ-aminolevulinic acid synthase* which has been partially purified as plant enzyme only from the purple bacterium *Rhodopseudomonas spheroides* so far; it is dependent on pyridoxal phosphate (p. 363 f) as coenzyme. With the loss of two molecules of water, two molecules of δ-aminolevulinic acid are condensed to yield the actual building block of the porphyrin system, porphobilinogen, comprising the N-containing pyrrol ring structure (p. 38 f). The reaction is catalyzed by *aminolevulinic acid dehydratase*.

It is still an open question whether in chlorophyll synthesis δ-aminolevulinic acid is formed in the chloroplasts. In this case the latter would have to be endowed with their own *succinate thiokinase* (p. 222) in order

Porphobilinogen

to ensure an adequate supply of succinyl-CoA; the occurrence of this enzyme in chloroplasts is still a matter of controversy. *Aminolevulinic acid dehydratase* has been demonstrated in *Euglena,* green plant cells and chloroplast preparations; it has been isolated and partially purified from callus cells.

The next reaction step resembles the formation of an oligo-compound from sugar or nucleotide units since four porphobilinogen molecules are assembled to form a complex tetrapyrrole compound, uroporphyrinogen III (Reaction 1). This differs from the theoretically expected structure of uroporphyrinogen I (cf. formula below) in that the substituents in position 7 and 8 have exchanged places. While type I of this compound appears only in cases of irregular porphyrin biosynthesis, types II and IV have not been found in

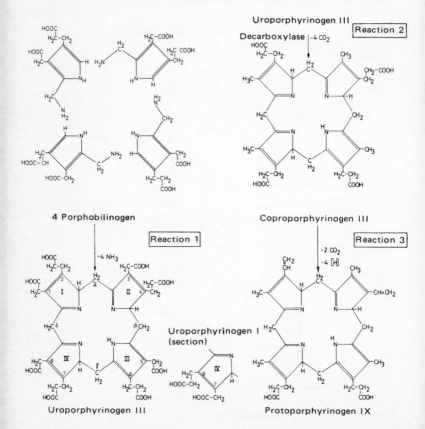

4 Porphobilinogen

Reaction 1

Uroporphyrinogen III

Uroporphyrinogen I (section)

Uroporphyrinogen III

Decarboxylase |-4 CO₂   Reaction 2

Coproporphyrinogen III

Reaction 3

Protoporphyrinogen IX

living cells so far (see p. 38 f for numbering of the substituted β-carbon atoms, and of the pyrrole rings in porphyrin).

The formation of uroporphyrinogen is a complex, enzyme-catalyzed process the details of which are not yet fully understood. Two enzymes are probably involved: the stepwise linkage of the porphobilinogen molecules is catalyzed by a *deaminase* (Reaction 1; see reaction scheme). With loss of NH_3 the aminomethyl group forms the link to the other pyrrole ring; the substituents at the β-carbon atoms are not involved. Accordingly, each pyrrole ring in the tetrapyrrole derivative formed carries an acetic acid residue and a propionic acid residue. The *deaminase* involved, *uroporphyrinogen-I synthetase,* has been isolated from spinach leaves where it is to some extent associated with the chloroplasts. It catalyzes the reaction:

$$4\ \text{Porphobilinogen} \xrightarrow{\text{"I-synthetase"}} \text{Uroporphyrinogen I} + 4\ NH_3$$

In the cell an *"isomerase"* prevents formation of uroporphyrinogen I in favor of uroporphyrinogen III. This enzyme has been isolated from wheat seedlings as *uroporphyrinogen-III cosynthetase.* By means of unknown reaction mechanism it introduces the fourth pyrrol ring "backwards", i.e. in the biologically "correct" position. The exclusive formation of uroporphyrinogen III in vivo requires close cooperation of the two enzymes; obviously a subsequent conversion of uroporphyrinogen I by the *cosynthetase* is not possible. Therefore, it is assumed that the *I-synthetase* is merely responsible for the synthesis of a di- or tripyrrole derivative which serves as substrate for the *cosynthetase.* The overall reaction may be summarized as follows:

$$4\ \text{Porphobilinogen} \xrightarrow[+\ \text{"Uroporphyrinogen-III cosynthetase"}]{\text{Deaminase (= "I-synthetase")}} \text{Uroporphyrinogen III} + 4\ NH_3$$

The individual reaction steps involved in the now following conversion of uroporphyrinogen III to protoporphyrin have still to be elucidated. A total of six decarboxylations is required, one COOH-group probably being removed from a side chain in each reaction step, since intermediates with 3 to 7 carboxyl groups have been demonstrated during the conversion of uroporphyrinogen III to protoporphyrin by a cell-free extract from *Chlorella.* One of these intermediates with 4 of the original 8 COOH-groups of uroporphyrinogen III has been identified as coproporphyrinogen III. This compound has a propionic acid residue and a methyl residue attached to each pyrrole ring as side chains (see structure p. 420). Their typical arrangement at ring IV is also found in protoporphyrin and in chlorophyll a.

Therefore, the role of an intermediate in the biosynthesis of these porphyrins may be assigned to coproporphyrinogen III. Of the four possible isomers types I and III are biologically important. Formation of these from the corresponding uroporphyrinogen compounds is catalyzed by *uroporphyrinogen decarboxylase* (Reaction 2 in the reaction scheme p. 420). Highly purified preparations of this enzyme from rabbit erythrocytes specifically split off CO_2 from the acetic acid residues of uroporphyrinogen III. Porphyrinogen compounds with 4 propionic acid residues and 3, 2 or 1 acetic acid residue appear as intermediates in the test mixture.

Out of the 15 isomers of protoporphyrinogen only one, namely IX, usually occurs in living cells. This compound differs from coproporphyrinogen in terms of the substituents in positions 2 and 4. Vinyl groups ($H_2C=CH-$) formed by oxidative decarboxylation compensate for the propionic acid residues (Reaction 3 in the reaction scheme p. 420). The enzyme involved is called *coproporphyrinogen decarboxylase* (or *coproporphyrinogen oxidase*); it has been demonstrated in cell-free extracts of *Euglena*, but has not yet been isolated from higher plants.

Protoporphyrinogen IX is converted to protoporphyrin IX by removal of hydrogen:

Protoporphyrinogen IX

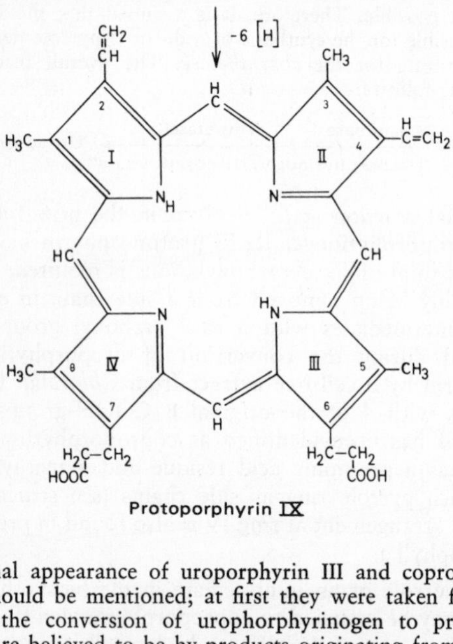

Protoporphyrin IX

The occasional appearance of uroporphyrin III and coproporphyrin III in the cell should be mentioned; at first they were taken for true intermediates in the conversion of urophorphyrinogen to protoporphyrin. Today they are believed to be by-products originating from spontaneous degradation of the true intermediates uroporphyrinogen III and copro-porphyrinogen III by oxidation (= removal of 6 H!).

Porphyrins in solution exhibit a specific absorption of radiation; hence they are easily identified by one of their most characteristic features: the absorption spectrum. The absorption maxima in the region of the "Soret peak", i.e. about 400 nm (p. 120) are of particular interest.

Fluorescence, another characteristic feature of porphyrins, may be used for identification purposes, too. The metal-free compounds exhibit an intensive red fluorescence in neutral or acid solution. Among the metallo-porphyrins the Mg-compounds such as the chlorophylls emit a red fluor-escent light; details were discussed previously (p. 101 ff).

Biosynthesis of Chlorophyll a

Studies on mutant strains of the green alga *Chlorella* (Granick and coworkers) provided first evidence that the pathways in biosynthesis of chlorophylls and of heme compounds are identical up to the stage of protoporphyrin.

The method used consisted in isolating and propagating cells in which the reaction chain of chlorophyll biosynthesis had been blocked at entirely different stages by mutation induced by previous exposure of the normal cells to ultraviolet radiation. Consequently, these cells could only grow heterotrophically in a glucose-containing medium. The main feature of the first mutant strain identified was a massive accumulation of protopor-phyrin in the cells; accordingly a function as intermediate in chlorophyll synthesis was ascribed to this compound. Further irradiation of this strain yielded a mutant the cells of which contained relatively large amounts of all the porphyrin and porphyrinogen compounds mentioned above. The results of labeling experiments with ^{14}C-glycine or ^{14}C-β-amino-levulinic acid in normal (= "wild-type") cells of *Chlorella* have to a certain extent confirmed the details already implied by the experiments with mutant strains: the tetrapyrrole structure of chlorophyll is formed by the same reaction mechanism as in animal cells.

The few details revealed so far concerning the conversion of proto-porphyrin to chlorophyll a come from studies on mutant strains of algae and photosynthetically active bacteria, which have lost their original ability to synthesize chlorophyll.

Three magnesium-containing compounds were originally isolated from *Chlorella* mutants: Mg-protoporphyrin, Mg-protoporphyrin monomethylester and Mg-2-vinylphaeoporphyrin a_5 (see formula). From their appearance it followed that Mg is introduced as central atom during a relatively early reaction step in the pathway of syn-thesis. It is unknown whether Mg-protoporphyrin is formed from protoporphyrin IX or from protoporphyrinogen IX by insertion of Mg; despite intensive efforts this conversion has not yet been ac-

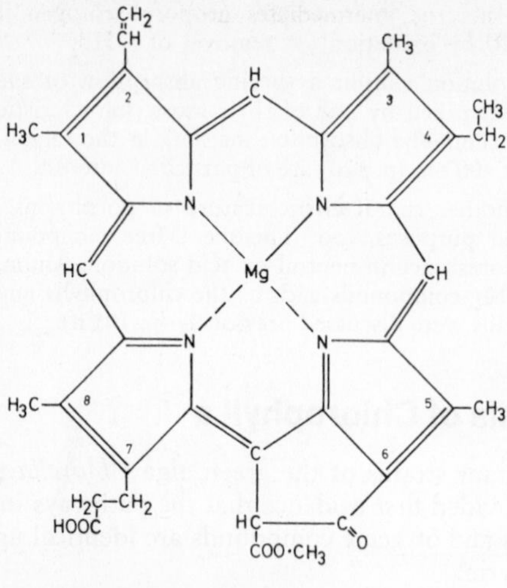

Mg-vinylphaeoporphyrin a₅

complished in cell-free systems. One may therefore assume that in vivo the substrate is bound to a protein or a lipoproteid, and that the introduction of Mg is catalyzed by a membrane-bound enzyme.

This view is supported by the finding that cell-free systems from greening plastids convert added δ-aminolevulinic acid to Mg-protoporphyrin only in the presence of membrane fractions in addition to a stroma fraction and several co-factors ($Mg^{2\oplus}$, $K^{\oplus}$, phosphate, coenzyme A, ATP, $NAD^{\oplus}$, CH_3OH). Probably all the reaction steps converting protoporphyrin IX to chlorophyll a proceed strictly membrane-bound; the enzymes involved may be assembled in a multienzyme complex (p. 406).

Mg-protoporphyrin monomethylester is the product of the first esterification which takes place at the carboxyl group of the pro- pionic acid side chain at position 6 prior to formation of the pen- tanone ring. The methyl group required is provided by the "pool" of C_1 fragments. In the "activated" form of S-adenosyl methionine (for formation see reaction scheme below). Feeding of methionine- 1-^{14}C to cells of *Rhodopseudomonas spheroides* or of ^{14}C-formate to *Chlorella* gives rise to specific incorporation of radiocarbon into the methylester group of bacteriochlorophyll and chlorophyll a, respectively. The enzyme involved, a *methyltransferase,* has been isolated and partially purified from chromatophore preparations (p. 69 f) of this purple bacterium and from chloroplasts of *Euglena;* it has been demonstrated in chloroplasts of higher plants, too. The following reaction is catalyzed in vitro:

"methyltransferase"

Mg-protoporphyrin + adenosyl ────────────────▶ Mg-protoporphyrin
             methionine   monomethylester
                                                 + adenosyl homocysteine

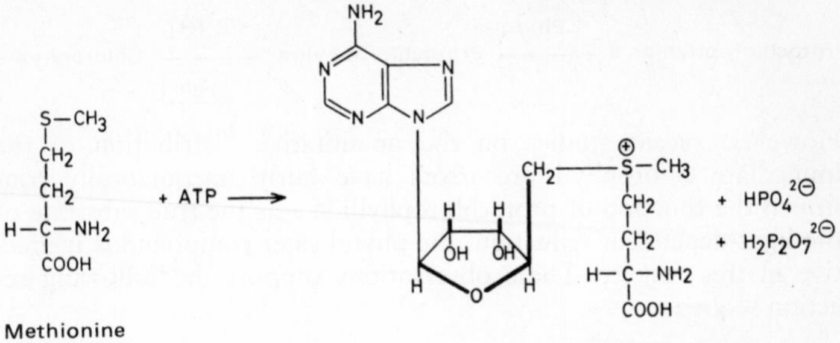

Methionine   + ATP ⟶

Adenosyl methionine, = "active methyl"

Mg-2,4-divinylphaeoporphyrin a_5, a compound formed by several unknown reaction steps from Mg-protoporphyrin monomethylester and differing from the latter by the additional structure of the iso-cyclic pentanone ring, probably acts as a precursor of Mg-2-vinyl-phaeoporphyrin a_5. Both compounds have vinyl groups in positions 2 and 4. Reduction of the one in position 4 by introduction of two hydrogen atoms gives rise to formation of Mg-2-vinylphaeoporphy-rin a_5 which is also termed "protochlorophyllide a". This com-pound is also a typical premature end product of chlorophyll syn-thesis in a *Chlorella* mutant. In etiolated leaves the product is attached to protein forming a complex: "protochlorophyllide holo-chrome" (see below). The pigment is probably only active as an intermediate of chlorophyll synthesis when bound to protein.

Protochlorophyllide a is converted by reduction to 7,8-dihydro-protochlorophyllide a or "chlorophyllide a". This step is light-de-pendent in many higher plant species; in some species, however, traces' of chlorophyll a are also formed in continuous darkness; in seedlings of conifers fairly large quantities are formed. Under hetero-trophic conditions algae are generally able to carry out this specific attachment of hydrogen to protochlorophyllide a and hence to ac-complish the biosynthesis of chlorophyll a in darkness as well as in light. Exceptions are *Cyanidium caldarium,* a few mutants of *Chlorella* and *Chlamydomonas* and the green flagellate *Euglena gracilis.*

With regard to the occurrence of protochlorophyll a in dark-grown seedlings it was assumed at first that the second esterification step, i.e. the attachment of phytol (p. 39 f) to the propionic acid residue in position 7, may occur at the stage of protochlorophyllide a. In this case the final step should be the light-dependent reduction of this compound to chlorophyll a:

$$\text{Protochlorophyllide a} \xrightarrow{\text{+ Phytol}} \text{Protochlorophyll a} \xrightarrow[\boxed{\text{light}}]{\text{+ 2 [H]}} \text{Chlorophyll a}$$

However, recent studies on the quantitative distribution of the immediate chlorophyll precursors have fairly unequivocally confirmed the function of protochlorophyllide a as the true substrate of the light-dependent reduction; the phytol ester compound is ineffective in this respect. These observations support the following reaction sequence:

$$\begin{array}{l}\text{Mg-2-vinylphaeoporphyrin } a_5 \xrightarrow[\boxed{\text{Light}}]{\text{+ 2 [H]}} \text{Chlorophyllide a} \xrightarrow[\boxed{\text{Darkness}}]{\text{+ Phytol}} \text{Chlorophyll a} \\ (= \text{"Protochlorophyllide a"})\end{array}$$

Phytol is an "isoprenoid" (p. 430 ff); the C_{20} compound geranylgeranyl diphosphate (p. 433 f) acts as precursor in its biosynthesis.

Photoreduction of protochlorophyllide a to chlorophyllide a also occurs in vitro in a cell-free system from etiolated leaves provided the structural association of pigment and protein, the "protochlorophyllide holochrome", has been preserved during preparation. The complex has been isolated and partially purified from dark-grown seedlings. Its molecular weight amounts to about 700 000; it exhibits a characteristic absorption maximum at about 635 nm, thereby differing from its conversion product, "chlorophyllide holochrome" which shows an absorption maximum at 675 nm. The complex probably also contains the specific enzyme activity catalyzing this conversion.

Since the higher alcohol phytol is practically insoluble in water the final steps in chlorophyll biosynthesis probably proceed in the lipoproteid regions of the chloroplast. It seems very likely that the attachment of phytol may be correlated with the final arrangement of the chlorophyll molecules in the structure of the thylakoid membrane (cf. p. 74 ff).

Uncertainty still prevails regarding the participation of *chlorophyllase* as an active enzyme in the esterification with phytol. In vitro the enzyme catalyzes only the hydrolytic cleavage of phytol from chlorophyll:

$$\text{Chlorophyll} \xrightarrow{\text{+ } H_2O} \text{Chlorophyllide + Phytol}$$

The observation that during illumination of etiolated leaves the activity of *chlorophyllase* markedly increases in correlation with the proceeding chlorophyll synthesis, indicates that the catalyzed reaction may also proceed in the opposite direction.

In cell-free extracts from isolated etioplasts chlorophyll a is synthesized from δ-aminolevulinic acid after addition of certain cofactors (p. 424) in the presence of O_2 when the reaction mixture is illuminated. Obviously, the previous illumination of the etioplasts in the intact tissue cells also decides whether the extracts prepared from these will form only chlorophyll a or chlorophyll b as well; the latter could only be demonstrated in the cell-free system after an extensive pre-illumination of etioplasts. Without this treatment the extracts obtained only catalyzed the synthesis of protochlorophyllide and protochlorophyllide ester (methylation in position 10?) from δ-aminolevulinic acid (Rebeiz and coworkers).

A complete synthesis of chlorophyll was achieved for the first time in the laboratory of R. B. Woodward (Nobel Prize 1965).

Synthesis of Chlorophylls b, c, d, and of Bacteriochlorophyll

The immediate precursor of chlorophyll b appears to be chlorophyll a or a special porphyrin which may form "a" or "b" at random.

Evidence for the first of these two possibilities was obtained in studies with cell-free systems (etioplasts and greening shoot tissues). In absence of δ-aminolevulinic acid only small amounts of chlorophyll a were converted to chlorophyll b. NADP⊕ appears to be required for the enzymatic oxidation of the methyl group in position 3. Phytol was transferred from chlorophyll a to chlorophyll b in a second reaction which indicates that the conversion observed proceeds at the stage of chlorophyllide a.

The pathway of biosynthesis of the unusual chlorophylls "c" and "d" is unknown. It is assumed that the pathway leading to chlorophyll c branches off at the stage of protochlorophyllide a.

Porphyrin synthesis proceeds by the same reaction mechanism in phototrophic bacteria as in other organisms. The similarity between biosynthesis of bacteriochlorophyll a and biosynthesis of chlorophyll a is inferred from the successfull isolation of similar or identical intermediates such as protochlorophyllide a or protochlorophyll a. Chlorophyllide a or chlorophyll a may even be precursors of bacteriochlorophyll.

Cells of *Rhodopseudomonas spheroides* form bacteriochlorophyll anaerobically in the light but not aerobically in darkness and accordingly exhibit autotrophic and heterotrophic metabolism, respectively. These observations provided a first answer to the important question: how does an organism or a cell regulate its porphyrin synthesis, so that an adequate formation of chlorophylls, cytochromes, phytochrome, vitamin B_{12} etc., is guaranteed in adaptation to the changing requirements of metabolism? In cells of the purple bacterium grown in darkness under aerobic conditions the activity of δ-*aminolevulinic synthase* is relatively low, but it increases rapidly when the cells are transferred to anaerobic conditions and illuminated. Synthesis of bacteriochlorophyll, however, is initiated only some time after the increase in enzyme activity. The cells dispose of two enzymes for conversion of coproporphyrinogen to protoporphyrinogen, one of which is activated by O_2, the other one only by anaerobiosis; depending on the environmental conditions only one of these is in the active state. Obviously the conversion of protoporphyrin to Mg-protoporphyrin has a significant control function since this reaction at the beginning of the separate pathway of bacteriochlorophyll synthesis is strongly inhibited by oxygen. However, no accumulation of precursors such as protoporphyrin or protoporphyrinogen etc. is observed which suggests that probably δ-aminolevulinic acid synthase and with it the starting reaction are simultaneously blocked under these conditions. It is still an open question whether this inhibitory effect is the result of an impaired formation of Mg-protoporphyrin or arises from a direct effect of O_2 on δ-aminolevulinic acid synthase. Probably both mechanisms are involved. Although induction and repression of enzyme activities (p. 446) appear at first sight to offer plausible explanation for the observed behavior of bacteriochlorophyll synthesis, doubts may arise in view of the velocity at which the changes proceed: a mere bubbling of air through an anaerobically growing light culture immediately blocks chlorophyll synthesis; it is restored as soon as aeration is stopped. A general change in the redox state of the cells is being discussed as possible reason for this behavior.

Formation of Iron Porphyrins

Since only few experimental data are available, the details of the reaction sequence employed in the formation of iron porphyrins in the cell are still rather obscure.

The synthesis of protoheme, "Fe-protoporphyrin IX", from protoporphyrin and Fe^{2+} is catalyzed by the enzyme system ferrochelatase (= porphyrin-iron-chelate enzyme), which was first partially purified from animal material. A corresponding preparation from rat liver promotes incorporation of Fe^{2+} or Co^{2+}, but not of Fe^{3+} into protoporphyrin IX and several other porphyrins except uroporphyrin and coproporphyrin. A more precise analysis of the enzyme system involved has revealed that an organism disposes not only of one ferrochelatase, but rather of a whole "family" of this enzyme. Its activity has been also demonstrated in extracts of the phototrophic bacterium Chromatium (strain D).

Incorporation of the metal atom into a porphyrin is measured by the resulting absorption change (p. 423).

The Fe-porphyrins which constitute the prosthetic groups in the cytochromes (= "porphyrinproteids") differ from one another in the nature of the side chains. Therefore, two different reaction mechanisms are assumed to exist in their formation: 1) the side chains in protoheme are changed by an enzymatic action, and 2) porphyrins other than protoporphyrin serve as precursors for the incorporation of $Fe^{2\oplus}$.

Since a group of *ferrochelatases* is obviously at the disposal of each cell, a certain variability in the synthesis of different kinds of iron porphyrins may result. The possible pathways are summarized in the diagram below:

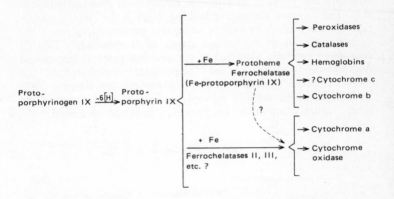

Phycobiliproteids

The biosynthesis of phycocyanins and phycoerythrins the chromophoric groups of which are chemically related to the porphyrins (p. 50), has not yet been elucidated. By analogy with the formation of bile pigments from hemoglobin in animal organisms it is assumed that the precursor is a hemoproteid. Initially, the α-methene bridge is probably replaced by oxygen, CO being released. This oxygen bridge is subsequently cleaved by reduction: the pyrrol rings I and II acquire an OH-group each. The concomitant opening of the original ring system of porphyrin at the same time releases the protein moiety and the central iron atom. The subsequent conversions of this open-chain tetrapyrrole compound still have to be clarified.

References

Machold, O., G. Scholz: Eisenhaushalt und Chlorophyll-Bildung bei höheren Pflanzen. Naturwissenschaften 56: 447, 1969

Price, C. A.: Iron compounds and plant nutrition. Ann. Rev. Plant Physiol. 19: 239, 1968

Rebeiz, C. A., P. A. Castelfranco: Proto-chlorophyll and chlorophyll biosynthesis in cell-free systems from higher plants. Ann. Rev. Plant Physiol. 24: 129, 1973

Vernon, L. P., G. R. Seely: The chlorophylls-physical, chemical, and biological properties. Academic Press, New York 1966

Carotinoids

Properties and structure of these compounds have been discussed in detail previously (p. 53 ff). Carotinoids are typical and widely distributed constituents of plant and animal cells, though they can be synthesized only by the former. They belong to the *isoprenoids,* a group comprising many compounds that are characteristic for plants: essential oils, resins, steroids and rubber. At first sight, they seem to lack common features, but a more precise evaluation of structure and pathway of synthesis reveals their chemical relationship. The common elementary building block is the "biological isoprenoid unit" which by various successive condensations forms the carbon skeletons of the various classes of "isoprenoids". This C_5 compound has been identified in the past few years as isopentenyl diphosphate (formerly called "isopentenyl pyrophosphate" or "active isoprene"). Thus the *isoprene hypothesis* postulated in 1922 by Ruzicka as the reaction principle underlying the biosynthesis of these "natural compounds" has been experimentally confirmed.

$= \; \rangle\!\!-\!\!=$ (short denotation)

$\rangle\!\!-\!\!-0\text{-}\textcircled{P}\text{-}\textcircled{P}$

Isoprene

"Active isoprene" = isopentenyl diphosphate

In contrast to the biosynthesis of macromolecules, biologically important compounds with quite different chemical structures and features are in this case formed from a single building block. First two or more isoprenoid units are always linked giving rise to unsaturated hydrocarbons with a multiple of C_5H_8. Subsequent oxidation, reduction or removal of carbon atoms contribute to the great variety of the isoprenoids: this is clearly indicated by the synthesis of carotinoids. We shall first briefly consider the formation of "active isoprene".

Biosynthesis of Isopentenyl Diphosphate

The precursor is "activated acetic acid", acetyl-CoA, whose key role as reactive precursor has been mentioned previously (p. 218). Two molecules react to form acetoacetyl-CoA; a third molecule is then added by condensation (see reaction scheme p. 431).

From the reaction product β-hydroxy-β-methylglutaryl-CoA coenzyme A is released with consumption of $NADP\text{-}H + H^{\oplus}$, the COOH group being reduced to a CH_2OH group. Mevalonic acid is formed, the key intermediate and immediate precursor of "active isoprene". It gives rise to formation of isopentenyl diphosphate in a complex reaction the details of which are still obscure; 3 moles of ATP are consumed, 1 mole of H_2O and 1 mole of CO_2 are released. Isopentenyl diphosphate is formally isoprene plus a pyrophosphoric acid residue (see above).

Mevalonic acid was originally discovered as a growth factor for a mutant of *Lactobacillus acidophilus* whose requirement for acetate is met by this compound. This was an important clue for the elucidation of the biosynthesis of mevalonic acid and of the isoprenoids as well. In higher plants mevalonic acid is one of those compounds which have a rapid turnover and consequently accumulate only in small quantities. Hence it has been possible to demonstrate and isolate this compound only after feeding the cells with acetate-2-^{14}C.

Several enzymes catalyzing the formation of isopentenyl diphosphate from acetyl-CoA are also active in higher plants. The results of labeling experiments carried out simultaneously with plant tissue slices indicate that the pathway of synthesis is the same in plant organisms and those animal organisms which form steroids.

Formation of the C$_{40}$-Polyisoprenoid Chain

In synthesis of the various isoprenoids and isoprenoid derivatives there is a common reaction chain with isopentenyl diphosphate as starting compound; repeated condensation of isopentenyl diphosphate building blocks yields the key compound of the individual classes of natural products. In carotinoid synthesis this function is assigned to a C_{20} compound ("C_{20}-terpenyl diphosphate", see reaction scheme p. 433) which is probably identical with geranylgeranyl diphosphate.

Initially a molecule of isopentenyl disphosphate is reversibly converted to the isomeric $\Delta^{2,3}$ compound dimethylallyl diphosphate (= "prenyl diphosphate" according to new nomenclature) by *isopentenyl diphosphate isomerase*:

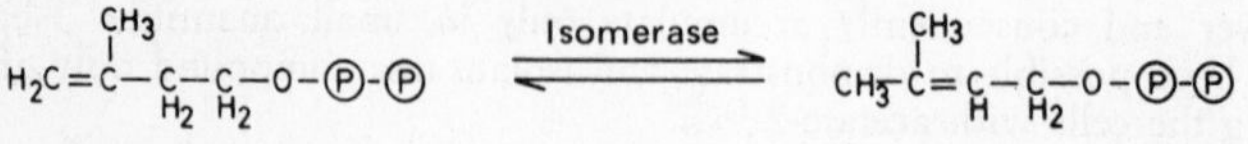

Isopentenyl diphosphate

Dimethylallyl diphosphate
= "prenyl diphosphate"

After cleavage of the pyrophosphate residue this compound reacts as a "carbonium cation" with the double bond of an isopentenyl diphosphate molecule yielding the dimeric compound geranyl diphosphate while a proton is released:

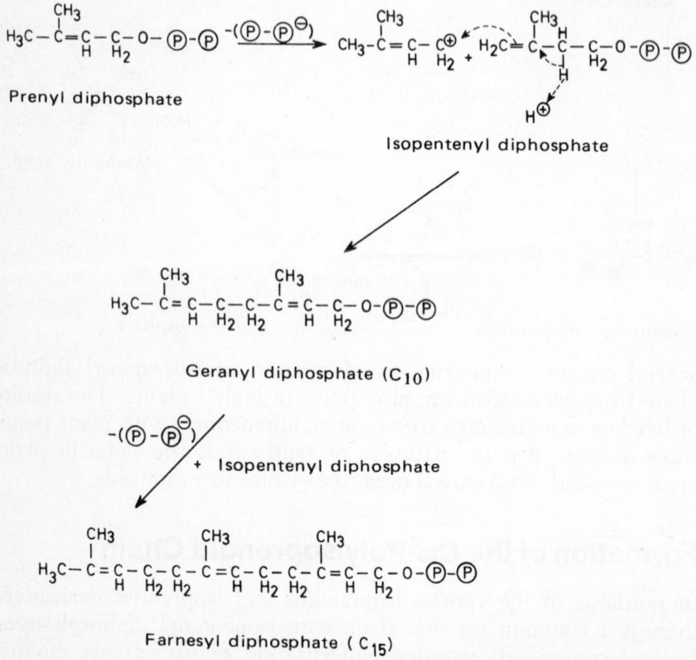

Prenyl diphosphate

Isopentenyl diphosphate

Geranyl diphosphate (C_{10})

+ Isopentenyl diphosphate

Farnesyl diphosphate (C_{15})

Geranyl diphosphate in turn via its carbonium ion condenses with an additional isoprenoid unit; farnesyl diphosphate is the reaction product (see reaction scheme).

This reaction sequence is characteristic for yeast and liver cells as well as for pea seedlings.

Elongation of the isoprene chain by an additional C_5 unit according to the reaction mechanism described above gives rise to formation of a C_{20}-compound, geranylgeranyl diphosphate, which probably represents the key compound in biosynthesis of carotinoids.

This view is supported by the fact that the tetra-isoprenoid is a characteristic constituent of organisms and tissues with a highly active biosynthesis of carotinoids. Its formation has been observed in cell-free extracts from pea seedlings and tomato fruit when mevalonic acid was present as substrate. The enzyme responsible for the formation of geranylgeranyl diphosphate from farnesyl diphosphate and isopentenyl diphosphate has been found in the plastids of tomatoes as well as in carrot roots and in the fungus *Phycomyces*.

The symmetry center of the skeleton of carotinoids consisting of 40 carbon atoms indicates that it is formed by a "head-to-head" condensation of two C_{20} building blocks. The first detectable reaction product, phytoene, contains a central double bond which is conjugated with those of the two neighboring isoprenoid units.

The reaction scheme below shows the formation of this C_{40} compound:

C_{20}-terpenyl diphosphate C_{20}-terpenyl diphosphate

new central bond! $+ 2\ \text{P-P}$

Phytoene (C_{40})

It has not yet been definitively proved that geranylgeranyl diphosphate is the immediate precursor of the carotinoids. Nevertheless, the ^{14}C-labeled compound is incorporated into β-carotene by a cell-free system from carrot plastids as is ^{14}C-farnesyl diphosphate.

Lycopersene has also been considered as a possible precursor of the C_{40} carotinoids; in contrast to phytoene, its molecule has a central single bond. Contrary to this consideration is the finding that lycopersene is not detectable as an intermediate in all the cells that synthesize carotinoids. According to the current concept phytoene is the key compound in synthesis of carotinoids. This is based on the fact that phytoene has always been found as an intermediate or, after blocking of the reaction chain, as a premature end product in studies of the synthesis of carotinoids performed in mutant strains of Neurospora, Rhodopseudomonas and Chlorella.

Conversion of Phytoene to Carotenes

The reaction sequence for the further conversion of phytoene to a typical carotene, lycopene, has been elucidated by analyzing the distribution of pigments in mutants of the tomato and of the green alga Chlorella. The unsaturated character of the molecule is gradually increased by a total of four oxidation (= dehydrogenation) steps (cf. reaction scheme p. 435); the number of double bonds increases from 9 to 13. At the same time the conjugated system is also enlarged step by step so that it finally comprises 11 double bonds. By these oxidative reaction steps the color of the individual intermediates is intensified (cf. p. 55). While phytoene with its three conjugated double bonds is colorless, neurosporene (discovered in a Neurospora mutant) has a weak yellow tint; the end product lycopene shows an intense yellowish-red color.

A limited cyclization occurs in several carotinoids; it takes place at the end of the molecular chain, at a relatively late stage in the process of synthesis.

It is not yet clear at which stage this terminal ring closure occurs. Recent findings suggest that both neurosporene and lycopene are possible stages; both may be converted to β-carotene by removal of two further hydrogen atoms; the closed ring structures at the end of the molecule contain one double bond each, conjugated to the last one in the molecular chain ("β-ionone structure"!). γ-Carotene probably assumes the role of an intermediate with only one closed ring structure (formula, p. 57).

The possibility of β-carotene being formed by an isomerization of α-carotene is also under discussion. The latter would formally be derived from ζ-carotene by removal of 4 H and subsequent cyclization at the chain ends.

Conversion of phytoene to lycopene

Xanthophylls

Nearly all the findings obtained up to now indicate that introduction of oxygen into the carotinoid structure – and, therefore, the biosynthesis of xanthophylls – takes place at a relatively late stage in the reaction sequence. This holds true for seedlings of higher plants and for *Chlorella* as well as for the phototrophic bacteria *Rhodopseudomonas* and *Chromatium*. Studies on a mutant strain of *Chlorella vulgaris* have provided some hints as to how a xanthophyll may be formed. Under heterotrophic conditions these cells synthesize the colorless phytoene in the dark; it is converted to colored carotinoids on illumination and in the absence of oxygen. When the cells are subsequently again placed in darkness without elimination of oxygen, the carotenes are converted to various xanthophylls. Similar reactions proceed in etiolated seedlings to which $^{14}CO_2$ is offered at the beginning of an illumination period: the specific activity of the xanthophylls formed in the developing chloroplasts is always lower than that of the carotenes.

Experiments in which the stable oxygen isotope ^{18}O was administered to cells of *Chlorella* as a gas or as $H_2^{18}O$ indicate that the oxygen of the OH-groups of lutein (p. 58 f; 3,3'-dihydroxy-α-carotene) originates from molecular oxygen and not from water. This certainly does not apply to the species of phototrophic bacteria exhibiting obligate anaerobic metabolism: they obtain the oxygen required for xanthophyll biosynthesis exclusively from water.

Conversion of individual xanthophylls depending on light and oxygen supply has been described for spinach leaves and cells of *Euglena gracilis*. In this way, antheraxanthin and violaxanthin may be formed from zeaxanthin, as shown by the following reaction scheme:

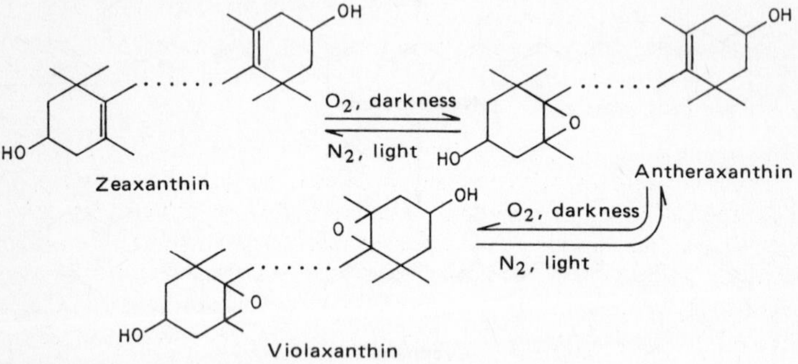

Conversion of the epoxy carotinoids

These reaction steps are reversible if the cells are illuminated in the absence of O_2 or in the presence of N_2.

The formation of "secondary carotinoids" in algae which may occur as a consequence of nutritional deficiencies has already been discussed in connection with nitrogen metabolism (p. 289).

References

Goodwin, T. W.: Natural substances formed biologically from mevalonic acid. Biochem. Soc. Symposia. Academic Press, New York 1969

Lynen, F.: Der Weg von der „aktivierten

Essigsäure" zu den Terpenen und Fettsäuren. Angew. Chemie 77: 929, 1965

Porter, J. W., D. G. Anderson: Biosynthesis of carotenes. Ann. Rev. Plant Physiol. 18: 197, 1967

Metabolic Regulation

In spite of the flux of many materials through various pathways the metabolism of a cell or organism proceeds in a harmonious and most precise manner, free of disorders and with great economy. A cyclic reaction system or a metabolic chain produces only as many intermediates or end products as the "operating metabolism" and the "building metabolism" can utilize. Enzymes are only activated or formed when required, i.e. when their specific substrate is available; they are inactivated or degraded when no substrate is present. Any interference with the interaction of the reaction sequences, for instance by external factors, is generally met with appropriate measures. Substrate and enzyme are obviously subject to certain controls which ensure harmony and order in metabolism.

The regulation of metabolic reactions has been briefly discussed in the preceding chapters of this book (p. 409). We know as yet little of the mechanisms concerned in higher plants as compared with those in microorganisms; the underlying principles, however, appear to be similar.

A first important device of regulation, which exerts control on the metabolic rate of whole reaction chains, consists of changing the concentration of metabolites, i.e. coenzymes, cosubstrates, intermediates etc. In this case, the enzymes involved generally do not change in activity or quantity; both of these features are affected, however, when regulation is exerted on enzymes directly. There are two possibilities for this: 1. The activity of the enzymes present is reduced or increased by substances reacting with the enzyme protein; in this way a rapid and reversible switching-on and switching-off of metabolic pathways is guaranteed (= "fine control"). 2. The synthesis of enzymes is initiated by *induction* or prevented by *repression;* the regulatory factors exert their control on the genetic material or on the transcription product. This mechanism is a relatively slow and coarse control device which permits changes in the amount or level of enzymes in response to demands made by a changing environment (= "coarse control").

This classification does not exclude the possibility of more than one of these control mechanisms being involved at a time in the regulation of a reaction chain or cycle. On the contrary: an essential metabolite, for example, may cooperate with a change in activity of one or several enzymes. Obviously only a close interaction of different regulatory principles ensures an efficient control of the diverse activities in metabolism.

To some extent the terminology has been taken over from comparable regulatory systems in technology. These serve as models for complex reaction pathways in metabolism which correspond to self-regulating systems applying the principle of *feedback control*. An intermediate or provisional end product exerts a negative feedback effect on an early reaction step in the catabolic system thus determining the turnover. Either the enzymes involved are influenced by this feedback effect or even the genes responsible for de-novo synthesis of particular enzymes.

Regulation by Metabolites

The important function of ADP in regulating the intensity of respiration and the ATP formation coupled to it has been briefly mentioned in discussing the respiratory chain (p. 235 ff). Since the amount of ADP available is limited the respiratory chain usually does not work at full capacity. More ADP becomes available only on increasing consumption of ATP, i.e. a high energy expenditure by metabolism; respiratory activity and oxidative phosphorylation will then rise in order to meet the mounting demand for ATP.

Lowering the activity of the respiratory chain also affects the citric acid cycle because the hydrogen provided by the latter is utilized only to a small extent. Over-production of reducing power and hence waste of valuable substrate is also prevented by regulation at the metabolite level. In this case, the key compound is oxalo-acetic acid which is consequently only formed in small amounts. Hence synthesis of citric acid slows down, and removal of acetyl-CoA from its "pool" is reduced.

Theoretically, $NAD^{\oplus}$ might also be involved as a limiting metabolite, like ADP, in the regulation of the citric acid cycle in mitochondria. Its activity would depend on the rate at which $NAD-H + H^{\oplus}$ delivers its hydrogen to the respiratory chain and is again at the disposal of the oxidative reactions of the cycle. This would lead to the same consequences as the regulation via oxaloacetic acid (see above). However, recent findings indicate that $NAD-H + H^{\oplus}$ is the effective metabolite which controls the citric acid cycle by interfering with the activity of *isocitrate dehydrogenase* (p. 221 f).

Another self-regulating system comparable with the respiratory chain is the pentose phosphate cycle (p. 255 ff). By interacting with the two oxidative reactions, $NADP^{\oplus}$ as a limiting factor controls the activity of the whole cycle (see p. 260): this activity is enhanced by reductive biosynthetic processes (amino acids, fatty acids) which remove the hydrogen rather rapidly thus restoring the oxidized state of the NADP molecules. On the other hand, a sluggish demand for hydrogen resulting from low anabolic activity has an inhibitory effect on the pentose phosphate cycle. Apart from this regulatory device via a metabolite there exists another one which exerts control by a modulation of enzyme activities similar to that in the citric acid cycle (p. 444).

The "Pasteur Effect" (p. 250) was originally also considered to be a metabolite regulation in which aerobic respiration and fermentation compete for ADP and orthophosphate, respectively. There are now several lines of evidence indicating that regulation primarily takes place at the level of the enzymes (for details, see p. 443 f).

Sub-division of the cell into compartments or reaction spaces (p. 5) plays an important role in metabolic regulation. By means of specifically located compartments synthesis and degradation of key compounds with common intermediates and metabolites can proceed independently via separate reaction sequences catalyzed by different enzymes (examples: starch, p. 178; fatty acids, p. 409). A key compound located in different compartments may become an important regulatory factor when it is exchanged between these compartments and influences the reactions concerned.

It is well known that synthesis of fatty acids takes place in the cytoplasm; the precursor acetyl-CoA, however, is produced in the mitochondria. In order to reach the specific reaction compartment acetyl-CoA is initially condensed with oxaloacetic acid, and the citric acid thus formed diffuses into the cytoplasm. This process accounts for the transport of acetyl-CoA across the mitochondrial membrane (cf. p. 203). In the cytoplasm cleavage to acetyl-CoA takes place catalyzed by the ATP-dependent *citrate lyase* (citric acid + ATP + HS-CoA → acetyl-CoA + ADP + P + oxaloacetic acid). It is now available for fatty acid synthesis. The rate of the latter will depend upon the amount of oxaloacetate which is available in mitochondria and upon the depletion of the pool of acetyl-CoA by respiration. Consequently, synthesis of fatty acids is possible only during repressed respiration. It is additionally controlled via the activity of several enzymes (p. 445). Compartments probably play the same important role in regulation of other pathways of biosynthesis outside the mitochondria which also require acetyl-CoA or intermediates of the citric acid cycle as precursors. Certainly the same is true for the reverse direction, namely the transport of glycerolphosphate, malic acid, glutamic acid and aspartic acid from the cytoplasm into the mitochondria, these compounds functioning as hydrogen carriers (p. 244 f).

For chloroplasts a "sucking effect" on 3-phosphoglyceric acid in the cytoplasm is under discussion which simultaneously inhibits glycolysis and the citric acid cycle.

Changes in Enzyme Activity as a Control Device

This control mechanism affects the activity of enzymes: by its stimulation or inhibition the rate of the reaction being catalyzed is adapted to the requirements of cellular metabolism. This type of control is more rapid and precisely pointed than metabolite regulation (see above) because it directly affects the key enzyme* of a metabolic sequence. One of the products formed – usually the end product** – stimulates or inhibits the activity of such an enzyme: as an "effector" it interacts with a second (= "regulatory") binding site and thereby influences, positively or negatively, the catalytic center, i.e. it results in either activation or inhibition of the enzyme. Only *allosteric enzymes* (p. 10) are able to react in this way; most of the key enzymes concerned are, in fact, of this type. This most important control device in metabolic regulation has therefore been termed "allosteric control" or "allosteric feedback".

An allosteric enzyme has two separate centers with different functions: a "catalytic center" for recognizing and binding of the substrate, and a "regulatory" (= allosteric) center with which an "effector", a low molecular weight substance may interact in a "sterically different" (name!) way (Fig. 101). This has been demonstrated with isolated enzymes in which the catalytic activity was fully preserved although the effector binding site had been experimentally eliminated. Reversible attachment of an effector to the native enzyme probably results in a change of the protein conformation, either completely blocking substrate binding to the catalytic center or else reducing or terminating its activity. The reaction kinetics indicate that a number of allosteric enzymes consist of closely cooperating subunits. Some of these may have an exclusively regulatory function, others an exclusively catalytic one. Positive effectors (= activators) are believed to react with a control center different from that for negative effectors (= inhibitors).

In spite of the greater significance of allosteric effects, it should not be overlooked that an effector may also influence enzyme activity by interacting with the catalytic center, provided that it has a structure very similar to that of the substrate which is specific for the enzyme concerned

---

* These enzymes often catalyze a highly exergonic and hence practically irreversible reaction step within a metabolic sequence of more or less reversible reactions. They have been recognized as "pacemaker enzymes", particularly in reaction sequences which are in a "dynamic steady state" (p. 7 f): glycolysis, citric acid cycle, respiratory chain.

** In this case the term "end product" does not signify a true end product of metabolism; it is the end (or key) product of a metabolic sequence or it is a metabolite (e.g. ATP) both of which are formed far removed metabolically from the key enzyme in the reaction chain – mostly several reaction steps away.

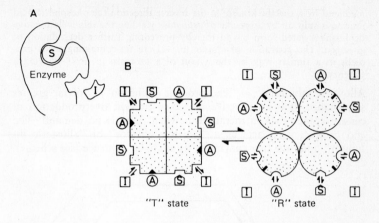

Fig. 101   Model of the structure and of the mode of action of allosteric enzymes. A: Location of two centers on the enzyme molecule: "catalytic center" with bound substrate (S) and "regulatory (=allosteric) center" with the binding site for an "effector" which may act as inhibitor (I) or activator (A). B: Substrate binding and effector activity in allosteric enzymes which consist of several subunits (after Monod, Wyman and Changeux); probably each of these subunits has a binding site for substrate (S), activator (A) and inhibitor (I). At least two conformations exist for the enzyme protein: "T" (= "tense") and "R" (= "relaxed"); they are in equilibrium with each other and are considered to have different grades of affinity for the allosteric effectors and the substrate molecules. A surplus of inhibitor keeps the enzyme in the inactive "T" state, bound substrate or activator, however, in the relaxed and active "R" state. Ultimately, the interaction between substrate and effectors determines the activity of the enzyme and hence the reaction to be catalyzed.

(hence "isosteric effect"). Substrate and effector may thus compete for the same binding site on the enzyme molecule, a phenomenon already familiar to us as "competitive inhibition" (e.g. inhibition of *succinic dehydrogenase* by malonic acid as a non-physiological effector, p. 224 f). The antagonism between substrate and effector determines the rate of the catalyzed reaction via the enzyme activity. The rate will increase in case of a surplus in substrate displacing the effector, and vice versa. This mechanism of self-regulation which is generally termed "product inhibition" has the same aim as allosteric control, i.e. to prevent accumulation of metabolic products. Product inhibition differs from allosteric control in that it can control only one reaction step, not the whole reaction chain.

Control exerted on enzyme activity may also be brought about by a modification of the enzyme structure, i.e. conversion of inactive enzyme precursors, "proenzymes", to their active form which often involves a specific control enzyme.

Thus, the inactive molecule form b of **glycogen phosphorylase** (p. 208) in muscle is converted to the active form a by phosphorylation with ATP mediated by a specific kinase. In the reverse direction the phosphoric acid residue is split off by a specific *phosphatase* yielding the relatively inactive dephosphorylated form b and thereby preventing further degradation of glycogen. The activation of *lipases* (p. 411) in fat-containing seeds proceeds in a similar way by the action of a *protease* (p. 397) cleaving a proenzyme.

Allosteric inhibition (= "end-product inhibition") is of far greater significance. It is an effective device to prevent overproduction of certain compounds in metabolism when there is no demand. The end product itself acts as a "negative effector" or "allosteric inhibitor" upon the first enzyme in its anabolic sequence (see scheme).

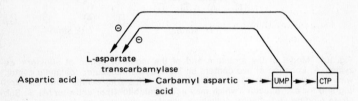

With the initial step inhibited the complete biosynthetic sequence is practically turned off thus avoiding a waste of precursors and an accumulation of intermediates. With decreasing amounts of end product due to rising demand the effector is released from the enzyme protein; hereby the starter enzyme is reactivated: the biosynthetic sequence starts working. Thus the system largely regulates itself. The binding constant of effector and enzyme as well as the enzyme concentration determine the efficiency and the fine tuning of the regulatory control.

This principle also underlies the regulation of pyrimidine nucleotide synthesis (p. 315 ff). If production exceeds the demand of nucleic acid synthesis, its most important "consumer", UMP, CTP and dCTP usually act as allosteric effectors and reversibly inactivate the allosteric enzyme *L-aspartate transcarbamylase* which catalyzes the initial reaction step in the biosynthetic sequence. This example is also interesting from another point of view. The initial step in pyrimidine nucleotide synthesis is formation of carbamyl aspartic acid. It is produced from aspartic acid and carbamylphosphate which are themselves intermediates of other synthetic sequences, e.g. that of amino acids (cf. p. 365). Such a branching point in metabolism creates a serious regulatory problem. In this case allosteric control again proves to be a very effective mechanism, as it acts on the first enzyme of the branching-off reaction chain behind the branching point, as demonstrated in our example. The chain can be turned off without impairing formation of the precursors or intermediates required for other pathways (see scheme).

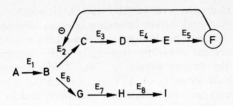

Another device of regulating branched metabolic pathways must be mentioned in this context which has so far been found only in microorganisms: formation of "isoenzymes" acting in parallel (p. 10; e.g. *aspartate kinases* in biosynthesis of lysine, methionine and isoleucine from the common precursor aspartic acid, p. 366 f).

The allosteric-controlled reactions considered so far may convey the impression that the effector acts exclusively as inhibitor. This is not the case; on the contrary, there are many reactions in which an effector exerts positive control, i.e. acts as "activator" (= "positive feedback"). The initial substrate or an intermediate may, for instance, speed up its own conversion by stimulating the activity of an allosteric enzyme operating several steps down the reaction chain. Moreover, the activity of individual enzymes may be positively affected by their own substrate or the end product of the reaction catalyzed (for examples, see below).

We shall now consider some important allosteric enzymes controlling various sites of metabolism as key enzymes.

Carbohydrate Metabolism

The processes of degradation, interconversion and synthesis of carbohydrates exert a strong and multiform influence on various reaction sites of metabolism. It is hence not surprising that these reactions in particular are subject to a complex control system ensuring that the various requirements are met.

Phosphofructokinase, the key enzyme of glycolysis (p. 209) is subject to allosteric control: ATP is not only utilized as cosubstrate but in higher concentrations acts as allosteric inhibitor. Since ADP, and to some extent also AMP, are positive effectors abolishing this inhibition, the ratio ATP: ADP (AMP) determines the activity of the enzyme and thus the flow of substrate through glycolysis. The respiratory chain usually yields plenty of ATP under aerobic conditions. Accordingly, this will choke glycolysis and lower consumption of carbohydrates by inhibiting the activity of *phosphofructokinase.* Vice versa, rising consumption of ATP by other areas of metabolism raises the level of ADP and AMP which now activate the key enzyme and hereby stimulate glucose degradation. This regulation is of great significance for those cells which meet their energy requirement either by respiration or by fermentation, depending on the oxygen supply (p. 246). Under anaerobic conditions only small amounts of ATP are formed via substrate chain phosphorylation (p. 211) which are, moreover, rapidly consumed. Consequently, ADP and AMP prevail and keep on activating *phosphofructokinase;* accordingly, the rate of glucose degradation is high in the absence of oxygen. When O_2 is available again, glucose consumption is immediately decreased to a low rate by the control mechanism described above. Thus, the principle underlying the "Pasteur effect" is an allosteric control of *phosphofructokinase.* A participation of metabolite regulation (p. 438), however, cannot be ruled out completely.

Pyruvate kinase of yeast is also inhibited by a surplus of ATP, but activated by fructose-1,6-biphosphate. It is synchronized with *phosphofructokinase:* the fructose-1,6-biphosphate arising from the activation of *phosphofructokinase* is metabolized at high rates by the subsequent reaction steps, particularly by the one catalyzed by *pyruvate kinase.* Vice versa, with a surplus of ATP these key enzymes at the beginning and the end of the metabolic sequence are inactivated.

These controls may also include reactions of the citric acid cycle as indicated by the finding that citric acid is also an allosteric inhibitor of *phosphofructokinase.* The respiratory chain is involved too, because the activities of *citrate synthase* and *isocitrate dehydrogenase* (p. 221 f) are inhibited by ATP as allosteric effector. An accumulation of intermediates of the citric acid cycle in yeast cells can thus be prevented by the control of at least three key enzymes. AMP or ADP as well as $NAD^\oplus$ consequently act as positive effectors of *isocitrate dehydrogenase* (p. 222).

Allosteric control is also exerted on the key enzyme of glycolysis by reaction chains outside the dissimilation process; these chains draw essential substrates from dissimilation. Thus intermediates of the pentose phosphate cycle (6-phosphogluconic acid, sedoheptulose-7-phosphate, erythrose-4-phosphate) act as allosteric inhibitors on *phosphohexose isomerase* which catalyzes the conversion of glucose-6-phosphate to fructose-6-phosphate. Under certain conditions this may eventually lead to a shifting of glucose degradation in favor of the pentose phosphate cycle and at the expense of glycolysis. The metabolite regulation of the former is thus supported by an allosteric control which additionally turns off the competing metabolic chain ("competitive side-chain regulation").

Vice versa intermediates from dissimilatory processes may assume the role of effectors in cases where they are key compounds of attached reaction chains. A significant example can be seen in the synthesis of starch and glycogen, where the key enzyme *pyrophosphorylase* (p. 170) is subjected to allosteric activation by 3-phosphoglyceric acid, phosphoenolpyruvic acid, fructose biphosphate (leaves) or pyruvic acid *(Rhodospirillum rubrum);* on the other hand, AMP and ADP bring about inhibition. So far the stimulating effect of glucose-6-phosphate on the specific *synthase* active in glycogen synthesis of the liver has not been definitively demonstrated in plant organisms.

Fatty Acid Biosynthesis

We are already well acquainted with the significance of regulation by metabolites and compartmentation in providing acetyl-CoA as extramitochondrial substrate for fatty acid synthesis (p. 439). The latter is additionally controlled by the citric acid released from mitochondria which, acting as an allosteric activator, affects *acetyl-CoA carboxylase,* the key enzyme of fatty acid synthesis. Hereby, the controls functioning within the mitochondria are supplemented (cf. p. 444): a high level of ATP, indicating a low activity of the citric acid cycle and the respiratory chain, inhibits further conversion of citric acid by *isocitrate dehydrogenase;* at the same time it is an important prerequisite for the synthesis of fatty acids. Consequently, there is also an allosteric inhibition of *acetyl-CoA carboxylase* by long-chain acyl-CoA compounds when these start to accumulate as a result of slow consumption by fat synthesis.

Calvin Cycle

In connection with photosynthetic CO_2 reduction to carbohydrate an allosteric sensitivity to AMP has been described for *phosphoribulokinase* (p. 152 ff). Complete exhaustion of ATP within the chloroplast or cell by CO_2 reduction is probably prevented by this control mechanism. The possible function of fructose-6-phosphate as an activator of *carboxydismutase* (p. 147 f) has also been discussed recently; on the other hand, fructose-1,6-biphosphate might act as an allosteric inhibitor of this enzyme.

Nucleotide Biosynthesis

Control of the synthesis of pyrimidine nucleotides has been discussed previously (p. 442). As regards the purine nucleotides it is known that the *transferase* which is involved in the initial reaction step leading to the synthesis of the ring structure (formation of 5-phosphoribosylamine, p. 319) is inhibited by the end products ATP and GTP as well as by AMP and GMP due to allosteric control.

Induction and Repression of Enzymes

The prominent features of this regulatory mechanism consist of 1) induction or repression of the synthesis of enzyme proteins; 2) control exerted on reactions acting on the genetic material during transcription; 3) enzymes controlled as groups. These regulatory principles were discovered first in those microorganisms which carry out an "adaptive" synthesis of new enzymes induced by the composition of the environment, i.e. by exogenous substrate. They are formed in case of need in addition to those which, as "constitutive" enzymes, belong to the basic pattern of the cell.

Generally, a compound in the nutrient medium which is suitable for energy production or for synthesis of cellular compounds is the "inducer". The most wellknown example is the induction of *β-galactosidase* by lactose in *E. coli*. This system has been studied in detail and has lead to the model of Jakob and Monod. According to this model enzyme induction is based on a specific control of transcription which is exerted on the activity of three types of genes: "regulatory gene", "operator gene" and "structural genes". The model offers a plausible explanation not only for enzyme induction but also for the reverse process of "repression", i.e. the switching-off of an enzyme synthesis in progress; this is an important control device in "building metabolism". Here a "repressor" in the form of a protein is required which is activated by an intermediate of low molecular weight, the "effector", and then blocks the formation of enzyme-specific m-RNA on the DNA. The de-novo synthesis of enzymes which are mostly identical with the complete set of a biosynthetic reaction chain is inhibited by this mechanism. It is not surprising that in many cases the end product of this chain acts as effector. For this reason it is referred to as "end product repression". In spite of the analogous effect of an end product – adaptation to a changed metabolic situation and prevention of overproduction of certain compounds – this must not be mixed up with the reversible "end product inhibition" (p. 442) by which merely the activity of key enzymes in reaction chains is changed. In end product repression, however, formation of all the enzymes involved in a reaction chain is switched off. Vice versa, the repressor is considered to be responsible for induction, too, in that it is inactivated by attachment of the inducing substrate, e.g. lactose. Now the block is removed from the DNA sequences carrying the information for *β-galactosidase* and the other enzymes of the reaction chain: de-novo synthesis of these enzymes will recommence ("de-repression"). Uncertainty still prevails as to whether this model, although confirmed by many findings, is also valid for eukaryotic cells. – Further details will be found in textbooks on genetics and micro-biology.

Higher Plants. It is far more difficult to demonstrate induction and repression experimentally in higher plants. In most cases genetic analysis and studies with mutants are not possible. The results obtained from examining enzyme activities, constitutive and in-duced ones, in an organ are sometimes ambiguous since the cells of the latter are of different types. A frequently employed method for demonstrating controlled enzyme synthesis in higher plants makes use of specific inhibitors of RNA and protein synthesis (p. 351 and p. 395). These are administered when a change in the internal or external conditions (see below) is accompanied by a sharp rise in the activity of a certain enzyme; suppression of this specific effect by the inhibitor indicates with some certainty that de-novo synthesis of the enzyme, i.e. an induction, had taken place.

With a change in the metabolic situation of such repressed cells, e.g. when the end product mentioned above is no longer available or is required again in relatively large amounts, "de-repression" will set in: all enzymes involved in the biosynthesis chain are now rapidly synthesized with high yields.

By means of this and other criteria induction and repression of enzymes have been demonstrated in higher plants, at least in several cases. In *Lilium longiflorum,* the thymidine which accumulates for a short time during the development of the pollen in the anthers obviously induces de-novo synthesis of *thymidine kinase* (thy-midine + ATP → thymidine-5′-phosphate + ADP); *thymidine kinase* may also be initiated by adding thymine to isolated buds, provided they are in the appropriate developmental stage. – The possible function of NH_3 as a repressor in the synthesis of *nitrate reductase* (p. 291) and *nitrogenase* (p. 296) has already been men-tioned. It cannot be ruled out completely, however, that enzyme molecules already present are perhaps inactivated. Enzyme induc-tion apparently occurs in fat-containing seeds during germination when the degradation of fat and the activity of the glyoxylic acid cycle (p. 226 f) are set in motion. Of all the enzymes involved *iso-citratase* (p. 227) is particularly affected.

The light-dependent induction of *phenylalanine ammonia lyase,* a key enzyme of secondary metabolism in plants, is probably control-led by the "phytochrome" system (Mohr and coworkers). – In free suspended callus cells of *Petroselinum sativum* blue radiation in-duces the formation of several enzymes of the flavonoid metabolism

(Hahlbrock). – The role of light and aerobiosis or anaerobiosis in controlling porphyrin biosynthesis in phototrophic bacteria has been discussed previously (p. 427 f).

Light is obviously an important factor in the induction of enzymes. De-novo synthesis of several enzymes of photosynthetic CO_2 reduction may take place during light-induced chloroplast development: *carboxydismutase, NADP-dependent phosphotriose dehydrogenase, aldolase* (cf. 157 f), and fructose-1,6-biphosphate splitting *phosphatase* (p. 150). Light-induced synthesis of enzymes also takes place in seedling tissues, in darkness-adapted cells of *Euglena gracilis* and in cells of *Nicotiana tabacum var. Samsun* (Bergmann).

Recent studies have shown that certain plant hormones may induce enzyme synthesis. After treatment with gibberellic acid the enzymes *α-amylase* (p. 205 f), *ribonuclease* (p. 351 f) and a *protease* were formed in cells of the aleurone layer of grains (barley, oats).

References

Atkinson, D. E.: Regulation of enzyme activity. Ann. Rev. Biochem. 35: 85, 1966

Betz, A.: Regulation der Enzymaktivität. Umschau 74: 297, 1974

Bergmann, L., A. Bälz: Der Einfluß von Farblicht auf Wachstum und Zusammensetzung pflanzlicher Gewebekulturen. Planta 70: 285, 1966

Bormann, E. J.: Regulationsvorgänge im glykolytischen und oxydativen Glucoseabbau. Biol. Rundschau 7: 193, 1969

Feierabend, J.: Regulationsvorgänge bei der Bildung von Photosyntheseenzymen. Umschau 67: 494, 1967

Glasziou, K. T.: Control of enzyme-formation and inactivation in plants. Ann. Rev. Plant Physiol. 20: 63, 1969

Hahlbrock, K., H. Ragg: Light-induced changes of enzyme activities in parsley cell suspension cultures. Arch. Biochem. Biophys. 166: 41, 1975

Marcus, A.: Enzyme induction in plants. Ann. Rev. Plant Physiol. 22: 313, 1971

Stadtman, E. R.: Allosteric regulation of enzyme activity. Advanc. Enzymol. 28: 41, 1966

Umbarger, H. E.: Regulation of amino acid metabolism. Ann. Rev. Biochem. 38: 323, 1969

Review Series

Advances in Botany (formerly „Fortschritte der Botanik") Ergebnisse der Biologie

Annual Review of Plant Physiology
Annual Review of Microbiology
Annual Review of Biochemistry

Journals

American Journal of Botany
Archive of Microbiology
Berichte der Deutschen Botanischen Gesellschaft
Biochemie und Physiologie der Pflanzen
Biochimica et Biophysica Acta
Canadian Journal of Botany
Journal of Experimental Botany
Journal of Molecular Biology
Photochemistry and Photobiology

Physiologia Plantarum
Phytochemistry
Plant and Cell Physiology
Plant Physiology
Planta
Planta medica
Plant Science Letters
Zeitschrift für Pflanzenphysiologie
Zeitschrift für Naturforschung

Index

A

Absorption
— of aromatic amino acids 358
— band 100
— change 97, 106, 134
— — light-induced 121, 134
— force 270, 272 ff, 277
— — of soil 272
— line 99
— of NAD-H and NADP-H 15
— spectrum 42 ff, 52 f, 56, 60 f, 85 ff, 102, 105, 120 f, 423
— — aqueous cell extracts 49, 62
— — of atoms 100
— — bacteriochlorophyll a 44
— — carotinoids 56
— — *chlorobium* chlorophyll 45
— — chloroplast fraction 60, 105
— — chromatophores 70 f
— — cytochrome 120 f
— — — reduced 121
— — — c and f 121
— — fucoxanthin 56
— — pheophytin 46
— — phycocyanins a. phycoerythrins 51 ff
— — porphyrins 423
— — spirilloxanthin 56
— — total pigment extract from leaves 56
Acer pseudoplatanus 187
Acetabularia mediter-ranea 95, 157, 172, 335
Acetaldehyde 248 ff, 252
— active 216, 368
Acetate consuming organisms 226 f
— as C-source 226 f
— kinase 298
Acetic acid 215 f, 225, 250 ff, 421
— — fermentation 252
Acetoacetyl-CoA 430 f

Acetobacter xylinum 185
α-Aceto-α-hydroxy-butyric acid 368 f
Acetolactic acid 369
Acetone 33, 102, 255
Acetyl-CoA carboxylase 405 f, 409, 445
— in chloroplasts 409
— as key enzyme of fatty acid synthesis 445
Acetyl-Coenzyme A 215 ff, 227 ff, 231 f, 245 f, 255, 368, 398 f, 404 ff, 412 ff, 430 f, 438 f, 445
— citric acid as transport metabolite 245
— pool 218, 232, 438
2-acetyl-desvinyl-3,4-dihydrochlorophyll cf. bacteriochlorophyll
Acetylene as substrate for nitrogenase 297
N-Acetylglucosamine 188 f
Acetylphosphate 298
Acetylserine 303 f
Achromobacterium 292
Acid, organic 164, 167, 200, 219, 250
Acidophilic plants 282
Acidophobic plants 282
Aconitase 221, 224 f
Actidion cf. cyclo-heximide
Actinomycetales 294
Actinomycin D 351
— peptide portion 376
Action spectrum of photosynthesis 85 ff, 105
Activation energy 9 ff, 90
Active absorption of ions 283 f
— transport 191 f, 269, 271
— — of molecules a. ions 271
Acyl carrier protein 409
— -CoA dehydrogenase 413

Acyl-CoA, desaturase 410
— -coenzyme A 413, 415
— phosphate 210
— -S-enzyme protein 148, 211
— thiokinase 413
— transfer in fatty acid biosynthesis 408
Adenine 13, 18 f, 313 f, 324 f, 339, 345 f
Adenosine 17 ff
— -3′,3′-diphosphate 217
— diphosphate (ADP) 18 ff, 126 ff, 181, 211, 222, 233, 242, 245 f, 297, 343, 438 f, 443 ff
— — as effector 443 f
— — glucose (ADP-glucose) 171, 180 f
— — — pyrophosphorylase 180
— — — -starch transgly-cosidase 181
— monophosphate (AMP) 18 ff, 195 f, 321, 443 ff
— — as allosteric inhibitor 445
— — as positive effector 443 f
— phosphorylsulfate kinase 302
— -5′-phosphate 321
— -5′-phosphorylsulfate (APS) 195 f, 302 ff
— triphosphate (ATP) 18 ff, 117, 119, 126 ff, 133, 136 ff, 155, 170 f, 180 f, 195 f, 199, 202 f, 204, 207 ff, 213 ff, 222, 225, 229 ff, 233, 240 ff, 249 f, 252, 254 ff, 260, 297 f, 301 ff, 306 f, 313, 315 ff, 319 ff, 331, 343, 366 f, 377, 386, 395, 405, 413, 431, 438 f, 443 ff
— — as allosteric inhibitor 443 ff
— — formation site 127 f, 136

472   Index